**ENVIRONMENTAL
SYSTEMS
ENGINEERING**

McGRAW-HILL SERIES IN WATER RESOURCES
AND ENVIRONMENTAL ENGINEERING

VEN TE CHOW, ROLF ELIASSEN, and RAY K. LINSLEY, Consulting Editors

CHANLETT: Environmental Protection
GRAF: Hydraulics of Sediment Transport
HALL AND DRACUP: Water Resources Systems Engineering
JAMES AND LEE: Economics of Water Resources Planning
LINSLEY AND FRANZINI: Water Resources Engineering
METCALF AND EDDY, INC.: Wastewater Engineering: Collection
    Treatment, Disposal
RICH: Environmental Systems Engineering
WALTON: Groundwater Resource Evaluation
WIENER: The Role of Water in Development: An Analysis of Principles
    of Comprehensive Planning

McGRAW-HILL BOOK COMPANY
New York
St. Louis
San Francisco
Düsseldorf
Johannesburg
Kuala Lumpur
London
Mexico
Montreal
New Delhi
Panama
Rio de Janeiro
Singapore
Sydney
Toronto

LINVIL G. RICH
College of Engineering
Clemson University

Environmental Systems Engineering

This book was set in Times Roman.
The editors were B. J. Clark and M. E. Margolies;
the production supervisor was Joan M. Oppenheimer.
The drawings were done by ANCO Technical Services.
The printer and binder was R. R. Donnelley & Sons Company.

Library of Congress Cataloging in Publication Data

Rich, Linvil Gene, 1921–
   Environmental systems engineering.

  (McGraw-Hill series in water resources & environmental engineering)
    1. Environmental engineering.   2. Sanitary engineering.   I. Title.
TD145.R49   620.8   73–4272
ISBN 0-07-052250-2

ENVIRONMENTAL
SYSTEMS
ENGINEERING

Copyright © 1973 by McGraw-Hill, Inc. All rights reserved.
Printed in the United States of America. No part of this publication may be reproduced,
stored in a retrieval system, or transmitted, in any form or by any means,
electronic, mechanical, photocopying, recording, or otherwise,
without the prior written permission of the publisher.

1234567890DODO79876543

To my mother, LILLIAN WATKINS RICH,
for her constant faith and support

CONTENTS

	Preface	xi
1	**Physical Phenomena**	1
1.1	Transport	1
1.2	Gas Transfer	7
1.3	Thermal Phenomena	13
1.4	Sedimentation	25
1.5	Continuous-Flow Models	31
2	**Chemical Phenomena**	39
2.1	Solution Equilibriums	39
2.2	Reaction Kinetics	45
2.3	Carbonate Equilibriums	49
2.4	Thermochemistry	58
2.5	Colloidal Behavior	61
3	**Biologic Phenomena**	70
3.1	Organic Materials	70
3.2	Microorganisms	76
3.3	Growth Kinetics	80

3.4	Biochemical Oxygen Demand	84
3.5	Anaerobic Decomposition	87
3.6	Photosynthesis	89
3.7	Food Chains	91

4 Ecological Systems — 99

4.1	Models	99
4.2	Analytic Solutions	102
4.3	Time-Domain Simulation	104
4.4	Continuous-Flow Microbiological Systems	106
4.5	Pesticide Concentration	111
4.6	Eutrophication	113

5 Natural Transport Systems — 138

5.1	Basic Models	138
5.2	Dissolved-Oxygen System	142
5.3	Streams	146
5.4	Estuaries	154
5.5	Transport in the Air Environment	165

6 Planning Factors — 170

6.1	Water-Quality Criteria and Standards	170
6.2	Air Pollution and Its Control	176
6.3	Radiological Health	181
6.4	Environmental Impact Statements	186
6.5	Population-Growth Models	194
6.6	Regional-Growth Model	197
6.7	Time-Capacity Expansion of Systems	208

7 Time Series — 214

7.1	Trend, Frequency, and Random Components	214
7.2	Time-Series Analysis	222
7.3	Synthetic Stream-Flow Sequences	236
7.4	Storage-Yield Relationships	243
7.5	Predicting Minimum Stream Flows	247

8 Management Systems — 254

8.1	Water-Quality Management	254
8.2	Solid-Waste Management	265
8.3	Waste-Water Reuse Systems	272

9 Engineered Transport Systems ... 280

- *9.1* Pipe Network Analysis ... 280
- *9.2* Water Distribution Systems ... 290
- *9.3* Open-Channel Flow ... 296
- *9.4* Domestic Waste-Water Collection Systems ... 307
- *9.5* Storm-Water Collection Systems ... 311

10 Water Treatment and Renovation Systems ... 318

- *10.1* Treatment Trains ... 318
- *10.2* Lagoon Systems ... 326
- *10.3* Individual Household Systems ... 331

11 Processes Used in Gross-Particulate Removal Trains ... 335

- *11.1* Screening Processes ... 335
- *11.2* Sedimentation Processes ... 336
- *11.3* Grit Chambers ... 343
- *11.4* Primary Sedimentation Basins ... 344
- *11.5* Flotation Processes ... 346

12 Processes Used in Suspended-Particulate Removal Trains ... 353

- *12.1* Activated-Sludge Processes ... 353
- *12.2* Trickling-Filter Process ... 366
- *12.3* Rapid-Sand-Filter Process ... 371

13 Processes Used in Dissolved-Materials Removal Trains ... 380

- *13.1* Aeration Processes ... 380
- *13.2* Carbon-Adsorption Processes ... 382
- *13.3* Chemical-Precipitation Processes ... 383
- *13.4* Ion-Exchange Processes ... 389
- *13.5* Membrane-Separation Processes ... 393
- *13.6* Disinfection Processes ... 396

14 Processes Used in Sludge Treatment Trains ... 404

- *14.1* Thickening Processes ... 404
- *14.2* Anaerobic Digestion ... 412
- *14.3* Conditioning Processes ... 418
- *14.4* Dewatering Processes ... 420
- *14.5* Drying and Incineration Processes ... 424

Appendix 427

A.1 Factors for Conversion to the International System (SI) Units 427
A.2 Nomogram for the Solution of Hazen-Williams Expression When $C = 100$ 430
A.3 Nomogram for the Solution of the Manning Formula ($n = 0.013$) 431
A.4 Hydraulic Elements of Circular Sections 432
A.5 Atomic Weights and Valences of Selected Chemical Elements 433
A.6 Algorithm Construction 434

Index 437

PREFACE

The title of this book, "Environmental Systems Engineering," is descriptive of the book's content, which includes topics from across the broad spectrum of man's environmental contacts. Although the water environment is considered in great detail, the reader is also introduced to the elements of air pollution and its control, solid-waste management, and radiological health. The stress placed on the water environment reflects the relative extent to which this environment lends itself to engineering control.

The book uses a systems approach in formulating and analyzing environmental phenomena, as well as in the selection and design of engineered facilities used for controlling the environment. In general, focus is placed on the system as a whole and how its components interact, as opposed to the components themselves. The mathematics of systems analysis and computer solutions is used extensively. Finally, the book was written with engineering objectives in mind.

This book is not a treatise or a design manual. Instead, it is a textbook written for engineering students at the junior and senior levels. Although many of the topics generally covered in water- and waste-water-treatment courses in civil engineering are included, the scope of the book is such that it can be used by students enrolled in other fields of engineering as well. The material dealt with is developed in a logical fashion, with earlier chapters laying a foundation for topics discussed subsequently. Time constraints imposed on the student are recognized, and for this reason no attempt has been made to cover all topics that could have been found in a book of this type. Only those topics generally recognized as being most important are included. Nonengineering students could use the book, but to do so would require the same mathematical proficiency expected of engineering students.

The organization of the book allows considerable flexibility in its use. Chapters 1 to 3 describe and formulate those phenomena that occur in environmental systems. Chapters 4 and 5 deal with natural environmental systems. Chapters 6 to 10 cover a variety of topics important to engineering systems, whereas Chapters 11 to 14 are devoted to treatment processes. The book was designed for a two-semester course in environmental systems engineering. However, it can be adapted to several types of single-semester courses, each with a different emphasis. For example, the book's contents can be grouped in the following chapter sets:

Natural environmental systems—Chapters 1 to 7
Water pollution control—Chapters 1 to 3, 5, 7, and 10; and Sections 6.1 and 8.1
Water and waste treatment systems—Chapters 1 to 3 and 10 to 14; and Sections 6.1 and 6.5 to 6.7

A modified engineering system of dimensions that is a combination of the absolute and gravitational systems is used. Recognizing that the student will be practicing engineering in an era when both metric and English units will be used, the author uses both. Notation is also varied, reflecting the fact that the topics covered in the book fall on both sides of the interface between the basic sciences and traditional engineering. Because the system of dimensions is uniform throughout the book, the student should experience no difficulty with either units or notation.

Solution procedures add yet another uniqueness to the book. Instead of using example problems consisting of equations solved by simple substitution of numerical values and routine mathematical manipulations, the author has relied on a more flexible and, he hopes, a more effective device for emphasizing principles of solution. The solution procedure may take the form of a computer algorithm, a description of ways to obtain data, a discussion of methods of application, or, simply, an elaboration on points brought out in the text. Such a device opens the door for the student, as opposed to solving his problem for him. Each chapter is followed by a selection of exercises to be solved by the student.

A large portion of the material presented in this book has been derived from the works of others. Their contribution is gratefully acknowledged. The author can take credit only for bringing the information together in an attempt to make it more accessible to the student.

Acknowledgment is made with thanks to B. H. Kornegay, W. E. Castro, J. C. Martin, T. M. Keinath, A. G. Law, and K. D. Tracy for reviewing certain portions of the book.

LINVIL G. RICH

1
PHYSICAL PHENOMENA

1.1 TRANSPORT

Transport in the environment takes place as a result of two phenomena—*advection* and *diffusion*. Although both phenomena are active in the natural environment, one or the other may predominate in any given situation.

Advection

Advection is that transport which results from fluid motion. Two types of time derivatives will be used here to describe such transport.

Consider a situation in which fluid is in flow in a given direction. Suppose that the concentration of some trace material is measured at a fixed reference point located internally and remote from the fluid boundaries. The time derivative describing the tracer concentration at the reference point is called the *partial time derivative* or *local derivative* $\partial c/\partial t$.

The other time derivative of interest here is called the *substantial time derivative* or the *derivative following the motion*. This derivative expresses the time rate of change in concentration of the tracer as it moves in the direction of flow and at the same velocity as the flowing fluid. The substantial time derivative consists of the time rate of change of concentration at a point together with the change due to the velocity field and is expressed in rectangular coordinates as

$$\frac{Dc}{Dt} = \frac{\partial c}{\partial t} + u_x \frac{\partial c}{\partial x} + u_y \frac{\partial c}{\partial y} + u_z \frac{\partial c}{\partial z} \qquad (1.1.1)$$

where D/Dt = operator

u_x, u_y, u_z = velocity components of flowing fluid

The substantial time derivative Dc/Dt describes, therefore, the time rate of change in concentration in an element of fluid as it travels with the flow to and away from the reference point. If there is no movement of material into or out of the element or if there is no generation or destruction of the material within the element, the concentration will remain the same, and Dc/Dt will be equal to 0.

Diffusion

Diffusion is a process in which a substance in solution or suspension (the *diffusing phase*) migrates in response to a concentration gradient through another substance (the *dispersing phase*). Diffusion is one of the most basic processes in nature and at the molecular level accounts for most of the transport that takes place. The diffusing phase can be a gas, solid, or liquid. The dispersing phase, which for our purpose constitutes an environmental medium, is most often either a liquid or a gas.

The basic relationships governing diffusion are called *Fick's laws of diffusion*. Fick's first law states simply that the rate of mass transport by diffusion across an element of area is proportional to the concentration gradient of the diffusing substance:

$$N_x = -D_m \frac{\partial c}{\partial x} \qquad (1.1.2)$$

where N_x = rate of mass transport in x direction across an element of area normal to x. $[FL^{-2}t^{-1}]$

$\partial c/\partial x$ = concentration gradient of diffusing phase. $[FL^{-4}]$

D_m = molecular diffusion coefficient. $[L^2 t^{-1}]$

Equation (1.1.2) can be used to derive Fick's second law.[1]

Consider Fig. 1.1.1. For the un-steady state, a mass balance of a dispersed phase diffusing along the x coordinate through a volumetric element of fluid is

$$\text{Accumulation} = \text{mass in} - \text{mass out}$$

$$\frac{1}{dA}\frac{\partial m}{\partial t} = \left(-D_m \frac{\partial c}{\partial x}\right)_1 - \left(-D_m \frac{\partial c}{\partial x}\right)_2 \quad (1.1.3)$$

The gradients at the two planes are related by

$$\left(-D_m \frac{\partial c}{\partial x}\right)_2 = \left(-D_m \frac{\partial c}{\partial x}\right)_1 + \left[-D_m \frac{\partial}{\partial x}\left(\frac{\partial c}{\partial x}\right) dx\right] \quad (1.1.4)$$

Substitution of Eq. (1.1.4) into Eq. (1.1.3) yields

$$\frac{1}{dA}\frac{1}{dx}\frac{\partial m}{\partial t} = \frac{\partial c}{\partial t} = D_m \frac{\partial}{\partial x}\left(\frac{\partial c}{\partial x}\right) = D_m \frac{\partial^2 c}{\partial x^2} \quad (1.1.5)$$

Equation (1.1.5) is called Fick's second law.

For the general three-dimensional case Eq. (1.1.5) can be written as

$$\frac{\partial c}{\partial t} = D_m \left(\frac{\partial^2 c}{\partial x^2} + \frac{\partial^2 c}{\partial y^2} + \frac{\partial^2 c}{\partial z^2}\right) = D_m \nabla^2 c \quad (1.1.6)$$

where ∇^2 = operator

The molecular diffusion coefficient is proportional to the absolute temperature and inversely proportional to the molecular weight of the diffusing phase and the viscosity of the dispersing phase.

---

[1] In the present text, a modified engineering system of dimensions is used that is a combination of the absolute and gravitational systems. In this system, seven dimensions will be used as follows:

Dimension	Symbol	Typical units	
		Metric	English
Time	[t]	Second	Second
Length	[L]	Meter	Foot
Mass	[M]	Gram-mass	Pound-mass
Force	[F]	Gram-force	Pound-force
Temperature	[T]	Degree Celsius	Degree Fahrenheit
Thermal energy	[H]	Calorie	British thermal unit
Mole	[m]	Gram-mole	Pound-mole

Factors for conversion to the International System (SI) of units are given in Table A.1. Although the SI system of units is based on a different system of dimensions than is used in this text, it provides a useful basis for conversion between units in common usage.

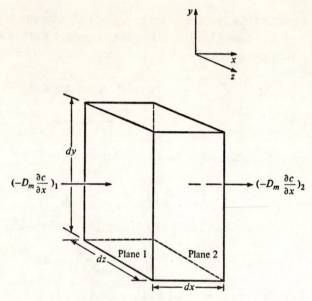

FIGURE 1.1.1
Definition sketch of diffusion into and out of an incremental volume.

Transport in the Environment

When both advection and diffusion occur, the effects of the two phenomena are additive and can be expressed mathematically by combining the expression for the substantial time derivative with Eq. (1.1.6):

$$\frac{Dc}{Dt} = D_m \nabla^2 c \qquad (1.1.7)$$

Equation (1.1.7) applies to situations in which no turbulence exists. However, the general form of the relationship has validity even when turbulence exists. In such cases, the instantaneous velocity terms u_x, u_y, and u_z can be replaced by temporal mean velocity terms $\bar{u}_x$, $\bar{u}_y$, and $\bar{u}_z$ plus the turbulent velocity fluctuations u'_x, u'_y, and u'_z. For the one-dimensional case, Eq. (1.1.7) becomes

$$\frac{\partial c}{\partial t} + (\bar{u} + u')\frac{\partial c}{\partial x} = D_m \frac{\partial^2 c}{\partial x^2} \qquad (1.1.8)$$

Often it is advantageous to use only the temporal mean velocity in the advective term. When this is done, the effect of the velocity fluctuations (which is actually advective) must be included. Normally, this is accomplished by including terms referred to as *turbulent diffusion coefficients*.[1]

$$\frac{\partial c}{\partial t} + \bar{u}\frac{\partial c}{\partial x} = (D_m + e)\frac{\partial^2 c}{\partial x^2} \quad (1.1.9)$$

where e = turbulent diffusion coefficient

For the typical environmental situation, Eq. (1.1.9) is modified so that the advective term is replaced by the cross-sectional average of the velocity U. Again, this is not accomplished simply by replacing $\bar{u}$ with U. Possible effects due to the lateral distribution of velocity are included in a dispersion term:

$$\frac{\partial c}{\partial t} + U\frac{\partial c}{\partial x} = E\frac{\partial^2 c}{\partial x^2} \quad (1.1.10)$$

where c = cross-sectional average of concentration. $[FL^{-3}]$

U = cross-sectional average of velocity. $[Lt^{-1}]$

E = longitudinal dispersion coefficient. $[L^2 t^{-1}]$

Expressions similar to Eq. (1.1.10) can be developed for two- and three-dimensional cases. It should be noted that although Eq. (1.1.10) is similar to Eq. (1.1.7) for the one-dimensional case, the dispersion term E for turbulent flow bears little relationship to the molecular diffusion coefficient D_m.

The use of Eq. (1.1.10) requires, for most environmental situations, an estimate of the value of E for the particular situation of concern. In some cases actual measurements can be made using tracer techniques.[2]

Diffusion and dispersion coefficients characteristic of various environments are indicated in Fig. 1.1.2.

Solution procedure 1.1.1 The differential equation developed for any particular situation must be solved subject to particular boundary conditions. For the case in which a conservative substance[3] in the amount W is released instantaneously at $t = 0$ and $x = 0$ into an environmental

---

[1] E. R. Holley, Unified View of Diffusion and Dispersion, *J. Hydraul. Div., ASCE*, vol. 95, no. Hy 2, pp. 621–631, 1969.
[2] H. B. Fischer, Dispersion Predictions in Natural Streams, *J. Sanit. Eng. Div., ASCE*, vol. 94, no. SA5, pp. 927–943, 1968.
[3] A substance that does not decay or degrade in the environment.

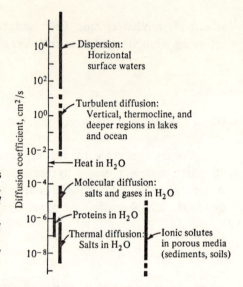

FIGURE 1.1.2
Diffusion and dispersion coefficients characteristic of various environments. (*After A. Lerman, Time to Chemical Steady-states in Lakes and Ocean, fig. 1, in "Nonequilibrium Systems in Natural Water Chemistry," Advances in Chemistry Series 106, American Chemical Society, Washington, 1971.*)

transport medium where there is one-dimensional dispersion and advection, Eq. (1.1.10) has the solution[1]

$$c = \frac{W}{S\sqrt{4\pi Et}} \exp\left[-\frac{(x - Ut)^2}{4Et}\right] \quad (1.1.11)$$

where W = weight of conservative substance. $[F]$

S = cross-sectional area normal to x axis. $[L^2]$

The concentration at any time t is spread out as a gaussian (bell-shaped) curve with the center moving downstream at velocity U.

For the same case but with no advection ($U = 0$), the solution is

$$c = \frac{W}{S\sqrt{4\pi Et}} \exp\left(-\frac{x^2}{4Et}\right) \quad (1.1.12)$$

For the case in which the substance is released continuously and at a constant rate, $\partial c/\partial t = 0$, and for all values of $x > 0$

$$c = \frac{W}{SU} \quad (1.1.13)$$

Theoretically, a steady-state condition such as is described in Eq. (1.1.13) would not be attained until $t = \infty$. Practically, the condition is

---

[1] W. E. Dobbins, Diffusion and Mixing, *J. Boston Soc. Eng., Civ. Eng. Ser.*, no. 114, pp. 108–128, 1963.

attained shortly after the front of the cloud of the conservative substance passes through the distance of concern.

The solution of Eq. (1.1.10) for the time-varying continuous release (or passage) of a conservative substance under conditions of steady, uniform stream flow can be conveniently obtained using numerical methods. ////

Solution procedure 1.1.2 Several methods for predicting the value of E from stream measurements have been reported. One such method[1] is based on the time change in variance of the measured time-concentration curve as the tracer cloud moves downstream:

$$E = \frac{\overline{U}^2}{2} \frac{\sigma_{t_2}^2 - \sigma_{t_1}^2}{\bar{t}_2 - \bar{t}_1} \quad (1.1.14)$$

where $\sigma_{t_1}^2, \sigma_{t_2}^2$ = variances of concentration-time curves at upstream and downstream stations, respectively

$\bar{t}_1, \bar{t}_2$ = mean time of passage of tracer cloud past each station

$\overline{U}$ = mean velocity of flow between stations

The mean travel time of a tracer cloud is rather indefinite. Some investigators have considered the mean travel time to be the time to the peak concentration. Others have used the time coinciding with the centroid of the tracer cloud. Also, the mean velocity of flow in Eq. (1.1.14) may not equal the discharge velocity Q/S. A more valid estimate of this variable can be determined from the mean time of flow $x/\bar{t}$. ////

1.2 GAS TRANSFER

When a liquid containing dissolved gases is brought into contact with an atmosphere of gas other than that with which it is in equilibrium, an exchange of gases takes place between the atmosphere and the solution. For gases of low or moderate solubility that do not react chemically with the solvent, the rate of change in concentration of dissolved gas in a liquid volume is

$$\frac{dc}{dt} = K_L \frac{A}{V}(c^* - c) \quad (1.2.1)$$

---

[1] H. B. Fischer, op. cit.

where c = concentration of dissolved gas in liquid. $[FL^{-3}]$

c^* = saturation concentration of dissolved gas. $[FL^{-3}]$

K_L = overall mass-transfer coefficient, liquid-phase basis. $[Lt^{-1}]$

A = surface area. $[L^2]$

V = liquid volume. $[L^3]$

Various models have been developed to explain the effect of liquid characteristics and behavior on the coefficient K_L. The *film model*[1] assumes that there is a stagnant liquid film at the liquid-gas interface in which steady-state molecular diffusion controls the rate of gas transfer.

$$\frac{dc}{dt} = \frac{D_m}{x} \frac{A}{V} (c^* - c) \qquad (1.2.2)$$

where D_m = molecular diffusion coefficient. $[L^2 t^{-1}]$

x = film thickness. $[L]$

Comparison of Eq. (1.2.2) with Eq. (1.2.1) reveals the relationship

$$K_L = \frac{D_m}{x} \qquad (1.2.3)$$

Suppose, for example, gas is being transferred to a liquid which is unsaturated with respect to the gas being transferred. First, through a combined process of mixing and diffusion the gas molecules are transported to the liquid-gas interface. At the interface, the gas is assumed to dissolve and be present in a concentration in equilibrium with the partial pressure of the gas in the gaseous phase. The dissolved gas then diffuses through the stagnant liquid film to the boundary between the film and the bulk-liquid phase, from where it is transported throughout the bulk-liquid phase by mixing. See Fig. 1.2.1. The same mechanism in reverse prevails when gases are released from a supersaturated solution. Since transfer by diffusion is slow compared with transfer by mixing, the rate of transfer from one phase to another is considered to be controlled by the stagnant film. The latter is thought of as persisting regardless of how much turbulence is present in the liquid, the turbulence serving only to reduce the film thickness. The film model, although questionable from a theoretical standpoint, is convenient to visualize.

---

[1] W. K. Lewis and W. G. Whitman, Principles of Gas Absorption, *Ind. Eng. Chem.*, vol. 16, p. 1215, 1924.

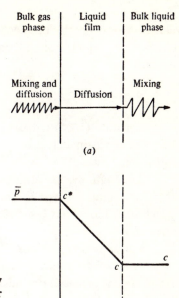

FIGURE 1.2.1
Film model for transfer of gases of low or moderate solubility. (a) Transfer mechanisms; (b) film gradient.

The *penetration model*[1] postulates that eddies orginating in the turbulent bulk of the liquid migrate to the gas-liquid interface, where they are exposed briefly to the gas before being displaced by other eddies arriving at the interface. During their brief residence (whose time is determined by a probability distribution function), the eddies through a non-steady-state diffusion process absorb molecules from the gas. Upon return of the eddies to the liquid bulk, the molecules are distributed by turbulence. According to the penetration model

$$K_L = \sqrt{D_m r} \qquad (1.2.4)$$

where r = surface renewal rate. $[t^{-1}]$

A comparison of Eq. (1.2.4) with Eq. (1.2.3) reveals a basic contradiction. The film theory holds that the coefficient K_L is proportional to the first power of the molecular diffusion coefficient, whereas for the penetration theory this relationship involves the square root of the coefficient.

The contradiction of these conclusions concerning K_L may have its resolution in the suggestion that the film may be assumed to maintain its existence in a statistical sense. According to the *film penetration model*,[2] the

---

[1] P. V. Danckwerts, Significance of Liquid-film Coefficients in Gas Absorption, *Ind. Eng. Chem.*, vol. 43, pp. 1460–1467, 1951.
[2] W. E. Dobbins, The Nature of the Oxygen Transfer Coefficient in Aeration Systems, in J. McCabe and W. W. Eckenfelder, Jr. (eds.), "Biological Treatment of Sewage and Industrial Wastes," pp. 141–148, Reinhold, New York, 1956.

film is considered to be always present, but its composition is continually changed by liquid from beneath the surface. The film penetration model predicts that

$$K_L = \sqrt{D_m r} \coth \sqrt{\frac{rx^2}{D_m}} \quad (1.2.5)$$

Equation (1.2.5) reduces to Eq. (1.2.4) as the hyperbolic cotangent term becomes approximately 3 or greater and approaches Eq. (1.2.3) as the renewal rate r approaches 0.

Although the penetration model holds promise in dealing with gas-transfer operations, application to present technology is limited. The film penetration model has been of value in correlating oxygen absorption data in streams.

For some applications, Eq. (1.2.1) is expressed in a different form. In most types of gas-transfer equipment, it is not possible to measure the interfacial area of contact. For this reason use is made of volumetric mass-transfer coefficients which combine the interfacial area per unit volume of equipment with the mass-transfer coefficient:

$$\frac{dc}{dt} = K_L a(c^* - c) \quad (1.2.6)$$

where $K_L a$ = overall volumetric mass-transfer coefficient, liquid-phase base. $[t^{-1}]$

When Eq. (1.2.1) is applied to oxygen transfer in stream reaeration, the equation appears as

$$\frac{dc}{dt} = k_2(c^* - c) = k_2 D \quad (1.2.7)$$

where k_2 = reaeration coefficient. $[t^{-1}]$

D = dissolved oxygen saturation deficit. $[FL^{-3}]$

Equilibrium Relationships

When a gas comes in contact with a liquid undersaturated with respect to a component of the gas, molecules of the latter will diffuse from the gas phase to the liquid phase. Concurrently, a portion of these molecules will return to the gas. The rates at which the molecules move from one phase to the other are governed by the concentrations of the diffusing component in the two phases. Ultimately, concentrations in both the liquid and gas will be such that the rate at which the molecules enter the liquid will equal the rate at which they leave.

Thereafter, the concentrations will remain constant, and the system is said to be *in equilibrium*.

The solubilities of gases (equilibrium concentrations) in liquids vary widely. For gases of low or moderate solubility that do not react chemically with the solvent, the quantity of gas which will dissolve at a given temperature can be determined by *Henry's law*:

$$c^* = K_H \bar{p} \qquad (1.2.8)$$

where $\bar{p}$ = partial pressure of gas, atm

c^* = concentration of gas in equilibrium with $\bar{p}$, mol/l

K_H = Henry's law constant, mol/(l)(atm)

Equation (1.2.8) can be used for partial pressures up to 1 atm. Values of Henry's law constant for several gases that are slightly soluble in water are listed in Table 1.2.1. According to *Dalton's law of partial pressures*, the partial pressure of each gas in a mixture of gases is proportional to its percentage by volume in the mixture.

The solubilities of the more soluble gases and those that react appreciably with the solvent do not behave as simple proportions of the partial pressures. Consequently, Henry's law does not express the equilibrium relationship for such gases. Instead, use must be made of experimental data in which the solubility of a gas has been determined as a function of temperature and pressure.

Table 1.2.1 HENRY'S LAW CONSTANTS FOR SEVERAL GASES SLIGHTLY SOLUBLE IN WATER*
$K_H \times 10^4$ mol/(l)(atm)

T, °C	Air	CO_2	CO	C_2H_6	H_2	H_2S	CH_4	NO	N_2	O_2
0	12.9	764	15.8	44.2	9.62	2,070	24.8	32.9	10.5	21.8
10	10.1	535	12.6	29.4	8.75	1,520	18.7	25.5	8.33	17.0
20	8.38	392	10.4	21.2	8.14	1,150	14.8	21.1	6.93	13.8
30	7.20	299	8.96	16.3	7.63	914	12.4	17.9	6.03	11.7
40	6.40	239	7.98	13.2	7.40	748	10.7	15.8	5.35	10.4
50	5.88	197	7.30	11.1	7.28	630	9.64	14.2	4.92	9.46
60	5.50	163	6.77	9.85	7.28	540	8.88	13.3	4.63	8.85
70	5.30		6.58	8.93	7.30	467	8.34	12.7	4.44	8.40
80	5.20		6.58	8.40	7.37	412	8.15	12.4	4.41	8.10
90	5.15		6.57	8.10	7.40	386	8.04	12.3	4.41	7.98
100	5.20		6.57	8.03	7.46	376	7.93	12.2	4.41	7.93

* Adapted from A. S. Foust et al., "Principles of Unit Operations," app. D-3, Wiley, New York, 1960.

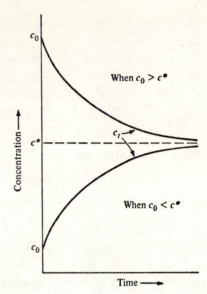

FIGURE 1.2.2
Gas concentration changes in a liquid as a function of time.

Solution procedure 1.2.1 When the concentration of the solute gas in the liquid remains relatively constant during gas transfer, transfer rates are estimated on the basis of steady-state conditions. For such conditions, Eq. (1.2.1) can be used directly. When the concentration changes with time, such a relationship can be derived by integrating the equation between the limits of time equal to 0 and t:

$$\int_{c_0}^{c_t} \frac{dc}{c^* - c} = K_L \frac{A}{V} \int_0^t dt \qquad (1.2.9)$$

$$\ln \frac{c^* - c_t}{c^* - c_0} = \ln \frac{D_t}{D_0} = -K_L \frac{A}{V} t \qquad (1.2.10)$$

$$D_t = D_0 \exp\left(-K_L \frac{A}{V} t\right) \qquad (1.2.11)$$

where D_0, D_t = saturation deficit initially and at time t. $[FL^{-3}]$

For a given situation, a plot of $\log D_t$ as a function of time will give a straight line, the slope of which will be equal to $-0.434 K_L A/V$.

Equation (1.2.10) when solved for the concentration at time t, c_t, yields an expression that gives a plot such as Fig. 1.2.2. ////

1.3 THERMAL PHENOMENA

Thermal phenomena play a major role in nature. Nowhere are the effects more profound than in water systems. Changes that occur either in the temperature of the water or in temperature-related water-quality constituents may affect in important ways the beneficial uses to be served by the water. The aquatic environment, including fish and the associated members of the food chain, is particularly sensitive to temperature. Thermal phenomena of central importance in the environment include the energy transport processes—conduction, convection, and radiation—and the energy transformation processes, of which evaporation will be discussed here.

The thermal energy content of a material consists of two types of energy: sensible heat and latent heat. *Sensible heat* is defined as that heat which when gained or lost by a body is reflected by a change in the temperature of the body. Energy of this nature is computed by multiplying the specific heat of the material by its temperature above some datum:

$$Q_s = c_p W T_0 \qquad (1.3.1)$$

where Q_s = sensible heat. $[H]$

c_p = specific heat at constant pressure. $[HF^{-1}T^{-1}]$

W = weight of material. $[F]$

T_0 = temperature of material above datum. $[T]$

Strictly speaking heat capacity should be used instead of the specific heat. *Heat capacity* is defined as the amount of heat required to raise the temperature of one unit of weight of a material by one degree of temperature. The *specific heat* is the ratio of the heat capacity of a material to the heat capacity of an equal weight of water at a reference temperature. For ordinary calculations, the two terms are used interchangeably. Both vary with temperature. At normal temperatures (0 to 100°C), the specific heat of water is approximately 1 cal/(g)(°C).

Latent heat is that heat required to bring about a change in state, no temperature change being involved. As most changes in state dealt with in heat calculations involve a transition between a liquid and a gas, concern commonly centers around the latent heat of vaporization. Latent-heat quantities may be computed from

$$Q_l = \lambda W \qquad (1.3.2)$$

where Q_l = latent heat. $[H]$

λ = unit latent heat. $[HF^{-1}]$

W = weight of material changing state. $[F]$

Like heat capacity and specific heat, latent heat is a function of the temperature. At normal temperatures (0 to 100°C), the unit latent heat of vaporization of water is approximated by $600 - 0.6T$ cal/g, where T is the temperature in degrees Celsius at which the change in state takes place.

In striking a thermal energy balance, it is necessary to consider the energy content of each component of the system. Since it is impossible to evaluate the total energy content of a material, a convenient temperature datum is selected, and the total energy is computed as the sum of the sensible- and latent-heat quantities above the datum. When thermal energy is absorbed into or lost from a system under constant pressure, the energy gain or loss is numerically equal to the *change in enthalpy*.

Conduction

The transfer of heat by *conduction* involves the transmission of energy through molecular motion, the molecules retaining their relative position and transferring only their momentum. The flow of heat from one end of a metal bar to the other provides an example of conductive heat transfer. The rate at which thermal energy is transported by conduction across a unit of area of a material is proportional to the temperature gradient:

$$B_x = -k \frac{\partial T}{\partial x} \quad (1.3.3)$$

where B_x = rate of thermal energy transport in x direction across unit area normal to x. $[Ht^{-1}L^{-2}]$

$\partial T / \partial x$ = temperature gradient in x direction. $[TL^{-1}]$

k = thermal conductivity. $[Ht^{-1}L^{-1}T^{-1}]$

Comparison of Eq. (1.3.3) with Eq. (1.1.2) reveals the similarity between conduction and diffusion. In fact, in aqueous systems thermal conductivity is commonly referred to as *thermal diffusion*.

For the un-steady state, an energy balance can be struck across an element of material in a manner similar to the mass balance expressed by Eqs. (1.1.3) to (1.1.5). This treatment yields

$$\frac{1}{\rho c_p \, dA \, dx} \frac{\partial Q_s}{\partial t} = \frac{\partial T}{\partial t} = \frac{k}{\rho c_p} \frac{\partial^2 T}{\partial x^2} \quad (1.3.4)$$

where ρ = density of conducting material. $[FL^{-3}]$

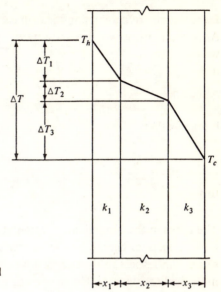

FIGURE 1.3.1
Temperature drop across compound barrier.

If the temperature gradient $\partial T/\partial x$ does not vary with time, the rate of energy transport through the material remains constant, and Eq. (1.3.3) can be written as

$$q = -k \frac{dT}{dx} \qquad (1.3.5)$$

which upon integration becomes

$$q = \frac{k}{x} \Delta T \qquad (1.3.6)$$

where $q =$ flow of thermal energy per unit time. $[Ht^{-1}L^{-2}]$

$\Delta T =$ temperature drop across material. $[T]$

The thermal conductivity is a property of the material through which the energy is being conducted and is found to be a function of its temperature. In general, the variation of conductivity is linear with temperature; hence, an arithmetic average of the conductivities at both sides of the material can be used in Eq. (1.3.6).

Figure 1.3.1 shows a compound barrier composed of three different thicknesses and types of materials. When thermal energy flows by conduction from one side of the barrier to the other, it must pass through each thickness.

Since the rate of conduction through each thickness is equal to the rate through each of the other two, at steady state

$$q = \frac{k_1 \Delta T_1}{x_1} = \frac{k_2 \Delta T_2}{x_2} = \frac{k_3 \Delta T_3}{x_3} \qquad (1.3.7)$$

Also, since

$$\Delta T = \Delta T_1 + \Delta T_2 + \Delta T_3 \qquad (1.3.8)$$

it follows that

$$\Delta T = q \left(\frac{x_1}{k_1} + \frac{x_2}{k_2} + \frac{x_3}{k_3} \right) \qquad (1.3.9)$$

and

$$q = \frac{\Delta T}{x_1/k_1 + x_2/k_2 + x_3/k_3} \qquad (1.3.10)$$

The terms in the denominator represent the individual resistances offered to the flow of thermal energy by the barrier materials. Since these materials are in series with respect to the energy flow, the total resistance equals the sum of the individual resistances. The temperature difference ΔT is the driving force which causes the energy to flow. An analogy can be drawn with Ohm's law governing the flow of electricity through a series of conductors:

$$\text{Current} = \frac{\text{potential difference}}{\text{total resistance}} = \frac{E}{R_1 + R_2 + R_3}$$

In aquatic systems, thermal transport through molecular motion is probably insignificant compared to the transport resulting from eddy currents. Consequently, to be fairly realistic in its representation, the thermal conductivity k for such systems should be, in fact, a dispersion term, much in the same way as E is in the mass transport equation [Eq. (1.1.10)].

Convection

The transmission of thermal energy by mixing or turbulence is called *convection*. Since the molecules of solids are relatively fixed and cannot move with respect to each other, convection is confined to fluids, where it accounts for a major portion of the thermal energy transported. Convection can be either forced or natural, depending on whether it is induced by stirring, agitation, or pumping, or by a gravitational field resulting from natural phenomena. Only natural convection will be considered here.

When considering convection in an air-water system, it is convenient to consider a fictitious viscous film of air as existing at the interface. The flow of thermal energy through the film can be thought of as a conductive process to which the film provides the major resistance to thermal transport. When the temperature gradient remains constant with time, convective heat transport can be represented by

$$q_h = h\, \Delta T \qquad (1.3.11)$$

where h = heat-transfer coefficient. $[Ht^{-1}L^{-2}T^{-1}]$

ΔT = temperature drop across film. $[T]$

Comparison of Eq. (1.3.11) with Eq. (1.3.6) will reveal that the heat-transfer coefficient h is analogous to the term k/x, which is the ratio of the thermal conductivity to the thickness of the material across which thermal energy is being conducted. Such an analogy suggests, therefore, that the heat-transfer coefficient includes not only the internal properties of the viscous air film but, also, its thickness. Any turbulence on the air side of the air film can be thought of as reducing the thickness of the film and, hence, increasing the value of the coefficient. The reader is reminded again that the film concept is a convenience and no evidence is available that such a film exists.

When meteorological data are used to evaluate thermal energy transport across an air-water interface, Eq. (1.3.11) is written as

$$q_h = CU_a(T_a - T_w) \qquad (1.3.12)$$

where q_h = thermal energy flux caused by convection. $[Ht^{-1}L^{-2}]$

U_a = wind velocity. $[Lt^{-1}]$

T_a = dry-bulb temperature of air. $[T]$

T_w = wet-bulb temperature of air. $[T]$

C = constant. $[HL^{-3}T^{-1}]$

Convective thermal energy flux can be either positive or negative depending on the temperature differential between the air and the water.

Evaporation

When water and air are in contact with each other, some of the water molecules have sufficient energy to break through the water surface and escape into the air as vapor. At the same time, some of the water molecules in the air penetrate the water surface to become part of the liquid phase. *Evaporation* is the net rate at which liquid water is transferred to the air. Figure 1.3.2 illustrates the evaporative process in a situation where undersaturated air is replaced continually across a water surface by more air, the air temperature and humidity being constant.

Assuming that the water surface is at the air temperature at the beginning, evaporation reduces initially the sensible heat of the water. At the same time, the resulting difference between the temperature of the air and that of the water

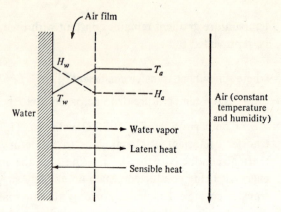

FIGURE 1.3.2
Evaporation from a free-water surface.

causes thermal energy to flow from the air to the water. As evaporation proceeds, the temperature difference becomes larger and larger until a gradient is reached where the flow of thermal energy from the air equals that which is released from the water through vaporization. The temperature, called the *wet-bulb temperature*, from there on will remain constant.

Under steady-state conditions the rate at which thermal energy is removed through evaporation can be found from

$$q_e = w_e \lambda_{T_w} \quad (1.3.13)$$

where q_e = thermal energy flux caused by evaporation. $[Ht^{-1}L^{-2}]$

w_e = weight of water evaporated per unit time. $[Ft^{-1}L^{-2}]$

λ_{T_w} = latent heat of vaporization at wet-bulb temperature. $[HF^{-1}]$

The rate at which water is evaporated can be determined from an expression developed for the concept in which the rate of diffusion of water vapor is controlled by a fictitious air film at the air-water interface:

$$w_e = k_G M_v (\bar{p}_w - \bar{p}_a) \quad (1.3.14)$$

where k_G = mass-transfer coefficient, gas phase. $[mt^{-1}L^{-2}]$

M_v = molecular weight of water. $[Fm^{-1}]$

$\bar{p}_w$ = vapor pressure of water at wet-bulb temperature

$\bar{p}_a$ = partial pressure of water vapor in air

Substituting Eq. (1.3.14) into Eq. (1.3.13) gives

$$q_e = k_G M_v \lambda_{T_w} (\bar{p}_w - \bar{p}_a) \quad (1.3.15)$$

The mass-transfer coefficient, like the heat-transfer coefficient, is influenced by turbulence in the air phase. Several semiempirical relationships have been developed relating the thermal energy flux resulting from evaporation to the wind velocity. These relationships have the form

$$q_e = C\lambda_{T_w} U_a(\bar{p}_w - \bar{p}_a) \quad (1.3.16)$$

where C = constant. $[HL^{-3}]$

U_a = wind velocity. $[Lt^{-1}]$

The constant is generally determined from water-loss studies for the particular situation of interest.

Radiation

Two types of radiation are of interest here: solar (or short-wave) radiation and long-wave radiation. The amount of solar radiation incident on a horizontal surface will vary depending upon the geographic location, the elevation, the season, and the meteorological conditions. Solar radiation intensity observations are made at a number of U.S. Weather Bureau stations throughout the United States. Energy values from the visible portion of solar radiation (4000 to 7000 Å) to be expected at various latitudes in the Northern Hemisphere during the year are presented in Table 1.3.1. Corrections to be made for elevation and cloudiness are indicated at the bottom of the table.

The intensity of short-wave radiation q_{rs} is reduced as it passes through an absorbing medium. The reduction in intensity can be predicted by Beer's law:

$$I = I_0 \exp(-ay) \quad (1.3.17)$$

where I_0 = initial intensity. $[Ht^{-1}]$

a = coefficient of absorption. $[L^{-1}]$

y = distance between points where intensities are I_0 and I. $[L]$

The coefficient of absorption a varies with the wavelength of the radiation and the nature of the absorbing medium. The presence of dissolved and suspended substances in the absorbing medium increases the value of the coefficient.

All materials at temperatures above absolute zero emit energy as long-wave electromagnetic radiation. The radiation, consisting of wavelengths predominately in the infrared region of the spectrum, travels in straight lines until it strikes other materials. Substances receiving radiation either reflect, transmit, or absorb it. That portion of the radiation absorbed is converted to thermal energy. Concurrently, the radiated body is emitting radiation of its own to materials it can "see." A continuous interchange of radiation exists between all bodies in "sight" of each other.

The energy emitted by a plane surface is given by the expression

$$q_{rl} = \varepsilon \sigma T_A^4 \qquad (1.3.18)$$

where q_{rl} = thermal energy flux caused by long-wave radiation. $[Ht^{-1}L^{-2}]$

ε = emissivity

σ = Stefan-Boltzmann constant. $[Ht^{-1}L^{-2}T^{-4}]$

T_A = absolute temperature of body. $[T]$

The emissivity ε is the ratio of the energy emitted by the surface in question to the energy emitted by the surface of a *blackbody*, a hypothetical material which absorbs 100 percent of the incident radiation. Since the value of q_{rl} is proportional to the fourth power of the absolute temperature, a small increase in surface temperature greatly increases the amount of radiation emitted.

Table 1.3.1 PROBABLE VALUES OF VISIBLE SOLAR ENERGY AS A FUNCTION OF LATITUDE AND MONTH*

Latitude		Jan.	Feb.	Mar.	Apr.	May	June	July	Aug.	Sept.	Oct.	Nov.	Dec.
0	max	255†	266	271	266	249	236	238	252	269	265	256	253
	min	210	219	206	188	182	103	137	167	207	203	202	195
10	max	223	244	264	271	270	262	265	266	266	248	228	225
	min	179	184	193	183	192	129	158	176	196	181	176	162
20	max	183	213	246	271	284	284	282	272	252	224	190	182
	min	134	140	168	170	194	148	172	177	176	150	138	120
30	max	136	176	218	261	290	296	289	271	231	192	148	126
	min	76	96	134	151	184	163	178	166	147	113	90	70
40	max	80	130	181	181	286	298	288	258	203	152	95	66
	min	30	53	95	125	162	173	172	147	112	72	42	24
50	max	28	70	141	210	271	297	280	236	166	100	40	26
	min	10	19	58	97	144	176	155	125	73	40	15	7
60	max	7	32	107	176	249	294	268	205	126	43	10	5
	min	2	4	33	79	132	174	144	100	38	26	3	1

* After W. J. Oswald and H. B. Gotaas, Photosynthesis in Sewage Treatment, table II, *ASCE*, vol. 81, separate no. 686, May 1955.
† Values of S in Langleys, cal/(cm²) (d)

Correction for cloudiness:

$$S_c = S_{\min} + r(S_{\max} - S_{\min})$$

where r = total hours sunshine/total possible hours sunshine

Correction for elevation up to 10,000 ft:

$$S_c = S(1 + 0.01e)$$

where e = elevation in feet

The net loss in long-wave thermal energy by a water surface is equal to the difference between the energy emitted by the surface and the energy absorbed from the atmosphere:

$$q_{rl} = \sigma(\varepsilon T_A^4 - \alpha T_{A_a}^4) \qquad (1.3.19)$$

where α = absorptivity of atmosphere

T_{A_a} = absolute temperature of air

Thermal Regimes

Two types of thermal regimes result from the interactions between surface waters and meteorological parameters: the completely mixed, or *homogeneous* regime, and the stratified, or *heterogeneous* regime. The homogeneous regime, which occurs in flowing streams, is characterized by the fact that no vertical temperature gradient exists. For such regimes, the thermal energy budget accounting for natural phenomena is

$$q_n = q_{rs} - q_e - q_{rl} \pm q_h \qquad (1.3.20)$$

where q_n = net energy flux passing air-water interface

q_{rs} = short-wave radiation flux passing through interface after absorption and scattering in atmosphere and reflection at interface

q_e = energy loss by evaporation

q_{rl} = net long-wave radiation flux between water and atmosphere

q_h = convective energy (sensible heat) flux between water and atmosphere

The value of the net energy flux can vary from positive to negative depending on the periodicity of certain terms in the equation. Methods for evaluating all the terms in the budget can be found elsewhere.[1,2]

The time rate change of temperature in a segment of a flowing stream can be formulated from a thermal energy balance struck across the segment.

---

[1] Water Resources Engineers, Inc., "Prediction of Thermal Energy Distribution in Streams and Reservoirs, Final Report," prepared for the Dep. of Fish and Game, state of Calif., 1968.

[2] J. C. Ward and S. Karaki, Evaluation of the Effect of Impoundment on Water Quality in Cheney Reservoir, *Sanit. Eng. Pap.* 4, Colorado State University, 1969.

Assuming a regime completely mixed in all directions,

Accumulation = energy in − energy out

$$V\gamma c_p \frac{dT}{dt} = \gamma c_p QT_0|_{\text{in}} + R - \gamma c_p QT_0|_{\text{out}} \pm q_n A \qquad (1.3.21)$$

where dT/dt = temperature change in segment. $[Tt^{-1}]$

V = volume of segment. $[L^3]$

Q = volumetric flow rate through segment. $[L^3 t^{-1}]$

A = air-water surface area. $[L^2]$

γ = specific weight of water. $[FL^{-3}]$

R = thermal discharge into segment. $[Ht^{-1}]$

When the time of flow in the segment is relatively short and there is no volumetric or thermal discharge into the segment, Eq. (1.3.21) reduces to

$$\frac{dT}{dt} = \frac{A}{V\gamma c_p} q_n \qquad (1.3.22)$$

indicating that temperature changes in the water are a function of meteorological conditions only. Any loss or gain of thermal energy through the stream bed is ignored in Eqs. (1.3.21) and (1.3.22).

In the absence of mixing, a stratified regime will occur as a result of the internal transfer of thermal energy from the air-water interface. The energy loss by evaporation q_e, the convective energy flux q_h, and the net long-wave radiation flux q_{rl} are essentially surface phenomena in the sense that they increase, or decrease, directly only the thermal energy of the water at the surface. The thermal energy of the water at the surface resulting from these phenomena is then transferred to the water below by conduction. On the other hand, short-wave radiation penetrates the surface water and increases directly the thermal energy of the underlying water. The net effect of thermal conductivity and short-wave radiation are temperature gradients which create a stratified regime in the water.

The mathematical expression for the time rate change of temperature at various depths in a stratified regime is

$$\frac{\partial T}{\partial t} = \frac{k}{\gamma c_p} \frac{\partial^2 T}{\partial y^2} + \frac{1}{\gamma c_p} \frac{\partial q_n}{\partial y} \qquad (1.3.23)$$

where y = depth below surface. $[L]$

The first term to the right of the equals sign is a thermal-conduction term commonly referred to as thermal diffusivity. The second term quantifies the

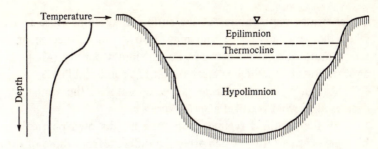

FIGURE 1.3.3
Summer thermostratification in lakes.

effect of the net energy flux passing the air-water interface. It must be kept in mind, however, that the only component of this flux that extends below the air-water interface is the short-wave radiation. Equation (1.3.23) can be solved by numerical integration.[1]

A stratified regime typical of that occurring in lakes located in the temperate zones of the world is illustrated by Fig. 1.3.3. Thermal stratification results in the formation of three zones, each with its own temperature characteristics. In the summer, the *epilimnion*, which is located next to the surface, is characterized by a relatively high and uniform water temperature. Because of the uniform temperature, the water density in the epilimnion is uniform, and the wind movement at the surface tends to keep the water in a mixed condition. The epilimnion is often referred to as the *zone of circulation*.

Below the epilimnion is located the *thermocline*, or *zone of transition*, characterized by a large temperature gradient. This zone is relatively small in vertical dimension and separates the epilimnion from the *hypolimnion*, or *zone of stagnation*. The latter is characterized by a relatively uniform temperature and the absence of significant circulation. In the stratified regime occurring in the summer, the coldest water, and hence the densest water, is in the hypolimnion, a situation imparting considerable stability to the regime. In early summer the thermocline is located close to the surface, but as the summer progresses, it migrates to greater depths.

In late fall or early winter, low air temperatures destroy stratification. As the air temperatures drop, the temperature of the water in the epilimnion drops to a level where the water density is greater than that in the hypolimnion. The unstable condition that results causes the *fall turnover*. This effect is generally triggered by wind movements and occurs rather abruptly. During the

---

[1] J. C. Sonnichsen, Jr. and C. A. Oster, Examination of the Thermocline, *J. Sanit. Eng. Div., ASCE*, vol. 96, no. SA2, pp. 353–364, 1970.

winter, water temperatures at the surface may come close to freezing (0°C), while the water at the bottom may drop to only a degree or two from the temperature at which water has its maximum density (4°C). In such cases, another stratified regime is created that has some stability.

As spring brings warmer air temperatures, the winter stratification is destroyed in what is called a *spring turnover*.

In the stratified regime, each zone has its own physical, chemical, and biologic characteristics which are quite different from those of the other zones. The properties of the epilimnion reflect its location with respect to the air-water interface and are greatly influenced by meteorological and atmospheric factors. On the other hand the hypolimnion has properties that reflect its isolation from such factors. Only during the time of turnover do the properties of the water become more or less uniform.

Note should be made here of the fact that thermal stratification on a seasonal basis does not ordinarily occur in ponds, shallow lakes, or bodies of water where the advective flow in and out is great enough to establish circulation patterns that destroy stratification.

Solution procedure 1.3.1 The steady-state form of Eq. (1.3.21) is useful in making energy balances in environmental situations. As an example, consider a cooling pond through which a waste water passes prior to its discharge into a stream. Assuming steady state, the thermal energy entering the pond must equal that which leaves:

$$\gamma c_p Q T_0|_{in} = \gamma c_p Q T_0|_{out} - q_n A \quad (1.3.24)$$

Since the temperature datum can arbitrarily be established at the temperature of the discharge from the pond, the first term to the right of the equals sign becomes 0. From an estimate of the average net energy flux passing the air-water interface, the required surface area of the pond can be determined.

Another use of Eq. (1.3.21) can be made in determining the temperature of water in a stream after receiving a discharge of a heated waste water. Assuming that mixing takes place rapidly and that steady-state conditions prevail, over a small section of the stream immediately below the discharge

$$\gamma c_p Q T_0|_{in} + R = \gamma c_p Q T_0|_{out} \quad (1.3.25)$$

Since we can express the waste discharge into the section as

$$R = \gamma c_p Q T_0|_{waste} \quad (1.3.26)$$

Eq. (1.3.25) can be reduced to

$$QT_0|_{in} + QT_0|_{waste} = QT_0|_{out} \quad (1.3.27)$$

Selecting 0°C as the temperature datum, Eq. (1.3.27) can be rearranged to solve for the temperature of the stream water after it has been mixed with the heated discharge:

$$T|_{out} = \frac{QT|_{in} + QT|_{waste}}{Q|_{in} + Q|_{waste}} \quad (1.3.28)$$

////

1.4 SEDIMENTATION

Sedimentation is a process in which a dispersed phase separates by gravity from a dispersing phase of a lighter density. In the environment, the dispersed phase of usual concern is a solid, whereas the dispersing phase is either air or water. Regardless of the nature of the phases, the same basic principles apply.

The classical laws of sedimentation apply to the settling of discrete, nonflocculating particles in dilute suspensions. Sedimentation of a particle from such a suspension is unhindered by the presence of other settling particles and is a function only of the properties of the fluid and the particle in question.

A discrete, nonflocculating particle settling through a quiescent fluid under the influence of gravity will accelerate until a velocity is reached (the *terminal settling velocity*) from which point on the acceleration is 0 and the velocity remains constant. In environmental systems, the terminal velocity is quickly reached.

A particle settling through a quiescent fluid at its terminal velocity is at balance between the effective gravitational force and a frictional force. For spherical particles, the effective gravitational force is

$$F_G = m \frac{g}{g_c} = \frac{\pi(\rho_s - \rho)D_p^3}{6} \frac{g}{g_c} \quad (1.4.1)$$

where F_G = effective gravitational force. $[F]$

m = mass of sphere. $[M]$

g = acceleration of gravity. $[Lt^{-2}]$

g_c = Newton's law conversion factor. $[LMF^{-1}t^{-2}]$

ρ_s, ρ = mass densities of particle and fluid. $[ML^{-3}]$

D_p = particle diameter. $[L]$

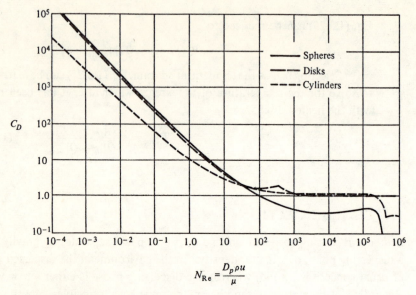

FIGURE 1.4.1
Drag coefficient for spheres, disks, and cylinders. [*By permission C. E. Lapple, from J. H. Perry (ed.), "Chemical Engineers Handbook," copyright, McGraw-Hill, New York,* 1950.]

The frictional force on a settling particle is a function of the roughness, size, shape, and velocity of the particle, as well as of the density and viscosity of the fluid. Experimentally, it has been found that the relationship is

$$F_F = \frac{C_D A_p \rho u^2}{2g_c} = C_D \frac{\pi D_p^2}{4} \frac{\rho u^2}{2g_c} \quad (1.4.2)$$

where F_F = frictional force. $[F]$

C_D = coefficient of drag

A_p = projected area of particle. $[L^2]$

u = terminal settling velocity of particle. $[Lt^{-1}]$

Equating the expressions given by Eq. (1.4.1) and (1.4.2) and solving for the terminal settling velocity yields

$$u = \left(\frac{4}{3} \frac{g}{C_D} \frac{\rho_s - \rho}{\rho} D_p \right)^{1/2} \quad (1.4.3)$$

The coefficient of drag C_D is a function of the Reynolds number N_{Re}. A correlation between the two is given in Fig. 1.4.1. It is to be noted that the

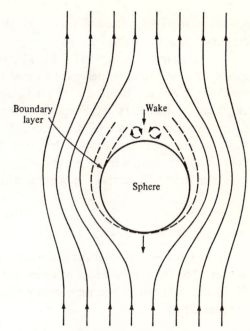

FIGURE 1.4.2
Fluid flow around a sphere.

correlation varies with the geometric shape of the particle. The relationship between C_D and N_{Re} can be interpreted in light of the flow characteristics exhibited by the fluid through which a particle is settling.

Consider Fig. 1.4.2. Here, a spherical particle is settling through a fluid. The solid lines, called *streamlines*, represent paths followed by fluid elements as they flow around the settling particle. At Reynolds numbers less than 0.1, in a region often referred to as the *Stoke's law* range, the fluid elements will pass smoothly over the surface of the sphere, leaving no wake. Resistance in this range of Reynolds numbers is composed almost entirely of *skin friction* (viscosity shear stresses). At larger values of the Reynolds number ($N_{Re} > 0.1$), the boundary layer separates from the rear surface of the sphere, enclosing a relatively stable wake. A transition occurs in the region between $N_{Re} = 0.1$ and 10^3, in which *form drag* contributes more and more to the total resistance on the particle. Form drag results from pressure differences caused by the acceleration of the fluid flowing around the sphere and from the high velocities of the turbulent eddies in the wake.

As the Reynolds number increases beyond 10^3, the point of boundary-layer separation moves to a position upstream from the midsection of the sphere, and the wake becomes unstable, shedding its contents into the main stream. From $N_{Re} = 10^3$ to 2.5×10^5, in a region often referred to as the *Newton's law* range, form drag accounts for almost all the drag on the particle, and the drag coefficient is roughly constant.

As N_{Re} increases beyond 2.5×10^5, the boundary layer becomes turbulent, first in the separated portion and then in that part which remains attached. When the boundary layer becomes completely turbulent, the point at which the boundary layer separates from the sphere shifts to a position downstream from the midsection. The shift is accompanied by a reduction in wake, which results in a drop in drag. The sketch in Fig. 1.4.2 illustrates the flow pattern that exists in the range of $N_{Re} = 0.1$ to 10^3.

The Reynolds number is computed with

$$N_{Re} = \frac{D_p \rho u}{\mu} \quad (1.4.4)$$

where μ = absolute viscosity. $[ML^{-1}t^{-1}]$

Of the two fluid properties ρ and μ, the latter is far more sensitive to temperature change. For the normal range of water temperature in the environment (0 to 30°C) the temperature-viscosity relationship for water can be expressed approximately by

$$\frac{\mu}{\mu_0} = \theta^{(T_0 - T)} \quad (1.4.5)$$

where μ, μ_0 = viscosities at temperatures T and T_0, respectively

θ = temperature coefficient[1]

Equation (1.4.3) can be used to determine the settling velocity given the diameter of the settling particle or, conversely, to determine the diameter from a knowledge of the settling velocity. Neither computation is straightforward, because both the velocity and diameter terms are included in the Reynolds number, which, in turn, is used to determine the coefficient of drag. This difficulty may be minimized by substituting the Reynolds number term for the velocity and diameter terms in Eq. (1.4.3) and rearranging to give two expressions, one of which has no velocity term and the other no diameter term:

$$C_D N_{Re}^2 = \frac{4}{3} g \frac{\rho(\rho_s - \rho)}{\mu^2} D_p^3 \quad (1.4.6)$$

$$\frac{C_D}{N_{Re}} = \frac{4}{3} g \frac{\mu(\rho_s - \rho)}{\rho^2} \frac{1}{u^3} \quad (1.4.7)$$

---

[1] When the temperatures T_0 and T are expressed in degrees Celsius, the temperature coefficient is equal to 1.029. At 10°C the viscosity of water is 1.310 cP [1 cP = 10^{-2} g/(cm)(s)]. Within the range of 0 to 30°C the density of water is approximately 1 g/cm³.

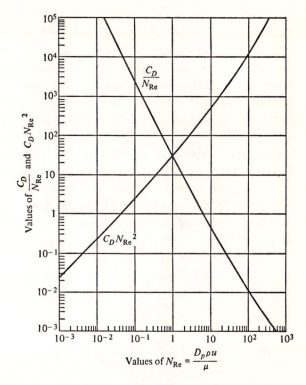

FIGURE 1.4.3
Functions of N_{Re} and C_D as a function of N_{Re}. (*Taken from T. R. Camp, Sedimentation and the Design of Settling Tanks, fig. 2, Trans. ASCE, vol.* 111, *p.* 898, 1946.)

The term on the left-hand side of the appropriate expression may be computed from given data, and the corresponding value of the Reynolds number found from Fig. 1.4.3. The Reynolds number is then used to compute either the velocity or the diameter, depending upon the particular parameter sought. The plot in Fig. 1.4.3 is a replot of the curve for spheres in Fig. 1.4.1.

Sedimentation in Streams

The preceding discussion dealt with the settling of discrete, nonflocculating particles in a quiescent fluid. In the environment, however, not all particulates settle as discrete particles. Some suspensions exhibit flocculating characteristics. Particles subsiding from such suspensions overtake and coalesce with smaller particles to form particles which settle at rates higher than did the parent particles. Furthermore, completely quiescent conditions rarely exist in nature.

As water flows or as turbulence is introduced through environmental factors such as wind, temperature changes, etc., velocity gradients are established within the body of the water which influence sedimentation rates. Although attempts have been made to quantify these effects, little success has been attained in formulating relationships that have been shown to have validity when applied to the natural environment.

Flow velocity exercises the greatest influence over sedimentation in flowing waters. From experience it is known that if flow velocity is greater than about 0.50 m/s, small sand particles are kept in suspension. Velocities as low as 0.20 m/s have been shown to keep organic solids in suspension. In fact, too-high velocities will scour previously settled deposits and re-form suspensions. For organic deposits, the velocity of scour is on the order of 0.30 to 0.50 m/s, depending on the compaction of the deposit.

When attention is focused on the rates at which suspended organic material is removed from a polluted river, use is sometimes made of the rate expression

$$\frac{dL}{dt} = k_3 L \qquad (1.4.8)$$

where L = concentration of suspended organic material. $[FL^{-3}]$

k_3 = sedimentation constant. $[t^{-1}]$

Equation (1.4.8) states that the rate of removal is proportional to the concentration remaining to be removed. The relationship is applied to that portion of the pollution load that is removed by either, or both, sedimentation or adsorption on bottom slimes.

Solution procedure 1.4.1 The engineer is often called upon to evaluate the potentialities for the accumulation of sludge deposits formed by organic materials settling out of a river. Time-of-passage curves provide a practical means for making this type of an evaluation. From information concerning the physical characteristics of the river channel, curves can be established showing the relationship between the cumulative time of passage and the flow distance. Such curves, plotted with English units, are shown in Fig. 1.4.4. The reaches of river where the flow velocity falls below the critical velocity (0.20 m/s) can be identified by the slope of the time-of-passage curve. Each flow has its own time-of-passage curve. A similar investigation using slopes of 0.30 to 0.50 m/s gives the velocity levels between which scour can be expected. The procedure lends itself to computer application. ////

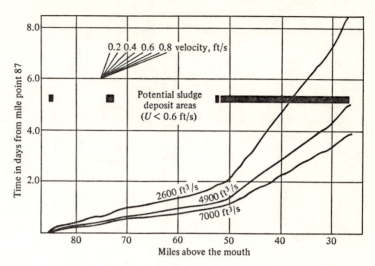

FIGURE 1.4.4
Time-of-passage curves for three runoff levels on the Miami River. (*Taken from C. J. Velz, Significance of Organic Sludge Deposits, fig. 2, "Oxygen Relationships in Streams," Tech. Rep. W58-2, Robert A. Taft Sanit. Eng. Cent., U.S. Public Health Serv., U.S. Dep. HEW, 1958.*)

1.5 CONTINUOUS-FLOW MODELS

Most environmental phenomena are time-dependent, and, as a consequence, such factors as residence time and time of flow are important in the consideration of environmental systems. The determination of residence time and time of flow in a system is facilitated by a study of flow models.

Conceptually, flow in continuous-flow systems approaches either *plug flow* or *completely mixed flow*.[1] Both of these conceptual models can be defined in terms of the behavior of individual increments of flow entering the system. In plug flow, increments of influent pass through the system and are discharged in the same sequence in which they enter. No mixing takes place. This type of flow is approximated in systems which are long and have small cross sections, such as pipes and rivers.

In completely mixed flow, increments of the influent upon entering the system are dispersed uniformly throughout the system. The fluid in the system is completely mixed, and its properties are uniform. The properties of the discharge stream are identical to those of the contents of the system. The completely mixed model is approximated in mixing vessels.

---

[1] P. V. Danckwerts, Continuous Flow Systems, *Chem. Eng. Sci.*, vol. 2, no. 1, 1953.

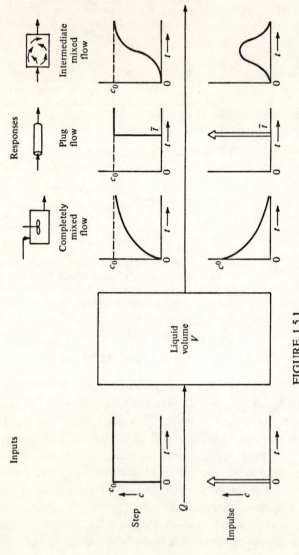

FIGURE 1.5.1
Time-domain, residence-time distribution curves for continuous-flow models.

It is to be emphasized that the two flow models, plug flow and completely mixed flow, are limiting cases. Actual flow patterns range in a broad spectrum between these idealized models.

Each model is characterized by the residence-time distribution of the elements constituting its discharge stream. The identification of such distributions is facilitated by the use of residence-time distribution (RTD) curves resulting in response to tracer inputs to the flow system. Time-domain RTD curves illustrating the response of the completely mixed flow, plug flow, and intermediate mixing flow models to step and impulse tracer inputs are shown in Fig. 1.5.1. Normalized RTD curves for the same system responses are shown in Fig. 1.5.2. Here in a dimensionless plot the ratio of the discharge concentration to the initial concentration is plotted as a function of the ratio of the time of observation to the mean residence time of the system, θ. For step inputs to all three flow systems, the initial concentration c_0 is the concentration of the tracer in the entering stream. For the impulse inputs, the initial concentration is the concentration of the tracer c^0 taken as if it were evenly distributed throughout the system. The normalized RTD curves resulting from step inputs are called *F curves*. Those from impulse inputs are commonly referred to as *C curves*.

The C curve which results from an impulse input to the completely mixed flow model has the mathematical description

$$C = \frac{c}{c^0} = e^{-\theta} = e^{-t/\bar{t}} \qquad (1.5.1)$$

The *F* and *C* curves are actually probability distribution curves, the former being cumulative distribution curves whereas the latter are their derivatives. Both types of curves are used to determine the fraction of the discharge stream which consists of elements having a residence time between θ and $\theta + d\theta$. For the C curve

$$\int_0^\infty C(\theta) \, d\theta = 1 \qquad (1.5.2)$$

or, expressed in the ordinary time domain,

$$\int_0^\infty \bar{t} C(t) \frac{dt}{\bar{t}} = \int_0^\infty C(t) \, dt = 1 \qquad (1.5.3)$$

It should be emphasized that these relationships between tracer curves and residence-time distributions hold strictly only in those cases where there is no back diffusion of the tracer in the input and discharge streams. However, they are useful for studies involving reactors, tanks, and small impoundments, such as lagoons and ponds.

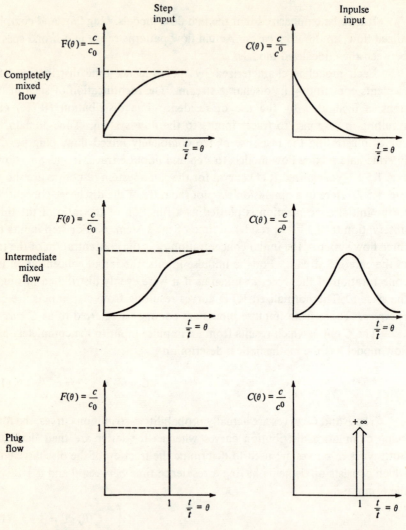

FIGURE 1.5.2
Normalized residence-time distribution curves for continuous-flow models.

The mean residence time can be determined from the first moment of the C curve about the origin:

$$\bar{t} = \frac{\text{moment}}{\text{area}} = \frac{\int_0^\infty tC(t)\,dt}{\int_0^\infty C(t)\,dt} = \frac{\int_0^\infty tc(t)\,dt}{\int_0^\infty c(t)\,dt}$$

$$\approx \frac{\sum_{i=0}^n t_i c_i \Delta t_i}{\sum_{i=0}^n c_i \Delta t_i} \qquad (1.5.4)$$

The measurement of the mean residence time of a system can be performed by using a stimulus-response technique employing a tracer injection into the input stream. Tracers employed for such purposes include colored dyes, radioactive compounds, and salt solutions. The tracer material can be injected as an instantaneous pulse (an impulse) and the response is measured with time. The observed data are used with Eq. (1.5.4) to calculate the mean residence time. When the time between samples Δt is constant, Eq. (1.5.4) can be reduced to

$$\bar{t} = \frac{\sum_{i=0}^n t_i c_i}{\sum_{i=0}^n c_i} \qquad (1.5.5)$$

Use of Eqs. (1.5.4) and (1.5.5) requires that the concentration return to 0; that is, $c_n = 0$.

The theoretical residence time in a continuous-flow system is equal to the ratio of the volume and the volumetric flow rate V/Q. When $\bar{t}/(V/Q) < 1$, short circuiting exists in the system. The ratio $\bar{t}/(V/Q)$ can be used as a measure of the degree of short circuiting.

EXERCISES [1]

1.1.1 A tracer material is released into a stream at point A continuously and at a steady rate. If the flow velocity in the stream is 0.3 m/s and the dispersion coefficient is 30 m²/s, what will be the concentration (in terms of the concentration at A) 1,500 m downstream from A? What will be the concentration if dispersion is negligible?

1.1.2 Ten kg of a tracer are released instantaneously into a stream at point A. What will be the profile of the tracer cloud 5,000 m downstream from point A if the flow velocity of the stream is 0.5 m/s, the dispersion coefficient is 50 m²/s, and the cross-sectional area at point A is 20 m²?

[1] The first and second digits in the exercise number refer to the section number.

1.1.3 A batch of liquid waste containing 170,000 kg of salt is to be pumped to the bottom of a lake 100 m deep. The lake has almost vertical sides and an average area normal to the vertical direction of 1 km². Using the information contained in Fig. 1.1.2, estimate what the concentration of salt will be at the surface of the lake 1 year later.

1.1.4 The following tracer data were collected at two locations during a river study.[1] From this information, estimate the coefficient of longitudinal dispersion for the reach of river between the two sampling points.

Section 4—2,400 m from release			Section 6—4,150 m from release		
Time	Time relative to first observation	Concentration	Time	Time relative to first observation	Concentration
11:49.0 A.M.	0.0	0.0	12:26.0 P.M.	37.0	0.0
11:52.0	3.0	0.26	12:31.0	42.0	0.07
11:55.0	6.0	0.67	12:36.0	47.0	0.22
11:58.0	9.0	0.95	12:41.0	52.0	0.40
12:00.0	11.0	1.09	12:45.0	56.0	0.50
12:02.0 P.M.	13.0	1.13	12:49.0	60.0	0.58
12:04.0	15.0	1.10	12:51.0	62.0	0.59
12:06.0	17.0	1.04	12:53.0	64.0	0.59
12:08.0	19.0	0.95	12:57.0	68.0	0.54
12:13.0	24.0	0.72	1:04.0	75.0	0.44
12:18.0	29.0	0.50	1:13.0	84.0	0.27
12:23.0	34.0	0.31	1:23.0	94.0	0.14
12:28.0	39.0	0.21	1:33.0	104.0	0.06
12:38.0	49.0	0.08	1:43.0	114.0	0.03
12:48.0	59.0	0.02	1:53.0	124.0	0.025
			2:03.0	134.0	0.02

1.2.1 At sea level, air contains 0.03 percent carbon dioxide by volume. What will be the saturation concentration of carbon dioxide (in mg/l) in water in contact with air at 1 atm and 20°C?

1.2.2 An outdoor spray aeration device is used to reduce the concentration of carbon dioxide in well water from 30 to 10 mg/l at a temperature of 20°C. The device creates a drop with an average diameter of 4 mm and provides an average exposure time of 3 s. What is the value of the mass-transfer coefficient?

1.2.3 An aeration device was found to increase the dissolved oxygen in water at a rate given by the following data:

Aeration time, min	0	10	20	30	40	50	60
Dissolved oxygen, mg/l	1.0	3.0	4.5	5.7	6.7	7.3	7.7

If the water was at a temperature of 20°C, what is the value of the overall volumetric mass-transfer coefficient $K_L a$?

---

[1] R. A. Godfrey and B. J. Frederick, Dispersion in Natural Streams, *U.S. Geological Survey Open-file Report*, 1963.

1.2.4 If pure oxygen were used in Exercise 1.2.3 instead of air, what would be the effect on the overall volumetric mass-transfer coefficient? What would be the effect on the rate of oxygen transfer?

1.3.1 How much thermal energy will be required to convert 1 l of water initially at 20°C to steam? How much will be required to evaporate 1 l of water at 20°C?

1.3.2 An insulation system consists of three layers of materials. The characteristics of these layers are as follows:

Material	Thickness, cm	Thermal conductivity at 150°C, cal/(h)(m)(°C)
Asbestos fiber	1	29
Glass fiber	2	6
Wood	4	8

What will be the thermal energy flow if a temperature drop of 100°C is maintained across the system? What percent change in energy flow will result if the thicknesses of the insulation layers are doubled?

1.3.3 The wet- and dry-bulb air temperatures over a water surface were determined to be 18 and 35°C, respectively. If the convective transport constant C is 1.6×10^4 cal/(m^3)(°C), estimate the thermal energy flux caused by convection when the average wind velocity is 0.2 m/s.

1.3.4 The water vapor pressure at a height of 2 m above a water surface was found to be 13.3 mbar, whereas the saturation vapor pressure of the air at the temperature of the water surface was determined to be 11.3 mbar. If the average wind velocity is 0.3 m/s, the latent heat of vaporization is 612 cal/g, and the thermal energy flux is -8 cal/(cm^2)(s), what is the value of the empirical evaporation constant?

1.3.5 A cooling pond is to be used to reduce the temperature of a waste water before it is discharged into a stream. The maximum temperature of the water in the pond is to be 30°C, at which temperature the average net thermal flux at the pond surface is estimated to be -5 cal/(cm^2)(h). If the waste flow is to be 100 m^3/h with a temperature of 60°C, how large a surface must the pond have? How deep must it be?

1.3.6 A waste water having a flow rate of 1 m^3/s and a temperature of 65°C is discharged into a stream with a flow rate of 10 m^3/s and a water temperature of 15°C. What will be the temperature in the stream below the point of discharge?

1.4.1 What is the subsidence velocity in water (15°C) of a spherical particle with a diameter of 0.1 mm and a specific gravity of 2.65? What percent increase in subsidence velocity will result if the water temperature is increased to 20°C?

1.4.2 A suspension of spherical particles, uniform in size with a particle diameter of 1 mm and a specific gravity of 2.65, is clarified in a vertical-flow sedimentation basin. At what rate can the clarified water be removed [in m^3/(s)(m^2)] from the top of the basin? Assume the temperature of the suspension to be 20°C.

1.4.3 The length and cross-sectional characteristics of a stream are recorded below. Using time-of-passage curves, such as those shown in Fig. 1.4.4, locate the potential sludge deposit points along the length of the stream for flows of 10, 20, and 30 m³/s.

Reach number	Length, km	Average cross section, m²
56	2.1	38
57	1.8	40
58	2.1	48
59	2.2	56
60	1.9	67
61	2.4	75
62	2.4	85
63	2.5	94
64	2.0	106
65	2.3	125
66	1.9	132
67	2.8	148
68	2.6	152
69	3.0	160
70	2.9	175

1.5.1 An impulse dose of a radiochemical tracer was injected into the input stream of a small mixing basin. Response data collected from the discharge were as follows:

Time, min	0.5	1.0	1.5	2.0	2.5	3.0	3.5	4.0	4.5	5.0	5.5	6.0	6.5	7.0
Net counts per min $\times 10^{-3}$	15	30	23	17	13	8	6	5	4	3	2	2	1	0

Determine the residence-time distribution and the mean residence time for the basin.

1.5.2 Using the data in Exercise 1.5.1, determine the residence-time distribution and the mean residence time if $\Delta t = 0.5$ min for the first four time periods and $\Delta t = 1$ min for the remaining periods.

2
CHEMICAL PHENOMENA

2.1 SOLUTION EQUILIBRIUMS

Most chemical reactions taking place in solutions are incomplete and involve the establishment of equilibriums between the reactants and products. Consider a situation wherein materials A and B react to form G and H. At equilibrium

$$mA + nB \rightleftharpoons pG + qH \qquad (2.1.1)$$

where A, B, G, and H = molecular or ionic species

m, n, p, and q = coefficients used to balance equation

From the *law of mass action*, it can be shown that

$$\frac{[a_G]^p [a_H]^q}{[a_A]^m [a_B]^n} = K \qquad (2.1.2)$$

where a_A, a_B, a_G, and a_H = activities of A, B, G, and H, mol/l

K = activity equilibrium constant

The value of the activity equilibrium constant is a characteristic of the specific equilibrium and will vary only with the temperature of the solution.

The activity of a solution is related to the concentration by

$$a = \gamma c \quad (2.1.3)$$

where c = concentration, mol/l

γ = activity coefficient

The value of the activity coefficient for most solutions of nonelectrolytes is close to unity. For solutions of electrolytes, the activity coefficient can be approximated using the expression

$$\log \gamma = -0.5 z^2 \frac{\sqrt{\mu}}{1 + \sqrt{\mu}} \quad (2.1.4)$$

where z = magnitude of ion's charge

μ = ionic strength

Equation (2.1.4) holds for solutions with ionic strengths up to 0.1 and for temperatures ranging from 0 to 40°C.

The ionic strength of a solution is defined as

$$\mu = 0.5 \sum_{i=1}^{n} c_i z_i^2 \quad (2.1.5)$$

where c_i = concentration of ith species of ion, mol/l

z_i = magnitude of charge on ith species of ion

When the ionic composition of water is unknown and the dissolved solids content is less than 500 mg/l, the ionic strength can be estimated with the empirical relationship[1]

$$\mu = 2.5 \times 10^{-5} c_s \quad (2.1.6)$$

where c_s = dissolved solids concentration, mg/l

When solutions are sufficiently dilute, that is, when the ionic strength is very low, the activity coefficient is approximately unity, and concentration terms can be used in Eq. (2.1.2).

Weak acids and weak bases in water dissociate in part to form ions. The equilibriums existing between the undissociated molecules and the ions are

$$H_m A_n \rightleftharpoons mH^+ + nA^- \quad (2.1.7)$$

and

$$B_m(OH)_n \rightleftharpoons mB^+ + nOH^- \quad (2.1.8)$$

where A^-, B^+ = anions and cations, respectively

---

[1] G. M. Fair and J. C. Geyer, "Water Supply and Waste-water Disposal," p. 468, Wiley, New York, 1954.

The equilibrium expressions for these dissociations are

$$\frac{[H^+]^m[A^-]^n}{[H_mA_n]} = K \quad (2.1.9)$$

and

$$\frac{[B^+]^m[OH^-]^n}{[B_m(OH)_n]} = K \quad (2.1.10)$$

The equilibrium constants in Eqs. (2.1.9) and (2.1.10) are called *ionization* or *dissociation constants*.

When dealing with the dissociation of water

$$HOH \rightleftharpoons H^+ + OH^- \quad (2.1.11)$$

a modified equilibrium constant is used:

$$[H^+][OH^-] = K[HOH] = K_w \quad (2.1.12)$$

where K_w = activity ion product of water

The activity of the undissociated water is so great as compared to the activities of the ions that it can be treated as a constant. In pure water or in any dilute solution at 25°C, the ion product of water is 1.00×10^{-14}.

Modified equilibrium expressions are also used to describe solubility relationships of slightly soluble salts. For example, consider the equilibrium existing in a saturated solution between the undissolved salt B_mA_n and its ions:

$$B_mA_n(s) \rightleftharpoons mB^+ + nA^- \quad (2.1.13)$$

The activity of the undissolved salt is considered constant, and the equilibrium expression is written as

$$[B^+]^m[A^-]^n = K[B_mA_n](s) = K_s \quad (2.1.14)$$

where K_s = activity solubility product

If the ionic activities in a solution are of a magnitude that the ion product in Eq. (2.1.14) (the term on the left-hand side of the equation) is less than the activity solubility product K_s, then the salt B_mA_n dissolves either until the total quantity of the salt in the solid state dissolves or until the ionic activities reach a magnitude where the ion product reaches K_s, whichever occurs first. On the other hand, if the ionic activities are such that the ion product is greater than K_s, the salt B_mA_n precipitates out of solution until the ion product and K_s become equal.

Consider a situation in which a solution initially is undersaturated with respect to the salt B_mA_n, that is, in which all the salt is dissolved and exists in an ionic state. If to this solution is added another solute C_pA_q containing a common ion A^-, the ion product in Eq. (2.1.14) $[B^+]^m[A^-]^n$ will be increased proportionately. If C_pA_q is added in quantities such that the activity of the common ion A^- increases the ion product until it equals the solubility product K_s, saturation occurs with respect to B_mA_n. Further addition of C_pA_q increases the activity of A^- and causes precipitation of B_mA_n with a corresponding decrease in the activity of B^+. This decrease in the solubility of an ionized salt as a result of adding one of the ionic constituents of the salt to the solution is called the *common ion effect*. The common ion effect is most demonstrable on slightly soluble salts. This effect plays a key role in the removal of ions from solution by the addition of chemicals.

The solubility concentrations of slightly soluble salts are increased by a phenomenon commonly referred to as a *secondary salt effect*. Attention is directed to Eq. (2.1.3). In concentrated solutions of electrolytes, the ionic strength is high, and, hence, the activity coefficient is low. Consequently, if for such solutions the activity equilibrium expression for the salt B_mA_n is to hold [Eq. (2.1.14)], then the solubility concentration has to increase.

When concentration terms are used in an equilibrium expression, a concentration equilibrium constant is substituted for the activity equilibrium constant:

$$\frac{[c_G]^p[c_H]^q}{[c_A]^m[c_B]^n} = \frac{(\gamma_A)^m(\gamma_B)^n}{(\gamma_G)^p(\gamma_H)^q} K = K' \qquad (2.1.15)$$

Ionic strength effects in the fresh-water environment are usually not great enough to be of practical significance. Furthermore, most of the equilibrium constants have not been evaluated with a high degree of accuracy. Consequently, concentration terms will be used for activities in most of the equations appearing in the sections that follow. It will be necessary to use activities instead of concentration terms only when applying the equations to brackish or seawater systems. In such cases, the appropriate value of γc can be used for the activity term. [See Eq. (2.1.3).]

Solution procedure 2.1.1

PROBLEM Calculate the activities of each ion in a solution containing 0.02 M $CaSO_4$ and 0.03 M $MgCl_2$.

SOLUTION (1) Compute the ionic strength of the solution using

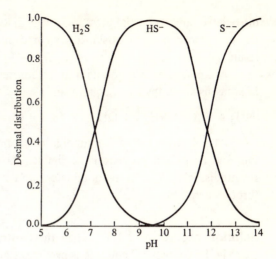

FIGURE 2.1.1
Distribution of hydrogen sulfide species as a function of pH (10^{-3} M solution at 25°C).

Eq. (2.1.5). (2) Compute the activity coefficients using Eq. (2.1.4). (3) Compute the activities using Eq. (2.1.3).

Ion	c	z	cz^2	$\log \gamma$	γ	a
Ca^{++}	0.02	+2	0.08	−0.582	0.26	0.0052
SO_4^{--}	0.02	−2	0.08	−0.582	0.26	0.0052
Mg^{++}	0.03	+2	0.12	−0.582	0.26	0.0078
Cl^-	0.06	−1	0.06	−0.145	0.72	0.0432
			0.34			

$\mu = 0.5(0.34) = 0.17$ ////

Solution procedure 2.1.2 Hydrogen sulfide solution equilibriums are of special interest to the environmentalist. Hydrogen sulfide in the gas phase gives rise to obnoxious odors. In solution, hydrogen sulfide dissociates in two steps to form hydrogen ions and sulfide ions:

$$H_2S \rightleftharpoons H^+ + HS^- \quad (K_1 = 6.3 \times 10^{-8} \text{ at } 25°C) \quad (2.1.16)$$

$$HS^- \rightleftharpoons H^+ + S^{--} \quad (K_2 = 1.3 \times 10^{-12} \text{ at } 25°C) \quad (2.1.17)$$

Figure 2.1.1 shows the distribution of the hydrogen sulfide species as a function of pH. At pH values greater than 8, hydrogen sulfide is dissociated, and its partial pressure is negligible. At pH values less than 8, the undissociated form exists in a significant concentration, and odors result.

Ammonia, both a pollutant and a nutrient in natural waters, exists in aqueous solutions as an equilibrium:

$$NH_3 + H_2O \rightleftharpoons NH_4^+ + OH^- \qquad (K = 1.65 \times 10^{-5} \text{ at } 25°C) \qquad (2.1.18)$$

At pH levels below 7, ammonia exists primarily as the ammonium ion, and its partial pressure is negligible. At pH values greater than 7, the undissociated form predominates, and ammonia can be removed through aeration. ////

Solution procedure 2.1.3 When thermostratification occurs in lakes (see Fig. 1.3.3), reduced conditions become established in the hypolimnion. Iron in the ferric state Fe(III) when introduced to such an environment is converted to ferrous iron Fe(II):

$$Fe^{3+} + e^- \rightarrow Fe^{++} \qquad (2.1.19)$$

When sulfur in the sulfate form is introduced to the reduced environment of the hypolimnion, it is converted through biologic activity to sulfide:

$$SO_4^{--} \xrightarrow{\text{microorganisms}} S^{--} \qquad (2.1.20)$$

A ferrous ion may react with the sulfide ion to form ferrous sulfide, a relatively insoluble compound:

$$Fe^{++} + S^{--} \rightleftharpoons FeS(s) \qquad (2.1.21)$$

$$[Fe^{++}][S^{--}] = K_s = 1 \times 10^{-19} \qquad (\text{at } 25°C) \qquad (2.1.22)$$

In addition, ferrous iron may react with any orthophosphate in the reduced environment to form ferrous phosphate, another insoluble compound:

$$Fe^{++} + PO_4^{3-} \rightleftharpoons Fe_3(PO_4)_2(s) \qquad (2.1.23)$$

$$[Fe^{++}]^3[PO_4^{3-}]^2 = K_s = 1.3 \times 10^{-30} \qquad (\text{at } 25°C)[1] \qquad (2.1.24)$$

Actually, ferrous phosphate appears to persist in reduced environments as *vivianite* $Fe_3(PO_4)_2 \cdot 8H_2O$.

As a result of the above reactions, iron as insoluble sulfide and phosphate may settle to the bottom of the hypolimnion and become

---

[1] P. C. Singer, Anaerobic Control of Phosphate by Ferrous Iron, *J. WPCF*, vol. 44, pp. 663–669, 1972.

incorporated in the bottom sediments. The latter thus becomes a sink for sulfur and phosphorus as well as for iron.

When thermostratification is destroyed by the fall or spring turnover, ferrous ions are oxidized by dissolved oxygen to the ferric state. At pH < 7.8, the ferric ions react with any orthophosphate ions present to form ferric phosphate:

$$Fe^{3+} + PO_4^{3-} \rightleftharpoons FePO_4(s) \qquad (2.1.25)$$

The ferric phosphate slowly hydrolyzes at a rate depending on the pH of the water to form ferric hydroxide. At pH > 7.8, the ferric ions hydrolyze directly to ferric hydroxide:

$$Fe^{3+} + 3OH^- \rightleftharpoons Fe(OH)_3(s) \qquad (2.1.26)$$

$$[Fe^{3+}][OH^-]^3 = K_s = 4 \times 10^{-38} \qquad (\text{at } 25°C) \qquad (2.1.27)$$

Ferric hydroxide, a very insoluble compound, may settle out of the water and become part of the bottom sediments. Ferric hydroxide and manganic hydroxide, a compound formed in a manner similar to that by which ferric hydroxide is formed, impart color to water and often give rise to consumer complaints when present in water supplies.

In environments in which oxygen is present, sulfides can be oxidized biologically to sulfates in a reaction similar to the reverse of Eq. (2.1.20).

////

2.2 REACTION KINETICS

The *reaction order* identifies the type of equation that expresses the rate at which a reaction occurs. Furthermore, it indicates the apparent mechanism of the reaction itself. Table 2.2.1 lists the rate equations and apparent mechanisms of a few irreversible reactions that demonstrate simple rate kinetics. The rate constant k is of key importance in defining the rate at which a reaction takes place. In *zero-order reactions*, the rate at which the concentration of a reactant (reactant A) decreases is constant and is independent of the reactant concentration. When the rate at which the concentration of a reactant decreases is found to be proportional to the concentration of the reactant remaining, the reaction is identified as *first order*. In *second-order reactions*, the rate of decrease is proportional either to the square of the concentration remaining, indicating that two molecules of the reactant react to form the products, or to the product of the concentration of A and some other reactant (reactant B). Higher reaction orders and fractional reaction orders are possible. However, most chemical reactions of significance in the environment occur at rates that can be approximated by the reaction orders listed in Table 2.2.1. Although no reaction ever goes to completion, many reactions can be considered to be essentially irreversible because of the large value of the equilibrium constant.

In a sequence of consecutive, irreversible first-order reactions such as

$$A \xrightarrow{k_1} R \xrightarrow{k_2} S \qquad (2.2.1)$$

the product of the first reaction step R reacts to form a product S. In these reactions, k_1 and k_2 are the reaction constants for the first and second reaction steps, respectively. The time rates of change in concentration for each of the materials in the reactions are as follows:

$$-\frac{d[A]}{dt} = k_1[A] \qquad (2.2.2)$$

$$-\frac{d[R]}{dt} = k_2[R] - k_1[A] \qquad (2.2.3)$$

$$\frac{d[S]}{dt} = k_2[R] \qquad (2.2.4)$$

These expressions can be integrated to yield

$$[A] = [A]_0 e^{-k_1 t} \qquad (2.2.5)$$

$$[R] = [A]_0 \frac{k_1}{k_2 - k_1} \cdot (e^{-k_1 t} - e^{-k_2 t}) \qquad (2.2.6)$$

$$[S] = [A]_0 \left[1 + \frac{k_2}{k_1 - k_2} e^{-k_1 t} + \frac{k_1}{k_2 - k_1} e^{-k_2 t} \right] \qquad (2.2.7)$$

Table 2.2.1 SIMPLE REACTION KINETICS

Reaction order	Rate equation	Apparent mechanism
Zero	$-\frac{d[A]}{dt} = k$	$A \rightarrow$ products
First	$-\frac{d[A]}{dt} = k[A]$	$A \rightarrow$ products
Second	$-\frac{d[A]}{dt} = k[A]^2$	$A + A \rightarrow$ products
Second	$-\frac{d[A]}{dt} = k[A][B]$	$A + B \rightarrow$ products

where $[A]_0$ = initial concentration of A, mol/l

k_1, k_2 = rate constants, s^{-1}

However, if $k_1 \ll k_2$, the sequence of reactions will behave kinetically in a manner similar to the single-step reaction

$$A \xrightarrow{k_1} S \qquad (2.2.8)$$

with the rate expression

$$-\frac{d[A]}{dt} = \frac{d[S]}{dt} = k_1[A] \qquad (2.2.9)$$

Thus, in a sequence of consecutive, irreversible reactions slow steps are rate-controlling.

When materials react in a reversible reaction such as the first-order reaction

$$[A] \underset{k_{-1}}{\overset{k_{+1}}{\rightleftarrows}} [R] \qquad (2.2.10)$$

the rate expressions must include the rates of both the forward and reverse reactions:

$$\frac{d[R]}{dt} = -\frac{d[A]}{dt} = k_{+1}[A] - k_{-1}[R] \qquad (2.2.11)$$

At equilibrium, $d[R]/dt = d[A]/dt = 0$, and

$$\frac{[R]}{[A]} = \frac{k_{+1}}{k_{-1}} = K \qquad (2.2.12)$$

where k_{+1}, k_{-1} = forward and reverse reaction rate constants, respectively, s^{-1}

K = equilibrium constant

Comparable relationships can be developed for other reaction orders.

In general, reaction rates are increased significantly by increases in temperature. Some rates will approximately double for each 10°C increase in temperature. The influence of temperature on the reaction constant is predicted by the *van't Hoff–Arrhenius equation*:

$$\frac{d(\ln k)}{dT} = \frac{E}{RT^2} \qquad (2.2.13)$$

where E = activation energy

R = gas constant

T = absolute temperature

Integrating Eq. (2.2.13) between limits T_0 and T yields

$$\ln \frac{k}{k_0} = \frac{E(T - T_0)}{RTT_0} \qquad (2.2.14)$$

where k, k_0 = rate constants at temperatures T and T_0, respectively

Temperatures in aquatic systems do not vary between wide limits. For such systems the product TT_0 does not change significantly, and the term E/RTT_0 can be considered to be a constant. Equation (2.2.14) can, therefore, be approximated by the exponential form

$$k = k_0 e^{C_k(T - T_0)} \qquad (2.2.15)$$

where C_k = temperature characteristic

If we use the expanded form of the term e^x, Eq. (2.2.15) becomes

$$k = k_0[1 + C_k(T - T_0) + C_k^2(T - T_0)^2 + \cdots] \qquad (2.2.16)$$

which has the approximate form

$$k = k_0[1 + C_k(T - T_0)] \qquad (2.2.17)$$

Often, the temperature dependence of reaction constants is represented by the empirical relationship

$$k = k_0 \Theta^{(T - T_0)} \qquad (2.2.18)$$

where Θ = temperature coefficient

The temperatures in the temperature-difference terms $T - T_0$ in Eqs. (2.2.17) and (2.2.18) can be expressed in degrees Celsius.

Solution procedure 2.2.1 Radioisotopes decay in a first-order reaction pattern. Rates of decay for such materials are expressed in terms of *half-lives*. The half-life $t_{1/2}$ is the time required for one-half of a pure radioisotope to undergo radioactive decay.

When $[A] = 0.5[A]_0$, Eq. (2.2.5) can be rewritten to give

$$\ln \frac{[A]_0}{0.5[A]_0} = \ln 2 = 0.693 = k t_{1/2} \qquad (2.2.19)$$

Whereupon

$$t_{1/2} = \frac{0.693}{k} \qquad (2.2.20)$$

////

Solution procedure 2.2.2 For a first-order reaction, the integrated form of the rate expression is given by Eq. (2.2.5). This expression can be written in the form

$$\log [A] = -\frac{k}{2.303} t + \log [A]_0 \quad (2.2.21)$$

and a plot of $\log [A]$ as a function of time t will yield a straight line, the slope of which is equal to $-(k/2.303)$. For such reactions, it is often convenient to modify Eq. (2.2.21):

$$\log ([A]_0 - [A]_R) = -\frac{k}{2.303} t + \log [A]_0 \quad (2.2.22)$$

where $[A]_R$ = concentration of A that has reacted by time t

For a second-order reaction in which A and B react, the integrated form of the rate expression can be written in the form

$$\log \frac{[B]_0([A]_0 - [A]_R)}{[A]_0([B]_0 - [B]_R)} = \frac{k([A]_0 - [B]_0)}{2.303} t \quad (2.2.23)$$

In a second-order reaction, $[A]_R$ and $[B]_R$ will be equal. A plot of the term on the left side of the equals sign in Eq. (2.2.23) as a function of t will yield a straight line, the slope of which will equal $k([A]_0 - [B]_0)/2.303$.

For a second-order reaction in which A reacts with A or for the special case where $[A]_0 = [B]_0$, the integrated form of the rate expression is

$$\frac{1}{[A]_0 - [A]_R} = kt + \frac{1}{[A]_0} \quad (2.2.24)$$

By plotting the term $1/([A]_0 - [A]_R)$ as a function of time, a straight line is obtained, the slope of which is equal to k.

For a zero-order reaction, the integrated rate expression is

$$[A] = -kt + C \quad (2.2.25)$$

where C = constant

A plot of $[A]$ as a function of t will yield a straight line with a slope equal to $-k$. ////

2.3 CARBONATE EQUILIBRIUMS

The carbonate equilibriums normally of interest when dealing with natural waters are presented in Fig. 2.3.1, along with the equilibrium expressions at 25°C. The influence of temperature on the equilibrium constants for the dis-

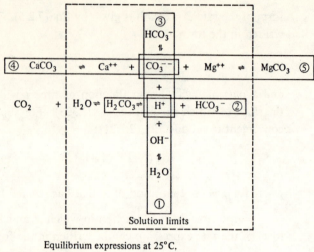

FIGURE 2.3.1
Carbonate equilibriums in H_2O saturated with $CaCO_3$, $MgCO_3$, and CO_2.

sociation of water, carbonic acid, and calcium carbonate is shown in Fig. 2.3.2. It is to be noted that the equilibriums are interrelated and that the hydrogen-ion concentration is involved directly in three of the equilibriums and indirectly in the other two. The effect that pH (logarithm of the reciprocal of the hydrogen-ion concentration in moles per liter) has upon the concentrations of the hydroxyl ions $[OH^-]$, normal carbonate ions $[CO_3^{--}]$, bicarbonate ions $[HCO_3^-]$, and dissolved carbon dioxide $[H_2CO_3^*]$ is explored in the discussion that follows.[1] Concentration terms will be considered to be expressed in moles per liter unless indicated otherwise.

---

[1] In aqueous systems the equilibrium concentration of carbonic acid in the undissociated form is small compared to that of dissolved carbon dioxide. For this reason, the sum of the concentrations is often referred to as the dissolved carbon dioxide.

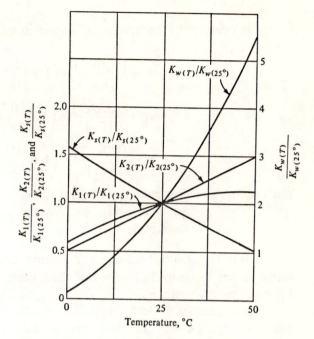

FIGURE 2.3.2
Variation of equilibrium constants with temperature.

The total concentration of the dissolved inorganic carbon species can be expressed by[1]

$$[C_T] = [H_2CO_3^*] + [HCO_3^-] + [CO_3^{--}] \quad (2.3.1)$$

where
$$[H_2CO_3^*] = [CO_2(aq)] + [H_2CO_3] \quad (2.3.2)$$

Equation (2.3.1) can be rearranged to give

$$\frac{[H_2CO_3^*]}{[C_T]} + \frac{[HCO_3^-]}{[C_T]} + \frac{[CO_3^{--}]}{[C_T]} = 1 \quad (2.3.3)$$

In Eq. (2.3.3) each of the terms to the left of the equals sign can be ex-

---

[1] W. Stumm, Chemistry of Natural Waters in Relation to Water Quality, *Proc. Symp. Environ. Meas., U.S. Public Health Serv. Publ.* 999-AP-15, U.S. Dep. HEW, 1964.

pressed in terms of the first and second dissociation constants K_1 and K_2 given in Fig. 2.3.1:

$$\alpha_0 = \frac{[H_2CO_3^*]}{[C_T]} = \left(1 + \frac{K_1}{[H^+]} + \frac{K_1 K_2}{[H^+]^2}\right)^{-1} \quad (2.3.4)$$

$$\alpha_1 = \frac{[HCO_3^-]}{[C_T]} = \left(1 + \frac{[H^+]}{K_1} + \frac{K_2}{[H^+]}\right)^{-1} \quad (2.3.5)$$

$$\alpha_2 = \frac{[CO_3^{--}]}{[C_T]} = \left(1 + \frac{[H^+]}{K_2} + \frac{[H^+]^2}{K_1 K_2}\right)^{-1} \quad (2.3.6)$$

where $\alpha_0, \alpha_1, \alpha_2$ = distribution coefficients for dissolved carbon dioxide (and carbonic acid), bicarbonate ion, and carbonate ion, respectively

It is to be noted that the sum of the distribution coefficients equals unity. Furthermore, the individual values of these coefficients are functions of the hydrogen-ion concentration.

For a situation in which water is unsaturated with respect to $CaCO_3$ but at equilibrium with the atmosphere or with a known concentration of dissolved carbon dioxide $[H_2CO_3^*]$, the hydrogen-ion concentration can be determined from an equation developed from an ion balance over the system.

In order to maintain electroneutrality,

$$[C] + [H^+] = [HCO_3^-] + 2[CO_3^{--}] + [OH^-] + [A] \quad (2.3.7)$$

where $[C]$ = concentration of positive charges held by balance of cations in system

$[A]$ = concentration of negative charges held by balance of anions in system

With rearrangement and substitution of the appropriate carbon dioxide equilibrium relationship for the bicarbonate and carbonate concentration terms, Eq. (2.3.7) can be written

$$[Z] = [C] - [A] = \frac{K_1[H_2CO_3^*]}{[H^+]} + \frac{2K_1 K_2[H_2CO_3^*]}{[H^+]^2} + \frac{K_w}{[H^+]} - [H^+] \quad (2.3.8)$$

By multiplying both sides of Eq. (2.3.8) by $[H^+]^2$ and by ignoring the term $[H^+]^3$ because of its relatively small magnitude at pH values normally encountered in the environment, Eq. (2.3.8) becomes the quadratic expression

$$[Z][H^+]^2 - (K_1[H_2CO_3^*] + K_w)[H^+] - 2K_1 K_2[H_2CO_3^*] = 0 \quad (2.3.9)$$

which can be solved for the hydrogen-ion concentration with

$$[\text{H}^+] = \frac{-b + \sqrt{b^2 - 4ac}}{2a} \quad (2.3.10)$$

where $a = [Z]$

$b = -(K_1[\text{H}_2\text{CO}_3^*] + K_w)$

$c = -2K_1 K_2 [\text{H}_2\text{CO}_3^*]$

The net concentration of positive charges $[Z]$ is equivalent to the *alkalinity* of the system. The alkalinity of a water is a measure of its capacity to neutralize acids. In the system under consideration, this capacity is established by the concentrations of bicarbonate, carbonate, and hydroxide ions.

When the system is in equilibrium with the atmosphere,

$$[\text{H}_2\text{CO}_3^*] = K_H \bar{p} \quad (2.3.11)$$

where K_H = Henry's law constant; for carbon dioxide it equals 3.92×10^{-2} mol/(l)(atm) (at 20°C)

$\bar{p}$ = partial pressure of carbon dioxide in atmosphere = 3×10^{-4} atm

For a situation in which water is unsaturated with respect to CaCO_3 and is not at equilibrium with the atmosphere or with a known concentration of dissolved carbon dioxide, the hydrogen-ion concentration is indeterminate. In such situations, both the alkalinity and the hydrogen-ion concentration in the system must be known in order to compute the total concentration of the dissolved inorganic carbon species and the values of the distribution coefficients.

Equation (2.3.7) expressed in terms of the distribution coefficients for bicarbonate and carbonate ions becomes

$$[Z] = \alpha_1 [C_T] + 2\alpha_2 [C_T] + \frac{K_w}{[\text{H}^+]} - [\text{H}^+] \quad (2.3.12)$$

Solving for $[C_T]$ gives

$$[C_T] = \frac{[Z]}{\alpha_1 + 2\alpha_2} - \frac{K_w/[\text{H}^+]}{\alpha_1 + 2\alpha_2} + \frac{[\text{H}^+]}{\alpha_1 + 2\alpha_2} \quad (2.3.13)$$

Figure 2.3.3 gives the concentrations of the alkalinity species, dissolved carbon dioxide, and hydroxide ion in a water with a total alkalinity of 200 mg/l at different pH values. Such curves can be computed with the use of Eqs. (2.3.4) to (2.3.6) and (2.3.13). The hydroxide concentrations are computed from the ion product of water.

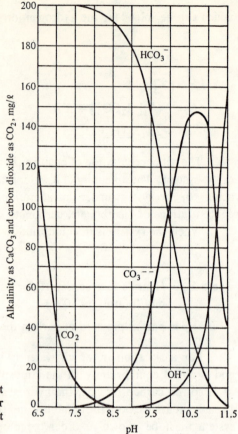

FIGURE 2.3.3
Alkalinity species and carbon dioxide at various pH values (calculated for water with a total alkalinity of 200 mg/l at 25°C).

For a situation in which water is saturated with respect to $CaCO_3$ and is not at equilibrium with the atmosphere or with a known concentration of dissolved carbon dioxide, the basic electroneutrality expression, Eq. (2.3.7), must be modified to include the calcium-ion concentration:

$$2[Ca^{++}] + [Z'] + [H^+] = [HCO_3^-] + 2[CO_3^{--}] + [OH^-] \quad (2.3.14)$$

where $[Z']$ = alkalinity minus calcium-ion concentration

Expressing Eq. (2.3.14) in terms of the appropriate distribution coefficients and the solubility product of $CaCO_3$ gives

$$2\frac{K_s}{\alpha_2[C_T]} + [Z'] + [H^+] = \alpha_1[C_T] + 2\alpha_2[C_T] + \frac{K_w}{[H^+]} \quad (2.3.15)$$

Upon rearrangement, Eq. (2.3.15) becomes

$$(\alpha_1 + 2\alpha_2)[C_T]^2 - \left([Z'] + [H^+] - \frac{K_w}{[H^+]}\right)[C_T] - 2\frac{K_s}{\alpha_2} = 0 \qquad (2.3.16)$$

which is a quadratic relationship having the solutions

$$[C_T] = \frac{-b + \sqrt{b^2 - 4ac}}{2a} \qquad (2.3.17)$$

where $a = \alpha_1 + 2\alpha_2$

$b = -([Z'] + [H^+] - K_w/[H^+])$

$c = -2K_s/\alpha_2$

Thus, for such systems the total concentration of dissolved inorganic carbon species may be computed from a knowledge of the pH of the system, the alkalinity, and the calcium-ion concentration. Expressions similar to Eqs. (2.3.14) to (2.3.17) can be developed for systems saturated with respect to magnesium carbonate.

For situations in which water is saturated with respect to $CaCO_3$ and is in equilibrium with the atmosphere or with a known concentration of dissolved carbon dioxide, an expression explicit for the hydrogen-ion concentration can be developed from the electroneutrality relationship, Eq. (2.3.14), by substituting for the ionic species their equivalents in terms of dissolved carbon dioxide concentration, equilibrium constants, and solubility constant for calcium carbonate. The expression takes the form of a quartic equation for which numerical methods of solution are available.

Up to this point, attention has been focused upon the equilibrium relationships existing in carbonate systems. Now attention is directed to the kinetics of such systems.

At pH ≤ 8, the reaction mechanism occurring in solution is

$$CO_2 + H_2O \underset{k_{-1}}{\overset{k_{+1}}{\rightleftharpoons}} H_2CO_3 \rightleftharpoons H^+ + HCO_3^- \rightleftharpoons 2H^+ + CO_3^{--} \qquad (2.3.18)$$

The dissociation of carbonic acid into bicarbonate ions and further into carbonate ions is a rapid process and may for all practical purposes be considered to be instantaneous. The hydration (and dehydration) of carbon dioxide proceeds at a finite rate. At normal temperatures, k_{+1} and k_{-1} have values of the order of 10^{-2} and 10 s^{-1}, respectively.

At pH ≥ 10, the dominating reaction mechanism is

$$CO_2 + OH^- \underset{k_{-1}}{\overset{k_{+1}}{\rightleftharpoons}} HCO_3^- \rightleftharpoons H^+ + CO_3^{--} \qquad (2.3.19)$$

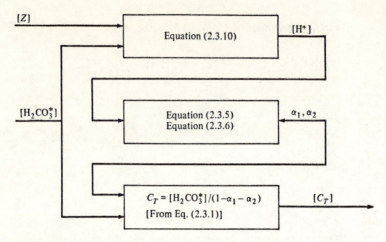

FIGURE 2.3.4
Information flow diagram for the determination of the total concentration of dissolved inorganic carbon in system unsaturated with respect to $CaCO_3$.

and k_{+1} and k_{-1} have values of approximately 10^4 l/(mol)(s) and 10^{-4} s^{-1}, respectively.

In summary, it appears that ordinarily the reactions in carbonate systems require a maximum of a few minutes to attain equilibriums.

Solution procedure 2.3.1 The carbonate system in natural waters provides an internal capacitance by which carbon dioxide is stored or withdrawn, more or less instantaneously, with the accumulation or reduction of carbon dioxide in solution. In biologically active waters, the concentration of dissolved carbon dioxide fluctuates over the diurnal cycle (daily cycle) in response to the net result of the utilization through photosynthesis and production by respiration. The effect of the carbonate system as a storage reservoir for carbon dioxide is important in modeling aquatic ecological systems such as will be discussed in Chap. 4.

At any given moment, the total concentration of the dissolved inorganic carbon in a system unsaturated with respect to $CaCO_3$ and in equilibrium with a known concentration of dissolved carbon dioxide (not necessarily in equilibrium with the atmosphere) can be computed in a sequence of steps indicated by the information flow diagram in Fig. 2.3.4. The average rate of withdrawal of inorganic carbon from the carbonate and bicarbonate ions at time $t + \Delta t$ can be approximated by

$$R = \frac{(\alpha_1[C_T] + \alpha_2[C_T])_t - (\alpha_1[C_T] + \alpha_2[C_T])_{t+\Delta t}}{\Delta t} \quad (2.3.20)$$

For these calculations, units of moles of carbon per liter of solution are convenient. ////

Solution procedure 2.3.2 Orthophosphates in solution exist in equilibriums similar to those of the carbonates. Such equilibriums are listed in Table 2.3.1 along with the equilibrium expressions at 25°C. The total concentration of the dissolved orthopolyphosphate species can be expressed by

$$[P_T] = [H_3PO_4] + [H_2PO_4^-] + [HPO_4^{--}] + [PO_4^{3-}] \quad (2.3.21)$$

which can be modified to yield

$$\frac{[H_3PO_4]}{[P_T]} + \frac{[H_2PO_4^-]}{[P_T]} + \frac{[HPO_4^{--}]}{[P_T]} + \frac{[PO_4^{3-}]}{[P_T]} = 1 \quad (2.3.22)$$

Each of the terms to the left of the equals sign in Eq. (2.3.22) can be expressed in terms of the appropriate dissociation constants found in Table 2.3.1:

$$\beta_0 = \frac{[H_3PO_4]}{[P_T]} = \left(1 + \frac{K_1}{[H^+]} + \frac{K_1 K_2}{[H^+]^2} + \frac{K_1 K_2 K_3}{[H^+]^3}\right)^{-1} \quad (2.3.23)$$

$$\beta_1 = \frac{[H_2PO_4^-]}{[P_T]} = \left(1 + \frac{[H^+]}{K_1} + \frac{K_2}{[H^+]} + \frac{K_2 K_3}{[H^+]^2}\right)^{-1} \quad (2.3.24)$$

$$\beta_2 = \frac{[HPO_4^{--}]}{[P_T]} = \left(1 + \frac{[H^+]^2}{K_1 K_2} + \frac{[H^+]}{K_2} + \frac{K_3}{[H^+]}\right)^{-1} \quad (2.3.25)$$

Table 2.3.1 POLYPHOSPHATE EQUILIBRIUMS

Equilibriums

1. $H_3PO_4 \rightleftharpoons H^+ + H_2PO_4^-$
2. $H_2PO_4^- \rightleftharpoons H^+ + HPO_4^{--}$
3. $HPO_4^{--} \rightleftharpoons H^+ + PO_4^{3-}$

Equilibrium expressions at 25°C

1. $\dfrac{[H^+][H_2PO_4^-]}{[H_3PO_4]} = K_1 = 7.52 \times 10^{-3}$

2. $\dfrac{[H^+][HPO_4^{--}]}{[H_2PO_4^-]} = K_2 = 6.32 \times 10^{-8}$

3. $\dfrac{[H^+][PO_4^{3-}]}{[HPO_4^{--}]} = K_3 = 4.8 \times 10^{-13}$

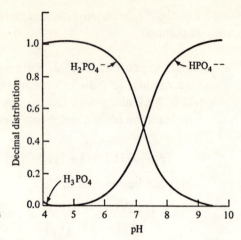

FIGURE 2.3.5
Distribution of orthophosphate species as a function of pH.

$$\beta_3 = \frac{[PO_4^{3-}]}{[P_T]} = \left(1 + \frac{[H^+]^3}{K_1 K_2 K_3} + \frac{[H^+]^2}{K_2 K_3} + \frac{[H^+]}{K_3}\right)^{-1} \quad (2.3.26)$$

where $\beta_0, \beta_1, \beta_2, \beta_3$ = distribution coefficients for phosphoric acid, dihydrogen phosphate, monohydrogen phosphate, and phosphate ions, respectively

The distribution of orthophosphate species in solution as a function of pH is shown in Fig. 2.3.5. ////

2.4 THERMOCHEMISTRY

Chemical reactions generally result in an absorption or an evolution of heat. When heat is absorbed or evolved from a system under constant pressure, the gain or loss in heat is numerically equal to the change in enthalpy. Absolute values of enthalpy are unknown. Consequently, enthalpy values are based on an arbitrary datum.

The quantity of heat absorbed or liberated in a chemical reaction is called the *heat of reaction* ΔH. Each heat of reaction must be referred to a particular temperature and to the particular physical states of both the reactants and products. By convention, positive values signify an absorption of heat, whereas negative values indicate that heat is liberated. For instance, in the reaction

$$CO_2(g) + H_2(g) \to CO(g) + H_2O(g) \qquad \Delta H_{25} = 9838 \text{ cal} \quad (2.4.1)$$

9838 cal are absorbed when the temperature is referenced to 25°C and both the reactants and products are in the gaseous state. However, when the water produced is written as being in the liquid state, $\Delta H_{25} = -681$ cal. The difference, 10,519 cal, represents the quantity of heat required to vaporize 1 mol of water. Although the heat of reaction is affected by pressure, the effect is usually small and is often neglected. Only in reactions involving gases at high pressure may the pressure effect be significant.

The heat of reaction can be determined in several ways. One method involves the use of the *heat of formation*. This quantity is defined as the heat of reaction when 1 mol of a compound is formed from its elements.

$$C(s) + O_2(g) \rightarrow CO_2(g) \qquad \Delta H_{f25} = 94{,}052 \text{ cal} \qquad (2.4.2)$$

The heat of reaction is computed by subtracting the heat of formation of the reactants from that of the products:

$$\Delta H = \sum (\Delta H_f)_P - \sum (\Delta H_f)_R \qquad (2.4.3)$$

For example, the heats of formation of the reactants and products of the reaction in Eq. (2.4.1) at 25°C are

$$\begin{array}{cccc} CO_2(g) + H_2(g) \rightarrow & CO(g) + H_2O(g) \\ -94{,}052 \quad\;\; 0 & -26{,}416 \quad -57{,}798 \end{array} \qquad (2.4.4)$$

It is to be noted that the heats of the elements in their most stable form at 1 atm pressure and the given temperature are taken as 0. The heat of reaction as computed by Eq. (2.4.3) is

$$\Delta H_{25} = (-26{,}416 - 57{,}798) - (-94{,}052) = 9838 \text{ cal}$$

The heat of reaction can be calculated from the *heats of combustion*, which is the heat evolved when 1 mol of a compound is completely oxidized. For the combustion of carbon monoxide

$$CO(g) + \tfrac{1}{2}O_2(g) \rightarrow CO_2(g) \qquad \Delta H_{c25} = -67{,}636 \text{ cal} \qquad (2.4.5)$$

The heat of reaction is computed by taking the difference between the heats of combustion of the products and reactants:

$$\Delta H = \sum (\Delta H_c)_R - \sum (\Delta H_c)_P \qquad (2.4.6)$$

For example, the heats of combustion of the reactants and products of the reaction in Eq. (2.4.1) at 25°C are

$$\begin{array}{cccc} CO_2(g) + & H_2(g) & \rightarrow & CO(g) + H_2O(g) \\ 0 & -57{,}798 & & -67{,}636 \quad\;\; 0 \end{array} \qquad (2.4.7)$$

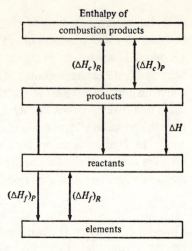

FIGURE 2.4.1
Enthalpy levels. (*After T. E. Corrigan, Chemical Engineering Fundamentals, Chemical Equilibrium*—1, *fig.* 1, *Chem. Eng., vol.* 61, *p.* 222, 1954.)

Since carbon dioxide and water are in the highest oxidation states possible, their heats of combustion are equal to 0. The heat of reaction as calculated by Eq. (2.4.6) is

$$\Delta H_{25} = (-57,798) - (-67,636) = 9838 \text{ cal}$$

The relationships between the heats of reaction, formation, and combustion are illustrated in Fig. 2.4.1. The heat of reaction can also be computed by combining the equations of reactions for which the heats of reactions are already known. For example, if the following formation reactions

$$C(s) + \tfrac{1}{2}O_2(g) \to CO(g) \qquad \Delta H_{f25} = -26,416 \text{ cal} \qquad (2.4.8)$$

$$C(s) + O_2(g) \to CO_2(g) \qquad \Delta H_{f25} = -94,052 \text{ cal} \qquad (2.4.2)$$

$$H_2(g) + \tfrac{1}{2}O_2(g) \to H_2O(g) \qquad \Delta H_{f25} = -57,798 \text{ cal} \qquad (2.4.9)$$

are rearranged and added together, then one has

$$C(s) + \tfrac{1}{2}O_2(g) \to CO(g) \qquad \Delta H_{f25} = -26,416 \text{ cal} \qquad (2.4.8)$$

$$CO_2(g) \to C(s) + O_2(g) \qquad \Delta H_{f25} = +94,052 \text{ cal} \qquad (2.4.2)$$

$$\underline{H_2(g) + \tfrac{1}{2}O_2(g) \to H_2O(g) \qquad \Delta H_{f25} = -57,798 \text{ cal} \qquad (2.4.9)}$$

$$CO_2(g) + H_2(g) \to CO(g) + H_2O(g) \qquad \Delta H_{25} = 9838 \text{ cal} \qquad (2.4.1)$$

The heat-of-reaction data found in the literature usually consist of the heats of formation and combustion of compounds in their *standard states*. The standard states for gases, liquids, and solids are the pure compounds at

1 atm pressure. Heats of combustion are generally presented for organic compounds, whereas the heats of formation are given for inorganic compounds.

The heat of reaction at any temperature can be computed from a knowledge of the heat of reaction at any other temperature:

$$\Delta H_{T_1} = \Delta H_{T_2} + \int_{T_2}^{T_1} \Delta c_p \, dT \quad (2.4.10)$$

where ΔH_{T_1}, ΔH_{T_2} = heats of reaction at temperatures T_1 and T_2

c_p = heat capacity or specific heat

and
$$\Delta c_p = \sum (c_p)_P - \sum (c_p)_R \quad (2.4.11)$$

It is often more convenient to use a mean value of the heat capacity between temperatures T_1 and T_2. In such cases, Eq. (2.4.10) can be written

$$\Delta H_{T_1} = \Delta H_{T_2} + \Delta \overline{c_p} (T_1 - T_2) \quad (2.4.12)$$

For example, the heat of reaction for the reaction in Eq. (2.4.1) at a temperature of 260°C can be computed as follows: From published data, the mean heat capacities of the products and reactants in Eq. (2.4.1) between the temperatures of 25 and 260°C are

$CO(g) = 7.1$ cal/(mol)(°C)

$H_2O(g) = 8.2$ cal/(mol)(°C)

$CO_2(g) = 9.9$ cal/(mol)(°C)

$H_2(g) = 7.0$ cal/(mol)(°C)

Then from Eq. (2.4.12)

$\Delta H_{260} = 9838 + (7.1 + 8.2 - 9.9 - 7.0)(260 - 25) = 9462$ cal

2.5 COLLOIDAL BEHAVIOR

The term *colloid* is used to describe a system in which particles of relatively small size (the *disperse phase*) are dispersed in a homogeneous medium (the *disperse medium*). Colloidal particles are larger than atoms and small molecules but are small enough to pass through the pores of ordinary filters. (See Fig. 2.5.1.) Arbitrarily, particles ranging in size from 1 nm (10^{-6} mm) to 1 μm (10^{-3} mm) are classified as colloidal. However, particles of even larger size do exhibit certain colloidal properties.

Several different types of colloid systems are possible. Table 2.5.1 lists eight known types with examples of each. Sols, emulsions, and aerosols are of

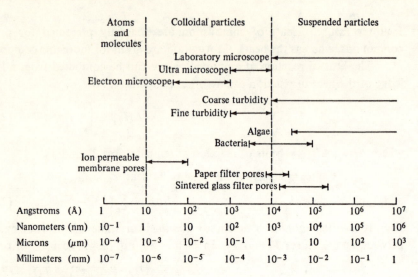

FIGURE 2.5.1
Classification of dispersed systems.

particular significance in the environment. Although the following discussion deals specifically with the properties of solids dispersed in liquids (sols), some of the properties discussed apply to emulsions and aerosols as well.

Colloid particles possess electrical properties that strongly influence their behavior. Charges located on particle surfaces establish an electrostatic field that is a major factor in determining the stability of the colloid system. These charges, called *primary charges*, can result from any one of three phenomena: (1) the ionization of reactive groups on the ends of the molecules constituting the particle structure, (2) the preferential adsorption of ions from the disperse

Table 2.5.1 TYPES OF COLLOID-DISPERSE SYSTEMS

Disperse medium	Disperse phase	Name	Example
Liquid	Solid	Sol	Clay turbidity in water
Liquid	Liquid	Emulsion	Oil in water
Liquid	Gas	Foam	Whipped cream
Gas	Solid	Aerosol	Dust, smoke
Gas	Liquid	Aerosol	Mist, fog
Solid	Solid		Colored glass
Solid	Liquid	Gel	Jelly
Solid	Gas		Pumice

medium, and (3) ion replacement in the ionic lattice structures located at the surface of the colloid.

Several types of reactive groups give rise to the primary charge. On hydrated silica colloids, the silanol groups —SiOH ionize to produce charges, the signs of which are dependent upon the pH of the system.

$$—SiOH_2^+ \underset{H^+}{\overset{OH^-}{\rightleftarrows}} —SiOH \underset{H^+}{\overset{OH^-}{\rightleftarrows}} —SiO^-$$

$$\text{pH} \xrightarrow{\quad\quad\quad\text{isoelectric point}\quad\quad\quad}$$

At a certain pH, called the *isoelectric point*, the charge becomes 0. Many other oxides and hydroxides behave in a similar manner.

The primary charge on organic colloids is often caused by the dissociation of the functional groups —COOH and —NH$_2$. Protein molecules and their degradation products contain carboxylic acid and amino groups on opposite ends of the molecule. In water, the amino group hydrolyzes, and one or both of the groups dissociate, depending upon the pH of the system. With the central molecular structure represented by the symbol R, the dissociation can be depicted by the expression:

$$R\begin{bmatrix}COOH \\ NH_3^+\end{bmatrix} \underset{H^+}{\overset{OH^-}{\rightleftarrows}} R\begin{bmatrix}COO^- \\ NH_3^+\end{bmatrix} \underset{H^+}{\overset{OH^-}{\rightleftarrows}} R\begin{bmatrix}COO^- \\ NH_3OH\end{bmatrix}$$

$$\text{pH} \xrightarrow{\quad\quad\quad\text{isoelectric point}\quad\quad\quad}$$

At the isoelectric point, both groups are ionized, and the net charge on the molecule is 0. An increase in pH from the isoelectric point depresses the ionization of the hydrated amino group and results in a net negative charge. A decrease in pH from the isoelectric point depresses the ionization of the carboxylic group, the result of which is molecules having a net positive charge. As indicated by the double arrows, the reactions are reversible.

The mechanism responsible for the preferential adsorption of ions is not clearly understood. However, such a mechanism is known to result in a primary charge on some clay and calcite colloids.

Other clay colloids receive their primary charge from ion replacement in the ionic lattice at the surface of the colloids. A net negative or positive charge is imparted to the colloid, depending on whether the replacement cation has a valence less than or greater than the cation being replaced.

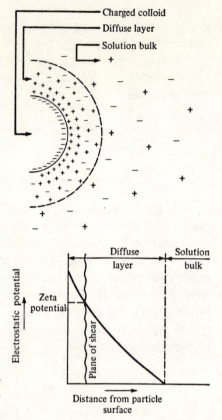

FIGURE 2.5.2
Schematic sketch of the double layer around a colloid particle and the electrostatic potential across the layer.

The primary charge on a colloid particle attracts solution ions of opposite charge. The counterions attracted to the colloid surround the latter in what is called the *diffuse layer*. The concentration of counterions in the diffuse layer varies from a relatively high concentration existing at the surface of the particle to that in the bulk of the solution. The charged colloid together with the diffuse layer constitutes what is commonly referred to as a *double layer*. A schematic sketch of a double layer is presented in Fig. 2.5.2. The primary charge is shown to be negative, whereas the counterions are positive.

Concentration differences between cation and anion species result in the establishment of electrostatic fields. If the bulk of the solution is considered to have zero potential, the potential at any point across the diffuse layer can be measured by the work required to transport a unit charge from the solution bulk to that particular point. The potentials through the electrostatic field of a charged colloid particle are illustrated in Fig. 2.5.2. The potential existing at

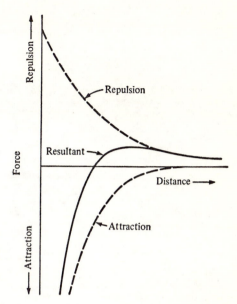

FIGURE 2.5.3
Interparticle forces as a function of interparticle distance. (*After K. J. Mysels, "Introduction to Colloid Chemistry," fig. 4-4a, p. 87, Interscience, New York, 1959.*)

the plane of shear is of particular importance. This plane forms a boundary between that portion of the solution around the particle that moves with the particle and the portion which can move independently of the particle. The plane of shear may occur close to the particle surface or, in the case of solvated colloids, at some distance from this surface. The potential at the shear plane, called the *zeta potential*, is responsible for the electrokinetic behavior exhibited by colloid particles.

Colloidal systems may be classified as stable or unstable. A colloidal system is said to be stable if the colloidal condition is more or less permanent. Stability depends upon the net resultant of the attraction and repulsion forces acting on the colloid particles. The attraction forces, called the *van der Waals forces*, are caused by the interaction of particle dipoles, either permanent or induced. The repulsion forces in a colloidal system are furnished by the zeta potential.

The interparticle forces acting on colloid particles as a function of interparticle distance are shown in Fig. 2.5.3. It is to be noted that the attraction (van der Waals) forces are effective only when interparticle distances are short.

The stability of colloids is also affected by *solvation*. The latter is caused by an affinity that certain functional groups present on colloidal surfaces have for water. These groups are water-soluble and, as such, attract and hold a layer of water firmly around the particle. Although the actual mechanism

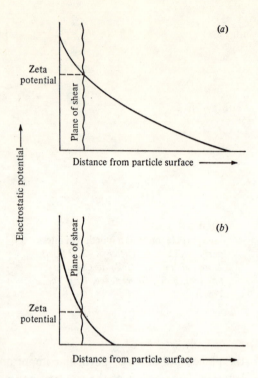

FIGURE 2.5.4
Reduction of colloid charge by compression of the diffuse layer with the addition of counterions. (a) Before addition of counterions; (b) after addition of counterions.

influencing stability is not clearly understood, it is convenient to consider the water of solvation as behaving as an elastic barrier to keep the colloidal particles apart.

The removal of colloid particles from suspensions is facilitated by destabilization. The objective in destabilization is the reduction of the repulsion forces to the extent that the forces of attraction prevail and the particles coalesce to form larger ones that are more easily separated from the system by gravity. Destabilization of colloidal systems may be accomplished in four different ways: (1) compression of the diffuse layer, (2) charge modification by ion adsorption, (3) interparticle bridging with organic polymers, and (4) enmeshment in precipitates.[1]

The addition of ions with a charge opposite to that of the primary charge serves to increase the concentration of counterions in the diffuse layer. As a result, the diffuse layer is compressed, and the potential at the plane of shear (zeta potential) is depressed. See Fig. 2.5.4. The valence of the counterion

---

[1] C. R. O'Melia, Coagulation and Flocculation, in W. J. Weber, "Physicochemical Processes for Water Quality Control," Interscience-Wiley, New York, 1972.

added is quite important. For example, a bivalent ion is about 50 times more effective than a monovalent ion, and a trivalent ion may be as much as 1,000 times more effective than a monovalent ion. The destabilization phenomenon observed in estuaries where large quantities of silt are released may be attributed, at least in part, to the effect that high salt concentrations have on the diffuse layer of colloids formed in nonmarine environments.

Some counterion species are adsorbed on the colloid surface. In such cases, the charge on the particle is modified. When modification results in reducing the charge, destabilization occurs. It is to be noted that destabilization through adsorption can occur at a much smaller concentration of the counterion than it does as a result of compression of the diffuse layer.

Organic polymers when placed in solution form long, branched molecules, each of which can be adsorbed on the surfaces of several colloid particles. This interparticle bridging results in an aggregation of polymer and colloids of a size that settles by gravity.

Certain metal salts, metal oxides, and metal hydroxides form precipitates in aqueous solutions. If colloids are present, they become enmeshed in the aggregating precipitates commonly referred to as *floc*. In fact, the colloids themselves may provide the nuclei around which the precipitates form. The precipitating agents are called *coagulants*. This method of colloid removal is common in water-treatment technology.

Theoretically, colloids can be destabilized by adjusting the pH of the colloid system to the isoelectric point. Practically, such a method finds little application because the cost is prohibitive.

EXERCISES

2.1.1 Using the law of mass action, express the dissociation constants of each of the following compounds in terms of equilibrium concentrations:
(a) H_2CO_3 (b) NH_4OH (c) $HOCl$
(d) H_2S (e) $NaCl$ (f) H_2O
(g) $CaCO_3$

2.1.2 Estimate the ionic strength of each of the following solutions:
(a) $0.10\ M\ CaSO_4$ (b) $0.05\ M\ HCl$
(c) $0.02\ M\ MgCl_2$ (d) $100\ mg/l\ CaCO_3$
(e) 200 mg/l dissolved solids

2.1.3 Calculate the activities of each ion in a solution containing $0.01\ M\ CaCl_2$, $0.01\ M\ H_2SO_4$, and $0.02\ M\ MgSO_4$.

2.1.4 Excess lime $Ca(OH)_2$ is sometimes used in water softening to obtain a more complete removal of Mg as $Mg(OH)_2$. Explain why.

2.1.5 Hydrogen sulfide odors evolve from many waste lagoons only at night. What can you conclude concerning the pH of these lagoons?

2.1.6 A waste with a pH of 6 is aerated to remove ammonia prior to the waste's discharge into a receiving stream. Will aeration be effective? Why?

2.2.1 Strontium 90 (Sr^{90}) is a radioisotope with a half-life of 29 years. How long does the radioisotope have to be stored to allow for 99.9 percent decay?

2.2.2 From the following kinetic data, determine the reaction orders and the rate constants for the reactions.

t, s	c_1, mol/l	c_2, mol/l
0	2.33	2.00
500	1.70	1.80
1,000	1.25	1.60
1,500	0.95	1.40
2,000	0.70	1.20

2.2.3 Two compounds in solution, both present initially in concentrations of 1 mol/l, react in a second-order reaction. If the reaction is 10 percent complete in 5 min, how long will it take for the reaction to be 90 percent complete?

2.2.4 Compound A reacts to form compound B in a first-order reaction with a rate constant equal to 0.5 d^{-1}. Concurrently, compound B reacts to form compound C in a first-order reaction with a rate constant of 0.8 h^{-1}. What will be the concentrations of the three compounds 24 h after the reactions are initiated if the initial concentration of A is 1 mol/l?

2.2.5 At 20°C a first-order reaction exhibits a rate constant of 0.1 d^{-1}, whereas at 30°C the same reaction exhibits a rate constant of 0.18 d^{-1}. Compute both the temperature characteristic and the temperature coefficient.

2.3.1 What is the pH of each of the following solutions?
(a) 10^{-5} M hydrogen ions
(b) 10^{-4} mg/l hydrogen ions
(c) 10^{-8} M hydroxyl ions
(d) 10^{-4} M HCl
(e) 10^{-4} mg/l HCl
(f) 10^{-5} mg/l NaOH

2.3.2 With the carbonate-equilibriums sketch in Fig. 2.3.1, explain the effect of increasing the partial pressure of carbon dioxide on the solubility of calcium carbonate.

2.3.3 What is the total concentration of the dissolved inorganic carbon species in water at equilibrium with the atmosphere at 20°C if the pH is 7.4? What are the bicarbonate- and carbonate-ion concentrations?

2.3.4 If water unsaturated with respect to $CaCO_3$ but at equilibrium with the atmosphere contains 200 mg/l alkalinity (as $CaCO_3$), what is the pH at 20°C? What is the pH if the partial pressure of CO_2 is increased tenfold?

2.3.5 A water saturated with respect to $CaCO_3$ and not at equilibrium with the atmosphere has an alkalinity of 600 mg/l (as $CaCO_3$), a calcium concentration of 400 mg/l (as $CaCO_3$), and a pH of 7.8. What is the total concentration of dissolved inorganic carbon species?

2.3.6 Write a computer program that will give a family of curves such as is shown in Fig. 2.3.2 for different values of alkalinity.

2.4.1 At 18°C the heats of formation of water as a gas and as a liquid are 57,826 and 68,387 cal/mol, respectively. Compute the heat of reaction when liquid water is converted to vapor. Compare this value with the latent heat of vaporization.

2.4.2 The heat of formation of methane gas $CH_4(g)$ at 25°C is 17,900 cal/mol. What is the heat of reaction when $CH_4(g)$ is oxidized to $CO_2(g)$ ($\Delta H_f = -94,052$) and $H_2O(g)$ ($\Delta H_f = -57,798$)?

2.4.3 The heat of combustion of ethane gas $C_2H_6(g)$ at 25°C and constant pressure is $-372,800$ cal/mol. What is the heat of reaction when $C_2H_6(g)$ is oxidized to $CO_2(g)$ and $H_2O(g)$?

2.4.4 Given the following reactions,

$$CO(g) + \tfrac{1}{2}O_2(g) \rightarrow CO_2(g) \quad \Delta H = -67,600 \text{ cal}$$
$$C(s) + O_2(g) \rightarrow CO_2(g) \quad \Delta H = -94,052 \text{ cal}$$

determine the heat of reaction for the following at the same temperature:

$$C(s) + \tfrac{1}{2}O_2(g) \rightarrow CO(g)$$

2.4.5 The heat of reaction at 25°C when $C_2H_6(g)$ is oxidized to CO_2 and H_2O is $-372,800$ cal/mol. What is the heat of reaction at 100°C if the mean heat capacities of the products and reactants are

$$C_2H_6(g) = 12.0 \text{ cal/(mol)(°C)}$$
$$O_2(g) = 7.0 \text{ cal/(mol)(°C)}$$
$$CO_2(g) = 10.0 \text{ cal/(mol)(°C)}$$
$$H_2O(l) = 18.0 \text{ cal/(mol)(°C)}$$

2.5.1 What causes the primary charge on colloids?

2.5.2 Explain the isoelectric-point phenomenon.

2.5.3 Discuss the double-layer phenomenon that exists around a colloid particle.

2.5.4 What is the zeta potential of a colloid, and what is its significance?

2.5.5 On what factors does the stability of a colloid depend?

2.5.6 What is solvation, and what is its effect on colloid stability?

2.5.7 What is the objective in the destabilization of a colloidal system?

2.5.8 In what four ways may a colloidal system be destabilized?

2.5.9 In the destabilization of a colloidal system with a negative zeta potential, which will be the most effective: $CaCl_2$, $Al_2(SO_4)_3$, or $NaCl$?

3
BIOLOGIC PHENOMENA

3.1 ORGANIC MATERIALS

The microbiological decomposition of materials of biologic origin is a natural phenomenon of considerable importance. Besides being a source of noxious odors, the decomposition of these natural organic materials often results in the depletion of the limited oxygen resources in surface waters. Consequently, it is important to know the composition of organic materials and to understand how several of the key chemical elements migrate between different forms of plant and animal life.

Organic materials of natural origin are composed to a large extent of a relatively few elements. These elements are:

1. Carbon (C)
2. Hydrogen (H)
3. Oxygen (O)
4. Nitrogen (N)
5. Phosphorus (P)
6. Sulfur (S)

The first three elements—carbon, hydrogen, and oxygen—are the building blocks out of which the major portion of the organic molecule is constructed. The elements nitrogen, phosphorus, and sulfur are found in smaller amounts. Other elements may be present in measurable and trace quantities.

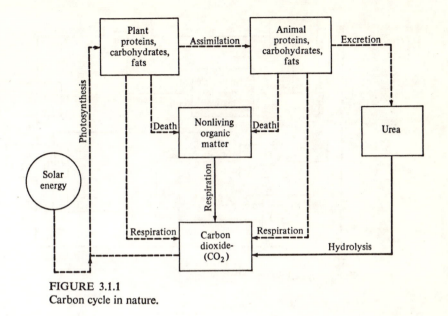

FIGURE 3.1.1
Carbon cycle in nature.

Attention is directed to the schematic diagram in Fig. 3.1.1. The diagram illustrates in broad outline the carbon transformations in nature. Beginning at the point labeled "Nonliving organic matter," carbon in the organic state is converted to carbon dioxide by the respiratory activities of microorganisms. Through photosynthesis the carbon in carbon dioxide is incorporated in plant materials. Some of the carbon is returned to the inorganic state through plant respiration, while the remainder ultimately ends up in nonliving organic matter or is assimilated by animal consumers. The carbon released from the animal before it dies either is converted directly to carbon dioxide through animal respiration or passes from the animal body in urea and other organic compounds, where it is converted by microorganisms to carbon dioxide.

A similar diagram for nitrogen is shown in Fig. 3.1.2. Nitrogen in nonliving organic matter is converted by microorganisms first to ammonia and then to nitrite and nitrate. Nitrogen either in the form of ammonia or nitrate is assimilated by the plants and incorporated in proteins. Animals, by consuming plants, utilize the nitrogen in their own protein structure. The major end product of nitrogen metabolism in most animals is urea. Urea nitrogen can be transformed through microbial activity to ammonia, nitrite, and nitrate. Denitrification and nitrogen fixation provide alternate routes in the scheme.

Figures 3.1.1 and 3.1.2 outline the carbon and nitrogen cycles in nature. Being true cycles, no inputs of these elements from external sources are indicated.

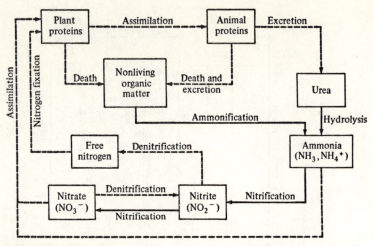

FIGURE 3.1.2
Nitrogen cycle in a typical ecosystem.

In typical ecosystems,[1] however, such inputs take place. Figure 3.1.3 illustrates the nitrogen transformations within the context of the aquatic ecosystem.

It is to be emphasized that carbon, nitrogen, and sulfur as well as the other elements in organic compounds follow a cyclic path, moving back and forth between the organic and inorganic states. The same general pattern is followed regardless of whether the ecosystem is a forest, a farm, or a polluted stream. As indicated in the figures, proteins, carbohydrates, and fats constitute the major groups of organic compounds. Fats and carbohydrates consist primarily of carbon, hydrogen, and oxygen, while proteins contain nitrogen and sulfur as well. From a practical point of view, however, it is advantageous (and even necessary, in most instances) to consider the composition of organic matter on an empirical basis $C_a H_b O_c N_d P_e S_f$.

Organics are high-energy-level compounds. Their formations from low-energy-level inorganics require an energy input, usually the sun:

$$\text{Inorganics} + energy \rightarrow \text{organics} \quad (3.1.1)$$

Organics are formed as constituent parts of living organisms. Once divorced from the living state, either through metabolism or death, organics are subject to decomposition and conversion back to low-energy-level inorganic compounds. The energy state varies with the type of organic compound.

---

[1] An *ecosystem* is any segment of nature in which there is an exchange of materials between organisms and their environment.

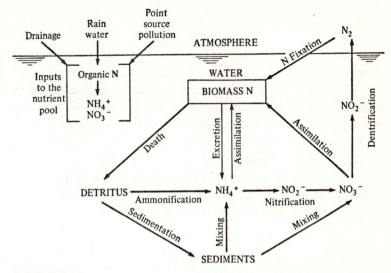

FIGURE 3.1.3
Nitrogen transformations in the aquatic ecosystem. (*Adapted from D. R. Keeney, The Fate of Nitrogen in Aquatic Ecosystems, fig. 1, Lit. Rev. 3, Water Res. Cent., University of Wis.*, 1972.)

The higher the energy state, the more reduced will be the compound. Inorganics from which organics are formed, or to which they revert, are materials in a low state of reduction (that is, a high state of oxidation).

An empirical relationship has been developed in which the unit heat of combustion is related to the degree of reduction of organic matter:

$$h = 127R + 400 \quad (3.1.2)$$

where h = unit heat of combustion, cal/g

R = value expressing degree of reduction

The R value for any type of organic matter can be computed from an expression developed from a rational consideration of the quantity of oxygen required for the complete oxidation of all carbon and hydrogen to carbon dioxide and water.[1]

$$R = \frac{100(2.66\alpha + 7.94\beta - \gamma)}{398.9} \quad (3.1.3)$$

where α, β, γ = percentages on ash-free weight basis of carbon, hydrogen, and oxygen, respectively

[1] H. A. Spoeher and H. W. Milner, The Chemical Composition of Chlorella; Effect of Environmental Conditions, *Plant Physiol.*, vol. 24, pp. 120–149, 1949.

As indicated above, the microbiological decomposition is of special concern to the environmental engineer. Such decomposition takes place either in the presence of oxygen (under *aerobic conditions*) or in its absence (under *anaerobic conditions*). When microbiological oxidation takes place, a demand is imposed upon the oxygen resources of the environment. In environments where the supply of oxygen is limited, oxygen depletion may occur with the subsequent development of nuisance conditions fostered by anaerobiosis. Such conditions constitute the focal point of the public's concern with the discharge of untreated organic waste materials.

The inorganic products resulting from the decomposition of organic materials containing carbon, nitrogen, and sulfur in aerobic and anaerobic environments are as follows:

	Aerobic environment	Anaerobic environment
Organic C	CO_2	CO_2, CH_4
Organic N	NO_3^-	NH_3
Organic S	SO_4^{--}	H_2S

The conversion of organic materials to inorganics requires a hydrogen acceptor. In an aerobic environment, microorganisms utilize molecular oxygen for this purpose, whereas in an anaerobic environment they utilize sulfates, nitrates, carbon dioxide, and organic compounds.

The quantity of oxygen required for the oxidation of organic material composed of the basic elements will be equal to the sum of the quantities required for the oxidation of each element minus that oxygen initially in the organic molecule. On an empirical-mole basis, the complete oxidation of organic matter can be expressed as

$$C_aH_bO_cN_dP_eS_f + (a + 0.25b + 1.25d + 1.25e + 1.5f - 0.5c)O_2$$
$$\rightarrow aCO_2 + (0.5b - 0.5d - 1.5e - f)H_2O + dNO_3^-$$
$$+ ePO_4^{3-} + fSO_4^{--} + (d + 3e + 2f)H^+ \qquad (3.1.4)$$

If nitrogen is not oxidized to nitrate, Eq. (3.1.4) must be modified:

$$C_aH_bO_cN_dP_eS_f + (a + 0.25b - 0.75d + 1.25e + 1.5f - 0.5c)O_2$$
$$\rightarrow aCO_2 + (0.5b - 1.5d - 1.5e - f)H_2O + dNH_3 + ePO_4^{3-}$$
$$+ fSO_4^{--} + (3e + 2f)H^+ \qquad (3.1.5)$$

In natural organic materials, the elements phosphorus and sulfur exist in small proportion with the other elements and, for this reason, can be neglected in most calculations. When oxidation of the organic material is incomplete, the oxygen requirement can be estimated from

$$C_aH_bO_cN_d + 0.5(ny + 2s + r - c)O_2$$
$$\rightarrow nC_wH_xO_yN_z + sCO_2 + rH_2O + (d - nz)NH_3 \quad (3.1.6)$$

where $r = 0.5[b - nx - 3(d - nz)]$

$s = a - nw$

The terms $C_aH_bO_cN_d$ and $C_wH_xO_yN_z$ represent on an empirical-mole basis the compositions of the organic material initially and at the conclusion of the process. An elemental analysis is required for an evaluation of the subscripts in these terms.

In an anaerobic process, the partial decomposition of an organic material can be represented by

$$C_aH_bO_cN_d \rightarrow nC_wH_xO_yN_z + mCH_4 + sCO_2 + rH_2O + (d - nz)NH_3 \quad (3.1.7)$$

where $s = a - nw - m$

$r = c - ny - 2s$

The ratio of oxidized materials to reduced materials in a system establishes an electric potential, called the *redox potential*:

$$E_h = \Psi \left(\ln \frac{\text{oxidants}}{\text{reductants}} \right) \quad (3.1.8)$$

Equations (3.1.4) to (3.1.7) formulate oxidation-reduction reactions. These reactions result in changing the ratio of oxidized materials to reduced materials in the environment in which the reactions occur, and, as a consequence, change the redox potential. Aerobic environments are generally characterized by redox values in the positive range, whereas anaerobic environments are characterized by negative values. It should be emphasized that the redox potential, like pH, is an intensity term and not a capacity term. Furthermore, the redox potential reflects the ratio of oxidant to reductant but not the concentration of either.

Solution procedure 3.1.1

PROBLEM Determine the empirical-mole composition of an organic material having an elemental composition of:

C   52.85 percent (dry weight)
H   6.48 percent (dry weight)
O   24.76 percent (dry weight)
N   15.12 percent (dry weight)

Table 3.1.1 COMPUTATION OF AN EMPIRICAL MOLE FROM THE ELEMENTAL COMPOSITION OF AN ORGANIC MATERIAL

Element	Atomic weight	Weight, %	Percent weight atomic weight	Ratio of atoms	Number of atoms in empirical mole
C	12	52.85	4.40	8.14	8
H	1	6.48	6.48	11.99	12
O	16	24.76	1.55	2.87	3
N	14	15.12	1.08	2.00	2

SOLUTION The percent compositions of the elements are divided by the appropriate atomic weights. The resulting ratios are converted proportionally using an even number for the lowest value (in this case, 2.00 for nitrogen), which yields an approximately even integer for all values. See Table 3.1.1. The empirical mole is found to be $C_8H_{12}O_3N_2$.

////

3.2 MICROORGANISMS

A wide variety of microorganisms are found in the environment. These include bacteria, molds, algae, protozoans, and small metazoans. The significance of microorganisms in environmental phenomena relates in large measure to their nutritional requirements. These requirements fall into four categories: (1) compounds that furnish the elements carbon and nitrogen with which cellular materials are constructed, (2) compounds used as sources of energy, (3) inorganic ions, and (4) growth factors, such as vitamins, etc.

Although microorganisms differ widely in many respects, the basic chemical composition of their protoplasm is quite uniform. Protoplasm is composed largely of proteins, but significant quantities of fats and carbohydrates may be in the cellular complex. The elemental composition of protoplasm can be expressed by the generalized relationship

$$C_aH_bO_cN_dP_eS_f$$

The elemental constituents in this relationship are the same as those found in the generalized expression for organic materials in Eqs. (3.1.4) and (3.1.5). The similarity between protoplasm and organic materials is not surprising when one considers the fact that most organics have their origin in living processes. The protoplasm in living organisms is often referred to as *biomass*.

Microorganisms that use carbon dioxide as a sole source of carbon are called *autotrophs*, whereas those that derive carbon from organic sources only are called *heterotrophs*. The term *facultative autotrophs* is applied to microorganisms that can use both carbon dioxide and organic compounds as carbon sources. Autotrophs utilize inorganic forms of nitrogen only. Heterotrophs may in some cases derive their nitrogen from inorganics, whereas in other cases they meet this requirement from organic materials.

Phosphorus and sulfur requirements of both autotrophs and heterotrophs may be met by inorganic phosphates and sulfates. Heterotrophs can obtain sulfur from organic compounds as well. Other ions important in nutrition, but in minute quantities, include magnesium, potassium, calcium, iron, manganese, copper, zinc, and cobalt. These ions are sometimes referred to as the *micronutrients*.

The need for certain growth factors, or *vitamins*, has been demonstrated in cultures of heterotrophs and many autotrophs. Such factors, though required in very small quantities, play important roles in life processes.

Living matter is distinguished from nonliving organic materials by its *metabolism*. Metabolism is defined as the chemical and physical processes continuously taking place in living organisms and cells, comprising those processes by which assimilated nutrients are built up into protoplasm (*anabolism*) and those by which protoplasm is used and broken down into simpler substances, with the release of energy (*catabolism*). The synthesis of protoplasm, an anabolic process, requires an input of energy:

$$\text{External substrate} + energy \rightarrow \text{cellular material} \quad (3.2.1)$$

The cellular material resulting from the synthesis process is energy-rich as compared to the substrate[1] utilized. Autotrophs obtain energy from synthesis either from inorganic compounds or, in the case of photosynthetic forms, from light. Heterotrophs derive their energy from organic sources only.

Organisms derive energy from organic materials in a process called *respiration*:

$$\text{External substrate} \rightarrow \text{metabolic products} + energy \quad (3.2.2)$$

Respiration is an oxidation process, requiring a hydrogen acceptor. Molecular oxygen serves as a hydrogen acceptor for aerobic organisms. Anaerobic organisms utilize sulfates, nitrates, carbon dioxide, and even organic compounds for such a purpose. The metabolic products resulting from respira-

---

[1] A substance or group of substances utilized as a source of nutrients for organisms.

tion are the inorganic compounds of carbon, nitrogen, and sulfur discussed in Sec. 3.1.

Synthesis, therefore, is *endergonic* (requires energy), whereas respiration is *exergonic* (releases energy). The two processes are complementary in metabolism. Although respiration can, and does, occur in the absence of synthesis, synthesis in heterotrophic organisms must always be accompanied by respiration. The energy transfer from respiration to synthesis is accomplished by a mechanism involving a chemical compound called *adenosine triphosphate* (ATP). This compound carries the energy to synthesis in the form of energy-rich phosphate bonds. With the destruction of the phosphate bond and the resultant release of energy to the synthesis process, the compound reverts to *adenosine diphosphate* (ADP), a low-energy compound. ADP is then available again for conversion to ATP by respiration. The ATP-ADP mechanism enables the chemical reaction comprising respiration and synthesis to be carried out at the relatively low temperatures characterizing metabolic processes. The ATP-ADP mechanism also points to the importance of phosphates to living organisms.

Energy is required also for the maintenance of life processes. These processes include transport, replacement of cellular parts, and all the other chemical and physical reactions that sustain life. The energy of maintenance can be obtained through the respiration process described by Eq. (3.2.2) or, in the absence of suitable external substrate, through the catabolic degradation of the internal cellular materials themselves. Such a process, called *endogenous respiration*, may be represented by

$$\text{Cellular material} \rightarrow \text{metabolic products} + energy \qquad (3.2.3)$$

The metabolic products and energy transfer mechanism are the same in endogenous respiration as in the respiration process described by Eq. (3.2.2). To avoid confusion, the process in Eq. (3.2.2) will be referred to hereafter as *primary respiration*.

Not all the energy released in respiration processes can be used for synthesis and/or maintenance. Some may be used in the motility of the organism. Furthermore, in conformance with the second law of thermodynamics, a portion of the energy is lost as heat. Figure 3.2.1 is a schematic diagram showing the release and utilization of energy in the metabolism of heterotrophic organisms.

The *ubiquity principle* has considerable significance in explaining the presence or absence of microorganisms in the environment. According to this principle, the species composition of the microorganism population present in any nonisolated environmental system reflects only the available nutrient supply and the prevailing environmental conditions and does not depend upon the special introduction of individual species. Microorganisms are ubiquitous in

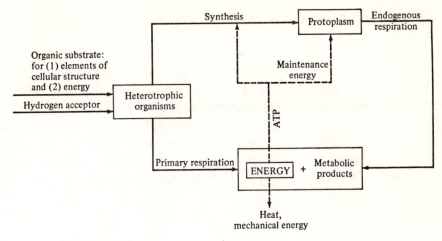

FIGURE 3.2.1
Release and utilization of energy in metabolism.

nature, and, for all practical purposes, it can be assumed that in such systems all species are present and their growth awaits only the availability of the right nutrients and the establishment of the proper environmental conditions. The use of inocula to introduce special species of microorganisms into an open system is redundant. Furthermore, as a corollary, if an inoculum is introduced into a system in which the nutrient supply and environmental conditions are unfavorable to growth, the species thus introduced will not become established and grow.

Solution procedure 3.2.1 The growth of microorganisms is influenced by the redox potential of their environment. In fact, the presence or

Table 3.2.1 CORRELATION OF REDOX POTENTIAL AND TYPE OF MICROORGANISM WITH REACTIONS TAKING PLACE IN LAKE WATERS AND SEDIMENTS*

Reaction	Redox potential range, mV	Microorganisms involved
1 Oxygen disappearance	+500–+350	Aerobes
2 Nitrate disappearance	+350–+100	Facultative anaerobes
Mn^{++} formation	below +400	
Fe^{++} formation	below +400	
3 Sulfide formation	below −150	Obligate anaerobes
Hydrogen, methane formation	below −150	

* Adapted from D. R. Keeney, The Fate of Nitrogen in Aquatic Ecosystems, table 9, *Lit. Rev.* 3, Water Resour. Cent., University of Wis., 1972.

absence of certain types of microorganisms in an environment can often be explained in light of the redox potential existing in that particular environment. Table 3.2.1 lists information correlating the redox potential and type of microorganism with certain reactions taking place in lake waters and sediments. Note should be made of the fact that odors resulting from sulfide formation are not expected to occur at redox potentials much higher than -150 mV. Also, methane formation is not expected at potentials higher than this value. It is to be emphasized that redox potentials are system specific. It is not possible to correlate redox potentials measured in one system with reactions occurring in another system. ////

3.3 GROWTH KINETICS

Given an environmental situation in which growth can occur, the concentration of biomass of an organism will tend to increase as the result of synthesis and decrease through endogenous respiration and death. Mathematically, the net change in concentration can be expressed as

$$\frac{dX}{dt} = \underset{\text{growth}}{\mu X} - \underset{\text{decay}}{k_d X} \quad (3.3.1)$$

where X = concentration of biomass. $[FL^{-3}]$

μ = specific growth rate. $[t^{-1}]$

k_d = specific decay rate. $[t^{-1}]$

When all nutrients are present and available to the organism in excess, the value of the specific growth rate remains constant for the particular set of nutrients, environmental conditions, and species of organism. When the concentration of one of the nutrients becomes growth limiting, however, the value of the specific growth rate decreases, and growth declines. The effect of the limiting nutrient on the specific growth rate is illustrated in Fig. 3.3.1. This hyperbolic relationship is known as the *Monod function*.[1]

$$\mu = \hat{\mu} \frac{S}{K_s + S} \quad (3.3.2)$$

where $\hat{\mu}$ = maximum specific growth rate. $[t^{-1}]$

S = limiting nutrient concentration. $[FL^{-3}]$

K_s = saturation constant; equal to nutrient concentration at one-half maximum growth rate. $[FL^{-3}]$

---

[1] J. Monod, *Recherches sur la croissance des cultures bactériennes*, Hermann & Cie, Paris, 1942.

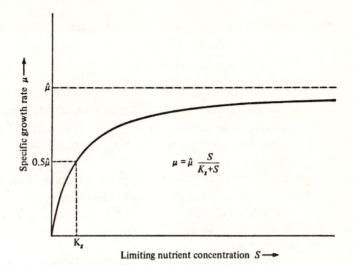

FIGURE 3.3.1
Specific growth rate as a function of the limiting nutrient concentration.

When all nutrients are present in excess, the specific growth rate μ approaches the maximum specific growth rate $\hat{\mu}$, and the growth term in Eq. (3.3.1) becomes a first-order reaction term $\hat{\mu}X$. However, when the concentration of the limiting nutrient becomes small compared with the value of the saturation constant $S \ll K_s$, the growth term becomes $(\hat{\mu}/K_s)SX$, a second-order reaction relationship.

The Monod function was first derived on an empirical basis. However, attempts have been made to relate it to the cellular mechanisms of transport and enzymatic utilization of the limiting nutrient. It is important to note that mathematical expressions other than the Monod function have been used to describe the decreasing-rate-of-increase relationship between the concentration of the limiting nutrient and the specific growth rate.

The maximum specific growth rate in the Monod function has been found to be a function not only of the inherent nature of the organism in question and the nutrients present in the growth medium but, also, of environmental factors, both physical and chemical. Temperature and pH exert profound effects. Salinity affects the growth rate through an influence on osmotic pressure. A wide variety of organic and inorganic toxic agents serve to reduce the growth rate. The temperature effect has been quantified by the relationship described by Eq. (2.2.18):

$$\hat{\mu} = \hat{\mu}_0 \Theta^{(T-T_0)} \qquad (3.3.3)$$

where $\hat{\mu}$, $\hat{\mu}_0$ = maximum specific growth rates at temperatures T and T_0, respectively. $[t^{-1}]$

Θ = temperature coefficient

From approximately 15 to 30°C, the temperature coefficient, based on thermodynamic information, has been estimated to have a value of 1.047. However, temperature coefficients ranging from 1.03 to 1.07 have been reported for microorganisms found in aerobic biologic treatment processes used in waste treatment.

A broad spectrum of microorganism species exhibits an optimal growth rate within the pH range of from 6 to 9. However, outside this range the growth rates of most of these microorganisms decrease rapidly. Most natural environmental systems, such as water and soil, have pH values at levels falling in the range of from 6 to 9.

The decay term in Eq. (3.3.1), as was indicated previously, is a function that expresses the rate at which the concentration of biomass decreases as the result of endogenous respiration and death. Empirically, this relationship has been found to follow first-order reaction kinetics. The value of the specific decay rate k_d is influenced by many of the factors influencing the specific growth rate.

The relative importance of each of the two terms on the right side of the equals sign in any situation will depend upon the nutrient-to-microorganism ratio and the existing environmental conditions. When the nutrient-to-microorganism ratio is relatively high, the first term predominates, and the second term may be insignificant. On the other hand, when the ratio is very low, the concentration change depends entirely on the decay term. Such behavior is revealed in batch-culture phenomena discussed in the following paragraphs. Adverse environmental conditions tend to magnify the decay influence at the expense of the growth term. Temperature influences the decay rate in much the same manner as it affects the growth rate.

Classic Growth Pattern

When discussing growth kinetics, use is generally made of batch-culture phenomena. In a batch-culture situation, there is no inflow of nutrients into the system or outflow of microorganisms from the system. The culture is isolated and totally dependent upon preceding events taking place within the system. Although batch phenomena normally do not apply to environmental

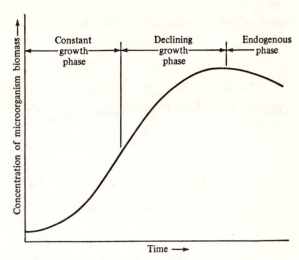

FIGURE 3.3.2
Classic growth pattern.

situations, batch-culture kinetics are useful in revealing certain characteristics of microbial growth patterns.

Attention is directed to Fig. 3.3.2. The curve, which illustrates the classic growth pattern exhibited by microorganisms in batch cultures, reveals that growth passes through at least three different phases. Initially, all nutrients are present in excess of the requirements of the microorganisms, and growth is unrestricted. During this period, called the *constant growth phase*, $\mu = \hat{\mu}$, and the concentration of microorganism biomass increases at an exponential rate; this phase is sometimes referred to as the *log growth phase*. Although organism decay through endogenous respiration and death takes place concurrently with growth, the maximum specific growth rate $\hat{\mu}$ is generally one to two orders of magnitude greater than the specific decay rate k_d, and the growth term in Eq. (3.3.1) masks the decay term during the constant growth phase.

At some concentration, one of the nutrients becomes growth-limiting, and the culture proceeds into the *declining growth phase*. During this phase, the specific growth rate decreases in response to the declining concentration of the limiting nutrient. When growth approaches cessation, the decay term in Eq. (3.3.1) finally predominates, and the concentration of biomass decreases in what is termed the *endogenous phase*. When microorganisms are introduced into a growth medium to which they are unacclimated, there occurs prior to the constant growth phase a *lag phase* in which the microorganisms become adjusted to the culture environment. The lag phase is not shown in Fig. 3.3.2.

Nutrient Utilization

Growth and nutrient utilization are directly related:

$$\frac{dS}{dt} = -\frac{1}{Y}\frac{dX}{dt} \quad (3.3.4)$$

where dS/dt = rate of nutrient utilization. $[FL^{-3}t^{-1}]$

dX/dt = rate of change in biomass concentration. $[FL^{-3}t^{-1}]$

Y = growth yield

The growth yield Y, which always is less than 1 on a mass basis, varies with the nutrient, organism species, and environmental conditions. Frequently, in an aerobic environment the value of the growth yield will be found in the neighborhood of 0.5. Although the growth yield may not remain constant even under a given set of conditions, for practical purposes it is generally considered to do so.

3.4 BIOCHEMICAL OXYGEN DEMAND

Most surface waters contain an indigenous flora of microorganisms. When a mixture of organic materials is discharged into such waters, the organic materials are utilized for growth by the heterotrophic microflora at a rate determined by Eq. (3.3.4):

$$\frac{dS}{dt} = -\frac{1}{\bar{Y}}\frac{dX}{dt} = -\frac{1}{\bar{Y}}\hat{\mu}\frac{S}{K_s - S}X \quad (3.4.1)$$

where S = concentration of organics. $[FL^{-3}t^{-1}]$

$\bar{Y}$ = average growth yield

Since the specific growth rate, even at low concentrations of the limiting nutrient, is generally much larger than the specific decay rate, a portion of the organics in the water is rapidly incorporated into the total biomass of the standing crop of microflora, and the remainder is converted into metabolic products through respiration. The period in which the organic materials are utilized in growth is followed by a relatively longer period in which the biomass wastes away in decay. In an aerobic environment, decay, being the result of

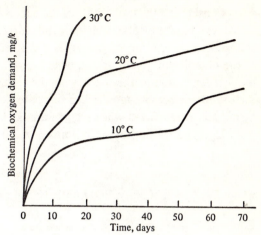

FIGURE 3.4.1
Biochemical oxygen demand as a function of time.

endogenous respiration and death, is accompanied by the utilization of the oxygen resources in the system. The quantity of oxygen utilized can be predicted by a stoichiometric relationship such as those represented by Eqs. (3.1.4) and (3.1.5). In the case of endogenous respiration, the organic-material term in the equations represents cellular biomass.

In the measurement of organic pollution in surface waters, it is common practice to measure concentrations in terms of the rate at which oxygen is utilized in the biologic decomposition of the organic materials. When organic materials are introduced to water containing a flora of microorganisms and oxygen, the dissolved oxygen concentration will vary with time in a relationship that generally appears to follow the curves in Fig. 3.4.1. These curves are compound curves, each composed of two different stages. The first stage, which at environmental temperatures extends over a period of at least several days, follows the first-order reaction expression

$$\frac{dL}{dt} = -k_1 L \qquad (3.4.2)$$

where L = first-stage biochemical oxygen demand (BOD). $[FL^{-3}]$

k_1 = deoxygenation constant. $[t^{-1}]$

It is to be noted that both Eq. (3.4.2) and the decay term in Eq. (3.3.1) are first order.

Considering the time frame of the curves in Fig. 3.4.1, one has to conclude that these biochemical-oxygen-demand curves for the most part reflect the endogenous respiration of protoplasm derived from the more rapid synthesis of the organics originally in the water. Synthesis may take place only during the first few hours at the most, with the result that oxygen uptake kinetics of such synthesis are obscured when data are determined on a daily basis. This hypothesis is supported when oxygen uptake is determined on an hourly basis.

Upon integration between limits of $t = 0$ and $t = t$, Eq. (3.4.2) becomes

$$L_t = L_0 e^{-k_1 t} \qquad (3.4.3)$$

where L_0, L_t = first-stage BOD initially and at time t, respectively. $[FL^{-3}]$

Equation (3.4.3) can also be expressed as

$$y = L_0(1 - e^{-k_1 t}) \qquad (3.4.4)$$

where $y = L_0 - L_t$ = BOD that has been exerted by time t. $[FL^{-3}]$

Equation (3.4.4) has wide use in water pollution control. In relating Eq. (3.4.4) to the BOD curves in Fig. 3.4.1 the term L_0 is considered to be equal to the asymptotic value approached by the first-stage portion of the curve. For this reason L_0 is commonly called the first-stage BOD and is designated by L.

The first stage of the BOD curve is interpreted as resulting from the BOD of the carbon in the organic materials being decomposed. The second stage is considered to reflect the oxygen utilized in the oxidation of the nitrogen compounds, a process called *nitrification*. However, because of the length of time generally elapsed before the onset of the nitrification stage, the latter in many cases has been considered as having little practical significance in the standard BOD test.

It is to be noted in Fig. 3.4.1 that both the first-stage BOD L and the deoxygenation constant k_1 are influenced by temperature. The temperature coefficients [see Eq. (2.2.18)] are often assumed to be 1.02 and 1.05 for L and k_1, respectively.

The 5-d 20°C BOD test is generally accepted as the standard BOD laboratory test. The deoxygenation constant k_1 varies widely with the type of organic material exerting the BOD. However, a value of 0.23 at 20°C is often used as an average for BOD exerted by organics with origins in domestic waste discharges.

Solution procedure 3.4.1 The use of Eq. (3.4.4) to predict the BOD of a particular water requires a knowledge of both L and k_1. These parameters can be estimated from laboratory data with a relatively simple graphic method.[1] The method is based on the similarity of the two expressions

$$1 - e^{-k_1 t} = k_1 t \left[1 - \frac{k_1 t}{2} + \frac{(k_1 t)^2}{6} - \frac{(k_1 t)^3}{24} + \cdots \right] \quad (3.4.5)$$

$$\text{and } k_1 t \left(1 + \frac{k_1 t}{6} \right)^{-3} = k_1 t \left[1 - \frac{k_1 t}{2} + \frac{(k_1 t)^2}{6} - \frac{(k_1 t)^3}{21.6} + \cdots \right] \quad (3.4.6)$$

Therefore, Eq. (3.4.4) can be approximated by

$$y = L k_1 t \left(1 + \frac{k_1 t}{6} \right)^{-3} \quad (3.4.7)$$

which can be rearranged into a straight-line form:

$$\left(\frac{t}{y} \right)^{1/3} = (k_1 L)^{-1/3} + \frac{k_1^{2/3}}{6 L^{1/3}} t \quad (3.4.8)$$

By plotting the term $(t/y)^{1/3}$ as the ordinate and t as the abscissa, the intercept of the straight line of best fit at $t = 0$ will give the value of $(k_1 L)^{-1/3}$, and the slope will be equal to the value of $k_1^{2/3}/6 L^{1/3}$. Simultaneous solution of the two expressions will yield k_1 and L. Experimental values of y used to determine these parameters should not exceed $0.9L$ in order to keep deviations from the approximating straight line from becoming significant. The straight line of best fit can be fitted by eye or by the method of least squares. ////

3.5 ANAEROBIC DECOMPOSITION

In an environment devoid of oxygen, organic materials are decomposed by bacteria in a process called *anaerobic decomposition*. Such an environment may become established within organic particulates, decaying masses of animal and vegetable origin, and aquatic systems in which all dissolved oxygen has been depleted. Figure 3.5.1 illustrates the stepwise nature of the process.

Complex organics such as fats, proteins, and carbohydrates are converted to soluble organic acids by a group of bacteria known collectively as *acid formers*. If conditions are favorable, a second group of bacteria called *methane formers* converts the organic acids to gaseous products, predominately methane

---

[1] H. A. Thomas, Jr., Graphical Determination of BOD Curve Constants, *Water and Sewage Works*, vol. 97, p. 123, 1950.

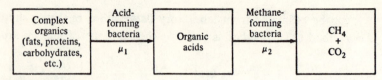

FIGURE 3.5.1
Anaerobic decomposition of organic materials.

and carbon dioxide. In both the acid-producing step and methane-producing step, bacterial cells result from growth.

The entire process may be represented by Eq. (3.1.7). Methane, carbon dioxide, and ammonia are the chief gaseous products of the process. These gases constitute approximately 95 to 98 percent of the gas evolved. The remaining volume is composed of hydrogen sulfide and hydrogen. Some organic materials such as lignin are resistant to the activity of both groups of bacteria and, hence, remain unaltered in an anaerobic environment.

The acid-forming bacteria thrive over a rather broad range of environmental conditions, whereas the methane producers are quite sensitive to such conditions and may not grow under circumstances readily adapted to by the acid formers. For this reason, the second step, the methane-forming step, is the rate-controlling step. Quite often in an environment where the growth rate of the methane bacteria is inhibited by adverse environmental conditions, organic acids will accumulate to reduce the pH to a level where almost all bacterial activity comes to a halt. That decomposition that does occur in such a situation is called *benthic decomposition*. The term is derived from the fact that the diffusion of organic materials at the bottom of rivers and lakes often is so slow that anaerobic decomposition is retarded to a very significant degree. The products of benthic decomposition, primarily organic acids, escape slowly from such deposits to the water above. When aerobic conditions prevail in the water above, the organic acids decompose aerobically, exerting a demand on the available oxygen resources of the water. The methane and carbon dioxide that do evolve from the deposit generally escape in bubble form to the atmosphere.

The growth of methane bacteria in sludges from domestic waste waters normally is inhibited at pH values less than 6.5 and by organic acid concentrations above 2,000 mg/l. In other wastes, growth has been observed at organic acid concentrations up to 10,000 mg/l. Furthermore, temperature exerts a profound effect on the activities of these bacteria. Such an effect can be predicted by the van't Hoff–Arrhenius relationship expressed in Eq. (2.2.18). At 27°C and above, growth increases gradually with increasing temperature, reaching an optimum at about 55°C. Below 27°C, growth and, hence, methane production decrease rapidly with decreasing temperature.

It should be noted that anaerobic metabolism is not as efficient as aerobic metabolism in producing cellular protoplasm. The energy mechanisms involved in the former are such that more substrate must be utilized to synthesize a unit mass of cell material than is the case in aerobic metabolism.

3.6 PHOTOSYNTHESIS

Life is created and sustained by the process known as photosynthesis. As shown in Fig. 3.1.1, through photosynthesis the carbon in carbon dioxide is incorporated in plant materials. Whereas some of the carbon is returned to the inorganic state through plant or microbial respiration, the remainder is assimilated by animal consumers. The energy from light radiation is locked into the structure of organic compounds to provide the fuel for sustenance of life, both plant and animal.

Of particular importance here is the photosynthesis that takes place in unicellular plants called *algae*. The growth and death of such organisms exerts a profound influence on the limited oxygen resources of surface waters.

Algal growth results from two interrelated processes—photosynthesis and secondary synthesis. In photosynthesis, carbohydrate and oxygen are produced from carbon dioxide and water through a reaction energized by illumination:

$$CO_2 + 2H_2O + energy \rightarrow (CH_2O) + H_2O + O_2 \qquad (3.6.1)$$

Some of the carbohydrate resulting from photosynthesis is converted to fats, proteins, and other organics required by the algal cell. A simplified diagram showing the overall process of photosynthesis and secondary synthesis is shown in Fig. 3.6.1. The cyclic pattern characterizing these metabolic processes is indicated by the circles. The stoichiometric relationships between carbon dioxide, oxygen, and algal cells can be represented by expressions that are the reverse of Eqs. (3.1.4) and (3.1.5).

The energy required for photosynthesis can be derived only from light having a wavelength falling between 4000 and 7000 Å. (See Fig. 3.6.2.) Light having wavelengths within this range is mostly visible and constitutes about 40 percent of the total energy of solar radiation.

The growth kinetics of algae are described by the same relationships discussed in Sec. 3.3. The specific growth rate, however, can be influenced by light intensity. That is, light intensity instead of a nutrient such as carbon

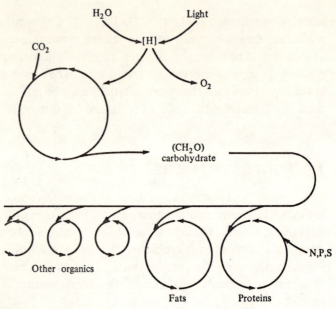

FIGURE 3.6.1
Simplified diagram of algal cell synthesis.

dioxide or phosphorus may be growth-limiting. For most algal species, light intensity is not growth-limiting above a value of about 2.5×10^4 ergs/(cm^2)(s) (600 fc). At a light intensity of about 10×10^2 ergs/(cm^2)(s)(24 fc), the so-called *compensation point*, growth and respiration are about equal. At intensities below the compensation point, the rate at which oxygen is utilized by respiration is greater than the rate at which it is produced by photosynthesis. At these low intensities, oxygen is drained from the aqueous environment, and the algae are in an endogenous phase.

Light penetration in an algal suspension can be approximated by the Beer-Lambert law:

$$I = I_0 \exp\left[-(\alpha + \beta X)z\right] \quad (3.6.2)$$

where I = light intensity at depth z. $[Ht^{-1}]$

I_0 = light intensity at water surface. $[Ht^{-1}]$

α, β = light absorption coefficients of water alone and of algal suspension, respectively. $[L^{-1}], [L^2 F^{-1}]$

X = concentration of algal suspension. $[FL^{-3}]$

z = depth below surface. $[L]$

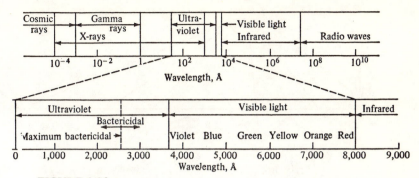

FIGURE 3.6.2
The electromagnetic spectrum.

The saturation intensity of 2.5×10^4 ergs/(cm^2)(s) is to be compared with 4.2×10^5 ergs/(cm^2)(s)(10,000 fc), the intensity of full sunlight. It is obvious, therefore, that the efficiency of light utilization in photosynthesis in natural environments is relatively low. The efficiency with which algae utilize solar radiation in the environment varies from approximately 2 to 9 percent with an average value of approximately 4 percent. That portion of the solar radiation that is not utilized in the formation of algal cells is converted to heat.

3.7 FOOD CHAINS

Any segment of nature in which there is an exchange of materials between organisms and their environment can be called an ecological system, or ecosystem. An ecosystem such as normally exists in a natural body of water consists of four groups of components: (1) abiotic substances, (2) producers, (3) consumers, and (4) decomposers.

The abiotic substances are the nonliving components of the system. This group includes water, carbon dioxide, detritus (nonliving organic and inorganic particulates), alkalinity, and all other inorganic ions.

The producer group comprises the photosynthetic organisms. These organisms are of two types: (1) rooted or large floating plants and (2) microscopic floating plants, usually algae, called *phytoplankton*. In large ponds and lakes, phytoplankton is generally more prominent than rooted vegetation.

The consumer group is made up of heterotrophic organisms, mostly animals, that feed on other organisms or organic detritus. The group includes the *herbivores* (zooplankton that feed on the producers) and the *carnivores*

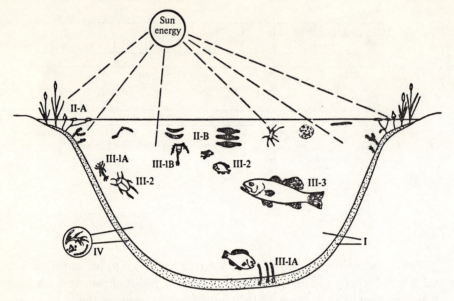

FIGURE 3.7.1
Diagram of the pond ecosystem. Basic units are as follows: I, abiotic substances—basic inorganic and organic compounds; IIA, producers—rooted vegetation; IIB, producers—phytoplankton; III-1A, primary consumers (herbivores)—bottom forms; III-1B, primary consumers (herbivores)—zooplankton; III-2, secondary consumers (carnivores); III-3, tertiary consumers (secondary carnivores); IV, decomposers—bacteria and fungi of decay. (*Taken from E. P. Odum, "Fundamentals of Ecology," 3rd ed., p. 13, fig. 2-2, Saunders, Philadelphia, 1971.*)

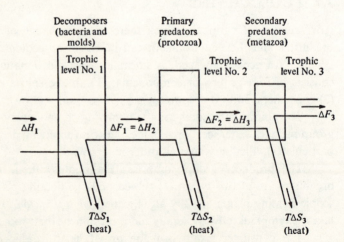

FIGURE 3.7.2
Diagram of energy flow through a decomposer food chain.

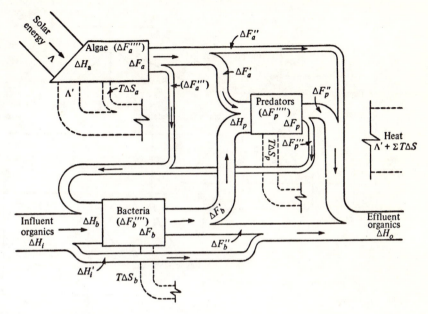

FIGURE 3.7.3
Energy flow through the biotic component of an oxidation pond.

(animals that feed on the herbivores or other carnivores). Often there are two or three groups of carnivores, with one or two groups feeding on a smaller group.

The decomposer group consists of *saprophytic* organisms, chiefly bacteria and fungi, which decompose the soluble organics and organic detritus. Figure 3.7.1 is a diagram of a typical pond ecosystem. The four groups of ecosystem components are identified in pictorial schematics.

The producers, consumers, and decomposers exist in a relationship called a *food chain*. A food chain is a sequence of prey-predator interactions in which a portion of the chemical energy initially in the protoplasm of the prey in the first interaction is passed on from the protoplasm of one organism to that of another.

A simplified diagram of energy flow through a food chain is presented in Fig. 3.7.2. According to the conservation principle in thermodynamics, the flow of energy through each *trophic level* (link in the food chain) equals the sum of the free energy available for growth and maintenance at that level and the entropy lost through respiration:

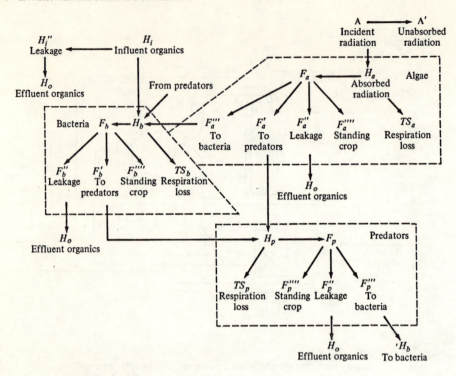

FIGURE 3.7.4
Energy sources and fates at time t.

$$\Delta H = \Delta F + T \Delta S \qquad (3.7.1)$$

where ΔH = enthalpy change

ΔF = free-energy change

ΔS = entropy change

T = temperature

It can be seen from Fig. 3.7.2 that of the initial energy assimilated by the decomposers ΔH_1 a fraction is lost in respiration $T \Delta S_1$ while the remainder ΔF_1 is utilized in growth and becomes available food energy ΔH_2 for the primary predators. A similar loss in energy takes place at subsequent levels.

Although there will normally be more than one food chain present in an ecosystem, all such food chains are interconnected in what is called a *food web*. A diagram of energy flow through a food web existing in a waste oxidation pond is presented in Fig. 3.7.3. Most of the energy entering the pond as solar radiation is shown as leaving the pond as heat or as being incorporated in bacterial and algal cells. The paths shown indicate the flow of energy through the system. For any specific aquatic ecosystem, the width of the path represents the relative rate of energy flow. The notation in Fig. 3.7.3 can be interpreted with the aid of Fig. 3.7.4. The latter is a diagram designating the energy sources and fates at time t.

EXERCISES

3.1.1 Discuss the carbon and nitrogen cycles in reference to the following ecosystems:
(a) An ocean (b) A city park
(c) A dairy farm (d) A forest

3.1.2 Arrange the following organic compounds in descending order of their heats of combustion:

Microorganisms	$C_5H_7NO_2$
Glucose	$C_6H_{12}O_6$
Glycerol	$C_3H_8O_3$
Acetic acid	$C_2H_4O_2$
Methane	CH_4
Carbon dioxide	CO_2

3.1.3 Following the discharge of a domestic waste, a river water ($T = 20°C$) contains 10 mg/l ammonia. If aerobic conditions are maintained, how much oxygen (in milligrams per liter) will be required to convert the ammonia to nitrate? How does this quantity compare with the saturation concentration of oxygen?

3.1.4 Determine the empirical molecular composition of an organic material with an elemental composition of $C = 54.5$ percent, $H = 6.0$ percent, $O = 27.1$ percent, and $N = 12.4$ percent.

3.1.5 One kg (dry weight) of organic material was partially decomposed microbiologically under aerobic conditions. Upon completion of the process, the organic material weighed only 200 g (dry weight). If the empirical molecular weight of the initial material was $C_{31}H_{50}O_{26}N$ and that of the final material $C_{11}H_{14}O_4N$, how many grams of oxygen were required by the process?

3.1.6 How much energy was released as heat in the decomposition process described in Exercise 3.1.5?

3.1.7 Ten kg (dry weight) of organic material with an empirical molecular weight of $C_8H_{12}O_3N_2$ are digested anaerobically. During the process, 39.2 l (STP) of carbon dioxide were evolved. At the end of the process, 4 kg (dry weight) of sludge solids with an empirical molecular weight of $C_5H_7O_2N$ remained. How much methane evolved during the process? How much energy was released?

3.1.8 What is the significance of redox potential in the environment?

3.2.1 Discuss the nutritional requirements of microorganisms.

3.2.2 What is meant by each of the following terms?
(a) Autotrophs (b) Biomass (c) Anabolism
(d) Heterotrophs (e) Metabolism (f) Synthesis
(g) Facultative autotrophs (h) Catabolism (i) Respiration
(j) Exergonic (k) Endogenous respiration

3.2.3 In the metabolic processes of microorganisms, what serves as a hydrogen acceptor in aerobic environments? In anaerobic environments?

3.2.4 Discuss the metabolic mechanism by which respiration is coupled to synthesis.

3.2.5 Discuss the fate of the energy released in microorganism respiration.

3.2.6 As an environment becomes anaerobic, in what order would one expect SO_4^{--}, NO_3, and O_2 to disappear? Would one expect CH_4 production prior to the disappearance of nitrates? Why?

3.3.1 With the use of a family of curves such as the one in Fig. 3.3.1, illustrate the effect of increasing the value of the saturation constant.

3.3.2 On what factors does the maximum specific growth rate in the Monod function depend?

3.3.3 Integrate Eq. (3.3.1). With the use of the integrated equation and with constant values of the specific growth rate and decay rate, plot the concentration of biomass as a function of time. Repeat using a decreasing value of the specific growth rate for successive points on the curve. Relate these curves to corresponding segments of the classic growth curve in Fig. 3.3.2.

3.3.4 A batch test was performed in the laboratory to determine the production of activated sludge solids when the latter is used to treat a waste. The data collected are tabulated below. Determine the growth yield.

Aeration time, h	COD* of waste, mg/l	Activated sludge solids, mg/l
0	1,200	2,000
1	940	2,068
2	720	2,118
3	540	2,163
4	430	2,183
5	360	2,196

*Chemical oxygen demand.

3.4.1 What relationship does the deoxygenation constant k_1 have with the specific growth rate μ?

3.4.2 What effect does temperature have on the first-stage BOD? On the deoxygenation constant?

3.4.3 What interpretation is given to the first-stage BOD? The second-stage BOD?

3.4.4 If the 5-d 20°C BOD of a river water is 10 mg/l, what will be the 3-d 25°C BOD? The 7-d 15°C BOD? Assume k_1 to be equal to 0.23 d^{-1} at 20°C.

3.4.5 Three hundred eighty m^3 of a domestic waste water containing no dissolved oxygen and having a 5-d 20°C BOD of 150 mg/l are discharged into a pond containing 5,680 m^3 of water saturated with oxygen (8.38 mg/l) and having a 5-d 20°C BOD of 1 mg/l. If the temperature of both the waste and the pond water is 25°C, what will be the dissolved oxygen in the pond 24 h after the introduction of the waste? Assume surface aeration to be negligible and the deoxygenation constant to be equal to 0.23 d^{-1} at 20°C.

3.4.6 The state water control board limits the daily waste discharge of an industry into a stream to 750 kg of first-stage 20°C BOD. As an economic method of treatment, the industry considers the possibility of permitting the waste to flow through a series of holding ponds connected by short baffled channels prior to discharge into the stream, the idea being that a portion of the waste's BOD will be exerted in the ponds and that the waste will become saturated with oxygen as it flows from one pond to another. The waste has a 5-d 20°C BOD of 150 mg/l, a deoxygenation constant of 0.20 d^{-1} at 20°C, and no settling solids; it is saturated with oxygen and has a constant flow of 63 × 10^{-3} m^3/s. Assuming the temperature of the waste in the ponds will be 15°C and that short circuiting and surface aeration in the ponds will be negligible, what will be the fewest number of ponds required, and how large will each have to be to permit the industry to discharge a waste of acceptable quality? The oxygen saturation concentration at 15°C is 10.15 mg/l.

3.4.7 From the following 20°C BOD data, estimate the values of the deoxygenation constant k_1 and the first-stage BOD.

Time, d	2	4	6	8
BOD, mg/l	178	280	334	364

3.5.1 Which draws most heavily on the oxygen resources of a stream, aerobic or anaerobic decomposition? Why?

3.5.2 Which of the following materials is the most resistant to anaerobic decomposition?
(a) Proteins   (b) Wood   (c) Fats
(d) Coal   (e) Carbohydrates   (f) Paper

3.5.3 Explain how an environment that results in benthic decomposition becomes established at the bottom of a flowing stream or a lake. What characterizes benthic decomposition?

3.5.4 Will more biomass be created from a unit weight of organic substrate in an anaerobic environment than will be the case in an aerobic environment? Why?

3.5.5 What gases characterize anaerobic decomposition?

3.6.1 Explain the difference between photosynthesis and secondary synthesis and how they are related.

3.6.2 One thousand kg of nitrogen are discharged into a lake with a volume of 1 million m^3. What will be the growth potential for algae if the empirical formula of the algal biomass is assumed to be $C_6H_{10}O_2N$?

3.6.3 Write the Monod functions for the case where light limits algal growth. Do the same for the cases where nitrogen, phosphorus, and carbon dioxide each limits growth.

3.6.4 What is the significance of the compensation point? Will it vary over a diurnal cycle?

3.6.5 At what depth will the compensation point be when the surface of the water is exposed to full sunlight and the algal concentration is 100 mg/l? Assume the light absorption coefficients of the water alone and of the algal suspension to be 2.95×10^{-3} cm^{-1} and 8.20×10^{-6} (mg/l)$^{-1}$ cm^{-1}, respectively.

3.6.6 Assuming that solar light intensity varies as a half-sine wave from 6 A.M. to 6 P.M. and that 4 percent of the energy falling incident to a pond is converted to algal cells, estimate the diurnal energy flux utilized in photosynthesis. How much is converted to heat? Assuming a biomass composition of $C_6H_{10}O_2N$, how much algae will be produced?

3.7.1 Define each of the following terms:

(a) Herbivores (b) Decomposers (c) Trophic level
(d) Carnivores (e) Phytoplankton (f) Food web

3.7.2 Construct an energy flow diagram for a food chain linking man with vegetation in a meadow.

3.7.3 What relationship does the energy flow through a food chain bear to the carbon and nitrogen cycles?

3.7.4 Resketch Figs. 3.7.3 and 3.7.4 for the following environmental modifications:
(a) No introduction of influent organics
(b) No effluent
(c) No sunlight

4

ECOLOGICAL SYSTEMS

4.1 MODELS

A *system* is a collection of components arranged and interconnected in such a way that when a change occurs with respect to one component the effect of the change is felt by the other components as well. The components, which themselves may be subsystems, may be physical, chemical, biologic, or a combination of all three. A distinguishing feature of systems is that they have one or more inputs and outputs. Inputs are sometimes referred to as *forcing functions*, and outputs as *responses*. As these terms imply, if the input is varied in a particular way, then the output will respond in its particular way. The dependence of the output upon the input is established by the properties of the system which collectively constitute the system's model.

Models describing the behavior of environmental systems are of three types: transport-phenomena models, empirical models, and population-balance models. These models are mathematical in form and embody, either deterministically or statistically, the characteristic properties of the system they represent. Transport-phenomena models are derived from the concepts of the balance of mass, energy, and momentum. They express the system's behavior in terms of deterministic mechanisms. Empirical models are developed from a statistical treatment of data derived either from the system in question or

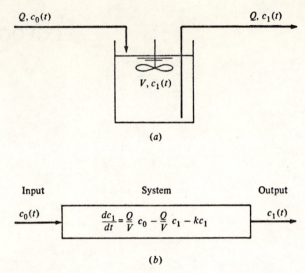

FIGURE 4.1.1
(a) System schematic; (b) information flow diagram for a continuous-flow completely mixed system.

similar systems. Population-balance models are developed from a consideration of residence-time distributions and other age distribution characteristics within systems. System models are only as accurate as the degree to which they represent the system being modeled. Considerable effort should be devoted to validating a model prior to its use in systems analysis and design.

A distinction of considerable importance in all considerations of systems is the one between a transient and steady-state output. If the input is changed suddenly from one constant value to another or from one periodic function to another, the output will go through a temporary adjustment (or transient) period before settling down to a new steady-state pattern. During *this* period the system is said to be exhibiting *dynamic behavior*.

The discussion that follows will deal with transport-phenomena models, specifically those models developed from the concepts of the conservation or balance of mass.

Attention is directed to Fig. 4.1.1a. Shown there is a system consisting of a tank through which an aqueous solution of a material flows. The volumetric rates of flow into and out of the tank are equal and constant. The tank contents are completely mixed, and while in the tank the concentration of the material in solution is reduced by a first-order reaction.

From a weight balance of the dissolved material across the tank one has

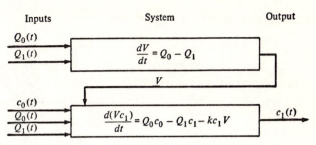

FIGURE 4.1.2
Information flow diagram for system in Fig. 4.1.1a when volumetric flow rates into and out of system are unequal and time-varying.

$$\text{Rate of accumulation} = \text{inflow} - \text{outflow} - \text{decay}$$

$$V\frac{dc_1}{dt} = Qc_0(t) - Qc_1 - kc_1 V \qquad (4.1.1)$$

where c_1 = concentration of material in tank and discharge. $[FL^{-3}]$

c_0 = concentration of material in inflow. $[FL^{-3}]$

Q = volumetric rate of flow into and out of tank. $[L^3 t^{-1}]$

V = volume of tank. $[L^3]$

k = first-order rate constant. $[t^{-1}]$

Equation (4.1.1) provides the model that describes the behavior of the system $c_1(t)$ in response to the time-varying input $c_0(t)$. Such a relationship is represented by the information flow diagram in Fig. 4.1.1b. The diagram indicates that by supplying the value of c_0 as a continuous function of time the value of $c_1(t)$ can be determined by a continuous integration of the model. Because Q, V, and k are constant, they, unlike $c_0(t)$, do not have to be supplied to the diagram.

If, instead of being equal, the volumetric rates of flow into and out of the system are unequal and are functions of time, the model of the system must be modified as follows: A flow balance struck across the system yields an expression for the time rate of volume change in the tank:

$$\frac{dV}{dt} = Q_0(t) - Q_1(t) \qquad (4.1.2)$$

The incorporation of this expression in the information flow diagram is shown in Fig. 4.1.2.

4.2 ANALYTIC SOLUTIONS

Equation (4.1.1) is a first-order, linear differential equation, one of the few types of differential equations for which there are convenient analytic solutions. This equation may be rearranged to give

$$\frac{dc_1}{dt} + ac_1 = \frac{Q}{V} c_0(t) \qquad (4.2.1)$$

where

$$a = \frac{Q}{V} + k \qquad (4.2.2)$$

Equation (4.2.1) has the form in which first-order, linear differential equations are often expressed. The reciprocal of Eq. (4.2.2) $1/a$ has the dimension of time and is referred to as the *time constant*. The term a and its reciprocal, the time constant, are a measure of the time required for the system described by the model to respond to a change in input concentration $c_0(t)$. The ratio Q/V is an important characteristic of any continuous-flow system. This ratio has the dimension of reciprocal time t^{-1} and is referred to as the *dilution rate* of the system. It is an expression of the number of volume changes occurring in the system per unit of time. The reciprocal of the ratio, called the *retention time*, is used to describe the time the fluid is theoretically held in the system.

The general solution to Eq. (4.2.1) is

$$c_1 = \frac{Q}{V} e^{-at} \int_0^t c_0(t) e^{at} \, dt + c_{1_i} e^{-at} \qquad (4.2.3)$$

where $c_{1_i} = c_1$ at time equals 0

The second term of Eq. (4.2.3) evaluates the initial condition of the system. The first term evaluates the concentration change brought about by the input $c_0(t)$. Solutions to the equation for three common forms of the input are presented in Table 4.2.1.

Table 4.2.1 SOLUTIONS TO EQ. (4.2.3) FOR VARIOUS INPUT FUNCTIONS*

Type of input	Type of $c_0(t)$	Concentration $c_1(t)$
Constant	c_0	$\frac{c_0}{a}\frac{Q}{V}(1 - e^{-at}) + c_{1_i}e^{-at}$
Linear	$c_0 \pm bt$	$\frac{c_0}{a}\frac{Q}{V}(1 - e^{-at}) \pm \frac{b}{a^2}\frac{Q}{V}(1 - e^{-at} - at) + c_{1_i}e^{-at}$
Exponential	$c_0 e^{\pm bt}$	$\frac{c_0}{a \pm b}\frac{Q}{V}(e^{\pm bt} - e^{-at}) + c_{1_i}e^{-at}$

* Adapted from D. J. O'Connor and J. A. Mueller, A Water Quality Model of Chlorides in Great Lakes, table 1, *J. Sanit. Eng. Div., ASCE*, vol. 96, August 1970.

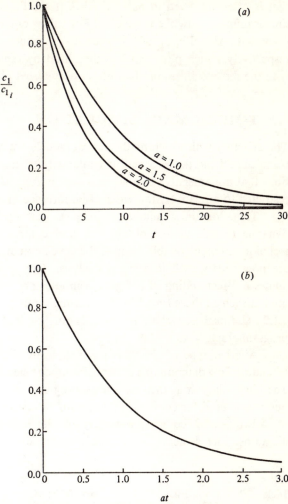

FIGURE 4.2.1
Plot of Eq. (4.2.4). (*a*) As a function of time t; (*b*) as a function of the dimensionless product at.

For the special case, where the inflow and outflow of the system are 0 [at $t \geq 0$: $Q = 0$, $c_0(t) = 0$], Eq. (4.2.3) reduces to the initial condition term

$$c_1 = c_{1_i} e^{-at} \qquad (4.2.4)$$

where c_1 = concentration of material in system

A plot of this expression is shown in Fig. 4.2.1*a* for three different values of *a*. It is to be noted that the expression describes a first-order decay relationship that is often observed in nature. Also, it should be noted that the response time is influenced by the value of a. For values of a equal to 1.0, 1.5, and 2.0, a fractional residual c_1/c_{1_i} of 0.2 is reached at values of t equal to 16, 11, and 8, respectively. Such plots can be normalized by replacing the time coordinate with the dimensionless product at. A normalized plot is shown in Fig. 4.2.1b.

4.3 TIME-DOMAIN SIMULATION

The difficulty with which mathematical models are solved by analytic methods is indicated from the information contained in Table 4.3.1. The types of equations that are amenable to analytic solution are usually of a trivial nature and restricted to a relatively few cases of importance in systems analysis.

Time-domain simulation techniques, however, provide means by which dynamic models almost of any size and complexity can be solved. These techniques, made possible through the development of the electronic computer, are now widely used in systems analysis. One type of simulation technique allows a direct coding of a digital computer program from the mathematical models such as those found in the information flow diagrams in Figs. 4.1.1 and 4.1.2. Use of these techniques requires only a minimal knowledge of computer programming.

Another type of simulation technique involves the reduction of the information flow diagram to a system of symbolic elements called a *block diagram*. The block diagram is then used for setting up the problem on an analog computer or is used for coding a program for simulation on a digital computer.

Table 4.3.2 contains a variety of elements commonly used in block diagrams. The elements found therein are not all-inclusive, nor are they specific

Table 4.3.1 CLASSIFICATION OF MATHEMATICAL PROBLEMS AND THEIR EASE OF SOLUTION BY ANALYTIC METHODS*

	Linear equations			Nonlinear equations		
Equation	One equation	Several equations	Many equations	One equation	Several equations	Many equations
Algebraic	Trivial	Easy	Essentially impossible	Very difficult	Very difficult	Impossible
Ordinary differential	Easy	Difficult	Essentially impossible	Very difficult	Impossible	Impossible
Partial differential	Difficult	Essentially impossible	Impossible	Impossible	Impossible	Impossible

* Courtesy of Electronic Associates, Inc.

for a particular simulation technique. However, they are general enough to translate readily into the language of any particular technique to be used. It is to be noted that each symbol contains two types of information: a capitalized alphabetic character identifying the type of element and a number

Table 4.3.2 BLOCK ELEMENTS USED IN TIME-DOMAIN SIMULATION PROGRAMMING

Type	Symbol	Description
Summer	Sn: $x, y, z \to x+y+z$	Element adds up to three quantities. Output is the algebraic sum of these quantities.
Integrator	In: $x \to \int x\,dt$	Element accepts only one input; its output is the integral of the input with respect to the independent variable.
Multiplier	Mn: $x, y \to xy$	Element accepts only two inputs. Output is the algebraic product of the inputs.
Divider	Dn: $x, y \to x/y$	Element accepts only two inputs. Output is the algebraic division of one input by the other.
Sign inverter	Nn: $x \to -x$	Element changes sign of input.
Gain	Gn: $x \to Px$	Element multiplies input by a constant P.
Constant	$Kn \to P$	Element introduces a constant P to the system.
Function generator	Fn: $x \to f(x)$	Element converts the input to some analytic or arbitrary function of the input.
Function switch	Rn: $x, y, z, P \to$ x if $P>0$, y if $P=0$, z if $P<0$	Element performs a variety of logical operations.

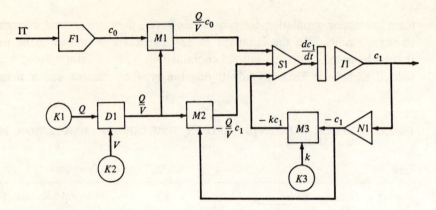

FIGURE 4.3.1
Block diagram for information flow diagram shown in Fig. 4.1.1b.

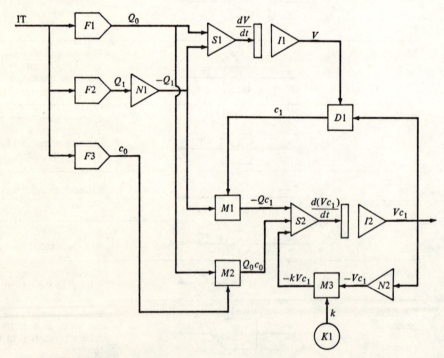

FIGURE 4.3.2
Block diagram for information flow diagram shown in Fig. 4.1.2.

identifying the specific element of its type. This information is used for coding the system for computer simulation. Block diagrams for the information flow diagrams in Figs. 4.1.1 and 4.1.2 are given in Figs. 4.3.1 and 4.3.2.

Up to this point, focus of attention has been on a single system, such as shown in Fig. 4.1.1. However, seldom does the engineer deal with a single system. In most instances, systems dealt with consist of many component subsystems. In such cases, the models are developed for the individual subsystem and incorporated into information flow diagrams. These diagrams are then used to construct an information flow diagram for the complete system. Although the system diagram is more complex than the diagrams of the component subsystems, the system diagram offers no particular problems in simulation. Given adequate computer capacity, large systems can be simulated with relative ease.

4.4 CONTINUOUS-FLOW MICROBIOLOGICAL SYSTEMS

The concepts of plug flow and completely mixed flow regimes have been discussed in Sec. 1.5, whereas growth kinetics have been discussed in Sec. 3.3. When a flow regime approximates the plug-flow model, batch kinetics can be applied to the growth behavior taking place in the system. The overall system can be thought of as consisting of a sequence of discrete batch subsystems, each deriving its initial conditions from some point source (which may or may not be continuous) but from that point on maintaining an isolation from adjacent subsystems. The plug-flow model and its implied batch-kinetics model have long been used to conceptualize the oxygen uptake mechanism observed in polluted streams.

In regimes that are considered as being completely mixed, such as might be expected to occur in ponds, short tanks, or relatively short reaches of a flowing stream, batch kinetics cannot be used directly to predict biologic behavior when there is continuous flow into and out of the system. In such systems there is generally a continuous input of nutrients into the system and a continuous discharge of microorganisms growing in the system and residual nutrients.

The working equations in continuous-flow completely mixed systems are derived from weight balances using the kinetic expressions in Eqs. (3.3.1) and (3.3.2):

Rate of accumulation = input − output + growth   − decay

$$V\frac{dX_1}{dt} = QX_0 - QX_1 + V\hat{\mu}\frac{S_1}{K_s + S_1}X_1 - Vk_dX_1 \quad (4.4.1)$$

where X_0, X_1 = concentration of microorganisms in input stream and in system, respectively. $[FL^{-3}]$

V = volume of system. $[L^3]$

Q = volumetric flow into and out of system. $[L^3 t^{-1}]$

For the change in limiting nutrient concentration

Rate of accumulation = input − output − assimilation

$$V\frac{dS_1}{dt} = QS_0 - QS_1 - \frac{V}{Y}\hat{\mu}\frac{S_1}{K_s + S_1}X_1 \qquad (4.4.2)$$

where S_0, S_1 = concentration of limiting nutrient in input stream and in system, respectively. $[FL^{-3}]$

Y = growth yield

It should be emphasized that in a continuous-flow completely mixed system the discharge characteristics are the same as those in the system. Hence, the terms X_1 and S_1 represent also the concentrations of microorganisms and limiting nutrient in the discharge stream.

Equations (4.4.1) and (4.4.2) are more conveniently expressed as

$$\frac{dX}{dt} = \frac{X_0}{\theta} - \frac{X_1}{\theta} + \hat{\mu}\frac{S_1}{K_s + S_1}X_1 - k_d X_1 \qquad (4.4.3)$$

and

$$\frac{dS_1}{dt} = \frac{S_0}{\theta} - \frac{S_1}{\theta} - \frac{1}{Y}\hat{\mu}\frac{S_1}{K_s - S_1}X_1 \qquad (4.4.4)$$

where θ = theoretical retention time. $[t]$

Equations (4.4.3) and (4.4.4) are the dynamic expressions for the growth of microorganisms and the concurrent assimilation of the limiting nutrient. Note should be made of the interrelationship between the two expressions. The dynamics of such systems can be studied by simulation either with analog computers or with digital computers using digital simulation languages. Simulation by both methods is facilitated by use of block diagrams such as the ones presented in Figs. 4.3.1 and 4.3.2.

At steady state, $dX_1/dt = dS_1/dt = 0$, and when no microorganisms are in the input, Eq. (4.4.3) reduces to

$$\frac{1}{\theta} = \hat{\mu}\frac{S_1}{K_s + S_1} - k_d = \mu - k_d \qquad (4.4.5)$$

and Eq. (4.4.4) reduces to

$$X_1 = \frac{Y(S_0 - S_1)}{1 + \theta k_d} \qquad (4.4.6)$$

The significance of Eq. (4.4.5) lies in the fact that the net growth rate (the difference between μ and k_d) is inversely proportional to the theoretical retention time of the system. Consequently, growth can be controlled by controlling the retention time. From Eq. (4.4.5), a relationship can be derived for the steady-state concentration of the limiting nutrient:

$$S_1 = \frac{K_s(1 + \theta k_d)}{\theta(\hat{\mu} - k_d) - 1} \quad (4.4.7)$$

If the value of k_d is small compared to $\hat{\mu}$, Eqs. (4.4.6) and (4.4.7) can be reduced still further to

$$X_1 = Y(S_0 - S_1) \quad (4.4.8)$$

and

$$S_1 = \frac{K_s}{\theta \hat{\mu} - 1} \quad (4.4.9)$$

Equations (4.4.8) and (4.4.9) predict the relationships illustrated in Fig. 4.4.1. These curves show the effects of retention time and concentration of limiting nutrient in the influent on the steady-state concentrations of microorganisms and limiting nutrient in the system and in the discharge. It is to be noted that for each level of concentration of nutrient in the influent S_0 there is a retention time below which the microorganisms are completely washed out of the system:

$$\theta_c = \frac{K_s + S_0}{\hat{\mu} S_0} \quad (4.4.10)$$

where θ_c = critical retention time. $[t]$

Also, it is to be noted that at any given retention time above the critical the limiting nutrient concentration in the discharge S_1 is independent of the nutrient concentration in the influent S_0.

The relationships that have been used in this section originally were derived in the laboratory using pure cultures of microorganisms and simple, well-defined substrates. However, subsequent investigations have indicated that these relationships apply reasonably well for continuous-flow completely mixed systems of mixed microorganism cultures using more complex substrates. Consequently, the principles discussed in the present section are being used more and more in the rational design of waste treatment processes.

Solution procedure 4.4.1 Use of the expressions developed in Sec. 4.4 for continuous-flow completely mixed microbiological systems depends upon a knowledge of the biologic coefficients $\hat{\mu}$, K_s, k_d, and Y. The values

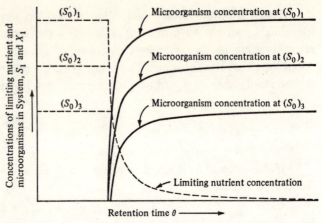

FIGURE 4.4.1
Steady-state relationships in a continuous-flow completely mixed system showing the effects of retention time and concentration of limiting nutrient in the influent on the concentrations of microorganisms and limiting nutrient in the system and in the discharge.

of these coefficients can be determined from data collected from laboratory-scale (or pilot-plant) continuous-flow completely mixed systems operating under steady-state conditions.

Equation (4.4.6) can be rearranged to give

$$\frac{S_0 - S_1}{X_1} = \frac{k_d}{Y}\theta + \frac{1}{Y} \quad (4.4.11)$$

A plot of the term $(S_0 - S_1)/X_1$ as a function of θ will yield a straight line, the slope of which will equal k_d/Y. The vertical intercept will be $1/Y$.

Equation (4.4.5) can be written as

$$\frac{1 + \theta k_d}{\theta} = \hat{\mu}\frac{S_1}{K_s + S_1} \quad (4.4.12)$$

which upon inversion and rearrangement becomes

$$\frac{\theta}{1 + \theta k_d} = \frac{K_s}{\hat{\mu}}\frac{1}{S_1} + \frac{1}{\hat{\mu}} \quad (4.4.13)$$

A plot of the term $\theta/(1 + \theta k_d)$ as a function of $1/S_1$ yields a straight line, the slope of which is $K_s/\hat{\mu}$. The vertical intercept is equal to $1/\hat{\mu}$.

When recycle of microorganisms is employed in the operation of the system such as is done in the completely mixed activated-sludge process discussed in Sec. 12.1, the term θ in Eqs. (4.4.11) and (4.4.13) is replaced by the biologic solids retention time θ_s, defined by Eq. (12.1.11).

It must be emphasized that each population of microorganisms, each substrate, and each set of environmental conditions will yield a particular set of values for the biologic coefficients.   ////

4.5 PESTICIDE CONCENTRATION

Pesticides embrace a wide variety of chemical compounds used in the control of undesirable forms of life. These compounds are used as insecticides, fungicides, and herbicides. Pesticides can be classified as nonpersistent, moderately persistent, persistent, and permanent, based on how long they last in the environment. Nonpersistent pesticides last from several days to about 12 weeks. Atrazine and 2,4-D are moderately persistent with lifetimes of 1 to 18 months. Most of the chlorinated hydrocarbons, such as DDT, aldrin, dieldrin, endrin, heptaclor, and toxaphene, are classified as persistent. DDT may persist in nature as long as 20 years. Pesticides containing mercury, lead, and arsenic are permanent.

Pesticides move through an ecosystem in many ways. Permanent pesticides are ingested by organisms and concentrated through the natural action of food chains as the organisms are consumed by others. Minute aquatic organisms and scavengers that live in water and bottom muds having pesticide concentrations of a few parts per billion can accumulate levels measured in a few parts per million—a thousandfold increase in concentration. Fish feeding on lower organisms can build up concentrations of pesticides in their visceral fat which may reach several thousand parts per million and levels in their edible flesh of hundreds of parts per million. While there is yet no evidence to indicate that pesticides presently in use cause carcinogenic effects in man, there is evidence that some pesticides can cause cancer in experimental mammals.[1]

Pesticides enter water through spraying, runoff, and waste discharges. Percolation through soil to groundwater and accidental dumping are minor sources. Some pesticides are adsorbed tightly to soil particles; therefore, pesticide pollution may result from silt and soil washed off agricultural lands into surface waters. Although a portion of the adsorbed pesticides is released to the water in solution and in suspended form, the remainder is incorporated in sediments. The sediments then continually release pesticides to the overlying water.

[1] Environmental Quality, *First Annu. Rep. Counc. Environ. Qual.*, pp. 130–140, 1970.

A model for pesticide concentration in nature can be developed using the *trophic-level concept* of representing the food-chain process in ecosystems.[1]

For any trophic level, the rate at which a pesticide is concentrated in the biomass of the organisms at that trophic level can be expressed as

$$\frac{d(Mc)}{dt} = R_{in} - r_d c - R_{ex} \qquad (4.5.1)$$

where M = trophic-level biomass. $[F]$

c = concentration of pesticide in biomass. $[FF^{-1}]$

R_{in} = rate at which pesticide is introduced to trophic level through ingestion of biomass from lower trophic levels. $[Ft^{-1}]$

r_d = rate at which biomass is removed from trophic level through death. $[Ft^{-1}]$

R_{ex} = rate at which pesticide is removed from trophic level through excretion. $[Ft^{-1}]$

Since persistent pesticides have the property of being resistant to metabolic destruction, a term representing such a fate is omitted from Eq. (4.5.1).

The population and, hence, the total biomass at any trophic level will fluctuate. However, if for present purposes the biomass is assumed to remain constant, then Eq. (4.5.1) can be rearranged and written as

$$\frac{dc}{dt} + \frac{r_d}{M} c = \frac{1}{M}(R_{in} - R_{ex}) \qquad (4.5.2)$$

As was indicated in Sec. 4.2, when a first-order, linear differential equation is expressed as it is in Eq. (4.5.2), the coefficient of the dependent variable, which in this case is the ratio r_d/M, and its reciprocal, the time constant, are a measure of the time required for the trophic level to respond to any disturbance in the system. The ratio in Eq. (4.5.2), however, has additional significance:

$$T = \frac{M}{r_d} = \frac{N\bar{m}}{n_d \bar{m}} = \frac{N}{n_d} \qquad (4.5.3)$$

where T = time constant. $[t]$

$\bar{m}$ = average biomass of each individual of trophic level. $[F]$

N = number of individuals

n_d = number of individuals dying per unit of time. $[t^{-1}]$

---

[1] Material for this section was derived from H. L. Harrison et al., Systems Studies of DDT Transport, *Science*, vol. 170, pp. 503–508, 1970.

The ratio, therefore, denotes the average life-span of members of the trophic level.

At steady state, Eq. (4.5.2) becomes

$$c = \frac{T}{M}(R_{in} - R_{ex}) \qquad (4.5.4)$$

Equation (4.5.4) reveals that at steady state the concentration of pesticide at any particular trophic level is (1) directly proportional to the average life-span of its members, (2) inversely proportional to the total biomass at that level, and (3) proportional to the net retention of pesticide in organisms constituting the trophic level. The net retention depends on the concentration of pesticide in the lower levels, the rate at which the members of the lower levels are ingested, and the amount of pesticide excreted. In general, as one moves up the ecosystem from one trophic level to the next higher one, the average life-spans increase, and the biomass in the trophic level decreases. Equation (4.5.4) provides an explanation for the observed concentration of pesticides in the higher trophic levels of the ecosystem.

4.6 EUTROPHICATION

The addition of certain materials called nutrients to natural waters often results in the growth of algae and higher aquatic plants, a process called *eutrophication*. Eutrophication is a natural process which can be greatly accelerated by those activities of man that introduce nutrients in the form of pollution. Accelerated eutrophication results in the production and accumulation of organic matter through the excessive growth of the algae and other plants. When the accumulation of vegetation decomposes, a condition occurs which fouls the air and results in the consumption of the dissolved oxygen so vital to fish and other animal life. Ultimately, eutrophication may result in the filling up and disappearance of lakes and estuaries.

It is often convenient to classify lakes as being either *oligotrophic* or *eutrophic*. Although these two terms lack exact definitions, it is generally agreed that oligotrophic lakes are relatively unproductive and receive only small amounts of plant nutrients, whereas eutrophic lakes are highly productive and receive nutrients in large quantities.

The total quantity of algae present in a body of water at a given time is referred to as the *standing crop*. This term should not be confused with *primary productivity*, which is a measure of the photosynthetic activity. Standing crop can be measured as suspended solids, by algal cell count, by cell volume, or by the quantity of chlorophyll present in the water. Primary productivity can be measured by comparing the oxygen evolution in a water sample exposed to light with that of one kept in the dark (light- and dark-bottle techniques), the rate of carbon 14 uptake, and the carbon dioxide consumption by diurnal pH fluctuations. Although primary productivity results in a standing crop, these two parameters of eutrophication are not coupled phenomena. For instance, at night the standing crop may be large, while primary productivity is 0.

Primary productivity is dependent upon solar radiation, temperature, and the concentration of plant nutrients available to the organisms. The effects of these factors are interrelated and difficult to assess individually. However, certain general observations can be made concerning each factor.

Light can be limiting to productivity in waters that are turbid or have high color. Light-limiting turbidities may be caused by high concentrations of inorganic materials (silt) or organic matter (sewage or other wastes). Such turbidity may also be caused by mutual shading of the algae themselves when present in a *bloom* (high concentration).

Algae appear to have a wide tolerance as far as temperature is concerned. Algal blooms have been observed in waters with temperatures near the freezing point. However, it appears that some of the more objectionable forms, such as the blue-green algae, prefer warmer water.

Algae and other aquatic plants require carbon, nitrogen, and phosphorus (the macronutrients) in relatively large quantities for growth. Other materials (the micronutrients) are needed in much smaller quantities. Although it has been shown that some species of algae can utilize the carbon either in the bicarbonate ion or in an organic molecule, in general the carbon requirement is met through the assimilation of carbon dioxide. Aquatic plants can derive nitrogen either from the nitrate form NO_3^- or from the ammonium ion NH_4^+. Some blue-green algae can utilize nitrogen gas N_2 through nitrogen fixation. The phosphorus requirement of algae is most readily met in the orthophosphate form PO_4^{3-}. In most environmental situations, it appears that either light or one of the macronutrients will be growth-limiting.

Nutrients are introduced into natural waters from a variety of sources. Table 4.6.1 shows the estimated sources of nitrogen and phosphorus in one natural body of water. Estimates for other natural waters will vary as to percent contribution of each source. Carbon can be introduced to natural waters in as great a variety of ways as can nitrogen and phosphorus. However, it appears that most carbon addition results from the discharge of domestic and industrial wastes and runoff, both urban and rural.

Eutrophication can have a significant effect on domestic, industrial, and recreational uses of water. Excessive growth of algae results in higher water-treatment costs to make the water potable. Even with proper treatment certain residual tastes and odors are imparted to the water by algae. Color, also, can result from large growths of algae.

Excessive growth of algae can seriously impair the aesthetic qualities of natural waters. Oxygen depletion, through algal respiration and decay, results, not only in fish kills, but also in a variety of insults to the visual and olfactory senses.

Although primary production can take place both in flowing and static waters, eutrophication is a phenomenon generally associated with relatively static waters or those heavily influenced by tidal action, such as estuaries. In static waters, such as lakes and reservoirs, eutrophication is greatly influenced by thermal stratification. Primary production in thermally stratified waters is confined to the well-mixed water of the epilimnion. In eutrophic waters, a large flux of dead algae settles from the epilimnion through the thermocline to the hypolimnion. Here the algal cells decompose, releasing plant nutrients to the relatively stagnant water. When thermal stratification is destroyed by a seasonal turnover, the nutrient-rich water of the hypolimnion is mixed with the water of the epilimnion to create conditions favorable to algal blooms.

A portion of the nutrients, soluble and insoluble, introduced to an aquatic system becomes a part of the bottom sediments. Such sediments may accumulate until disturbed by nature or man-made processes. When disturbed, the entrapped nutrients are released and again become available to the plant life in the system.

Table 4.6.1 ESTIMATED NUTRIENT SOURCES FOR LAKE MENDOTA*

Source	Annual contributions, lb		Estimated contributions, %	
	Nitrogen	Phosphorus	Nitrogen	Phosphorus
Municipal and industrial waste water	47,000 (total)	17,000 (total)	10	36
Urban runoff	30,300 (soluble)	8,100 (soluble)	6	17
Rural runoff	52,000 (soluble)	20,000 (soluble)	11	42
Precipitation on lake surface	97,000	140–7,600	20	2
Groundwater	250,000	600	52	2
Nitrogen fixation	2,000		0.4	
Marsh drainage	?	?		
Total	478,300	47,000		

* After G. F. Lee et al., "Report on the Nutrient Sources of Lake Mendota," Lake Mendota Probl. Comm., Madison, Wis., 1966.

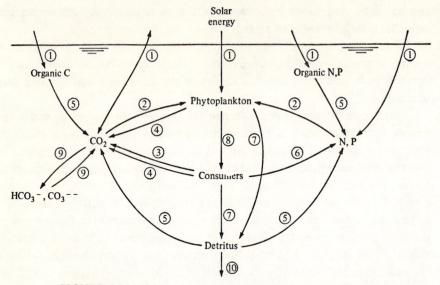

FIGURE 4.6.1
An aquatic ecosystem. (1) External inputs; (2) photosynthesis; (3) primary respiration; (4) endogenous respiration; (5) biologic decomposition; (6) excretion; (7) death; (8) grazing; (9) equilibrium; (10) sedimentation.

Modeling and simulation provide tools with which eutrophication can be studied quantitatively. Although many factors and interrelationships are still obscure, sufficient information is known to construct a model for an aquatic ecosystem, the simulation of which will provide an insight into the factors influencing eutrophication.

Attention is directed to Fig. 4.6.1. Shown there is a sketch of an aquatic ecological system. For tractability in modeling, the sketch does not include all factors influencing eutrophication or all the results. For instance, the dissolved-oxygen subsystem, which is heavily influenced by eutrophication processes, is omitted. Only those subsystems are included which contribute most directly to the growth of algae. The inclusion of additional subsystems results in a more complete, but complex, model. The community of algal species in the system is grouped together in the subsystem labeled "phytoplankton." The animal species constituting the other trophic levels in the system are also grouped together under the label of "consumers." For present purposes the consumer group includes the entire food chain in the system, ranging from minute crustaceans, feeding on algae up to and including the fish life present. The "detritus" subsystem is composed of the remains of dead algae and consumers. Not included, but of considerable importance to many ecosystems, are the decomposers. This group, consisting of bacteria and other fungi, is responsible for the decomposition of nonliving organic material. Decomposition is included in the ecosystem in Fig. 4.6.1, but only as a term influencing the behavior of the other subsystems.

The elements carbon, nitrogen, and phosphorus are shown to be present both in organic and inorganic forms. The inorganic form of nitrogen N includes nitrogen in nitrates NO_3^-, nitrites NO_2^-, and ammonium ions NH_4^+.

In the development of a model for the ecosystem shown in Fig. 4.6.1, a column of water will be considered of unit cross-sectional area and extending from the water surface through the depth of the epilimnion. It will be assumed (1) that the water in this volumetric element is in a completely mixed condition, (2) that the only vertical transport of material in the element to the water below the epilimnion is through sedimentation of detritus, (3) that no vertical transport from the water below to the element takes place, and (4) that horizontal hydrodynamic transport into and out of the element can be ignored. Such assumptions describe in a convenient way a situation that occurs in stratified lakes during the summer season.

For an ecological system contained in a volumetric element of this nature, models of the component subsystems take the general form of

$$\frac{dc}{dt} = \sum \text{sinks} - \sum \text{sources} \qquad (4.6.1)$$

The expression is to be compared to the model in Fig. 4.1.1. Shown there are the two horizontal hydrodynamic transport terms $(Q/V)c_0$ and $(Q/V)c_1$, which are ignored in Eq. (4.6.1), and the single sink term kc_1.[1]

The dynamic models for each of the nine subsystems shown in Fig. 4.6.1 are listed in Table 4.6.2. The individual terms are defined in the glossary found in Table 4.6.3.

Phytoplankton System

As the photosynthetic component of the ecosystem, the phytoplankton system is of central importance. The rate of change in biomass concentration of the phytoplankton group can be expressed as

$$\frac{dX_p}{dt} = R(2) - R(4) - R(7) - R(8) \qquad (4.6.2)$$

---

[1] The model developed in this section was derived from a model discussed in Water Resources Engineers, Inc., "A Proposed Ecologic Model For a Eutrophying Environment" prepared for the Federal Water Pollution Control Administration, 1968.

The order in which the individual terms occur at the right of the equals sign of Eq. (4.6.2) corresponds to the order in which the terms appear in Table 4.6.2. The numbers appearing in the parentheses identify the terms with respect to the metabolic or environmental process involved. The key to these processes is given in Fig. 4.6.1. In the discussions which follow, attention will be paid only to those terms posing special problems in the modeling process.

Table 4.6.2 MODELS OF SUBSYSTEMS PRESENT IN THE AQUATIC ECOLOGICAL SYSTEM REPRESENTED IN FIG. 4.6.1

1 Phytoplankton

$$\frac{dX_p}{dt} = \mu_p X_p - k_{rp} X_p - k_{mp} X_p - \frac{\mu_c X_c}{Y_{c/p}}$$

2 Consumers

$$\frac{dX_c}{dt} = \mu_c X_c - k_{rc} X_c - k_{mc} X_c$$

3 Detritus

$$\frac{dC_D}{dt} = k_{mp} X_p + k_{mc} X_c - k_{dD} C_D - k_{sD} C_D$$

4 Carbon dioxide

$$\frac{dC_{CO_2}}{dt} = k_{doc} C_{OC} + \frac{1 - Y_{c/p}}{Y_{c/p}} \mu_c X_c + k_{rp} X_p + k_{rc} X_c$$

$$+ k_{dD} C_D + K_L \frac{A}{V}(C_s - C_{CO_2}) + f(C_{CO_2}) - \mu_p X_p$$

5 Inorganic nitrogen

$$\frac{dC_N}{dt} = k_{doN} C_{ON} + k_{eN} X_c + k_{dDN} C_D + L_N - \gamma_{pN} \mu_p X_p$$

6 Inorganic phosphorus

$$\frac{dC_P}{dt} = k_{doP} C_{OP} + k_{eP} X_c + k_{dDP} C_D + L_P - \gamma_{pP} \mu_p X_p$$

7 Organic carbon

$$\frac{dC_{OC}}{dt} = L_{OC} - k_{doc} C_{OC}$$

8 Organic nitrogen

$$\frac{dC_{ON}}{dt} = L_{ON} - k_{doN} C_{ON}$$

9 Organic phosphorus

$$\frac{dC_{OP}}{dt} = L_{OP} - k_{doP} C_{OP}$$

The growth rate of the phytoplankton can be expressed as

$$R(2) = \mu_p X_p \qquad (4.6.3)$$

As was discussed in Sec. 3.3, the specific growth rate μ_p varies as a function of the concentration of the limiting nutrient. At a given temperature, the specific growth rate is often expressed as

$$\mu_p = \hat{\mu}_p \frac{S}{K_s + S} \qquad (4.6.4)$$

In nature, however, it is possible that under certain conditions two or more

Table 4.6.3 GLOSSARY OF TERMS USED IN TABLE 4.6.2

X_p = concentration of phytoplankton biomass. $[FL^{-3}]$
X_c = concentration of consumer biomass. $[FL^{-3}]$
C_D = concentration of detritus. $[FL^{-3}]$
C_{CO_2} = concentration of free carbon dioxide. $[FL^{-3}]$
C_N = concentration of inorganic nitrogen. $[FL^{-3}]$
C_P = concentration of inorganic phosphorus. $[FL^{-3}]$
C_{OC} = concentration of organic carbon. $[FL^{-3}]$
C_{ON} = concentration of organic nitrogen. $[FL^{-3}]$
C_{OP} = concentration of organic phosphorus. $[FL^{-3}]$
L_{OC} = rate of addition of organic carbon from external sources. $[FL^{-3}t^{-1}]$
L_N = rate of addition of inorganic nitrogen from external sources. $[FL^{-3}t^{-1}]$
L_{ON} = rate of addition of organic nitrogen from external sources. $[FL^{-3}t^{-1}]$
L_P = rate of addition of inorganic phosphorus from external sources. $[FL^{-3}t^{-1}]$
L_{OP} = rate of addition of organic phosphorus from external sources. $[FL^{-3}t^{-1}]$
μ_p = specific growth rate of phytoplankton. $[t^{-1}]$
μ_c = specific growth rate of consumers. $[t^{-1}]$
k_{rp} = specific reaction rate for endogenous respiration of phytoplankton. $[t^{-1}]$
k_{rc} = specific reaction rate for endogenous respiration of consumers. $[t^{-1}]$
k_{mp} = specific reaction rate for mortality of phytoplankton. $[t^{-1}]$
k_{mc} = specific reaction rate for mortality of consumers. $[t^{-1}]$
k_{dD} = specific reaction rate for decomposition of detritus. $[t^{-1}]$
k_{sD} = specific reaction rate for detritus sedimentation. $[t^{-1}]$
k_{dOC} = specific reaction rate for biologic release of carbon. $[t^{-1}]$
k_{dON} = specific reaction rate for biologic release of nitrogen. $[t^{-1}]$
k_{dOP} = specific reaction rate for biologic release of phosphorus. $[t^{-1}]$
k_{dDN} = specific reaction rate for biologic release of nitrogen from detritus. $[t^{-1}]$
k_{dDP} = specific reaction rate for biologic release of phosphorus from detritus. $[t^{-1}]$
k_{eN} = specific reaction rate for consumer excretion of nitrogen. $[t^{-1}]$
k_{eP} = specific reaction rate for consumer excretion of phosphorus. $[t^{-1}]$
$Y_{c/p}$ = yield coefficient converting phytoplankton biomass to consumer biomass
γ_{pN} = ratio of nitrogen to carbon uptake by phytoplankton during photosynthesis
γ_{pP} = ratio of phosphorus to carbon uptake by phytoplankton during photosynthesis
K_L = transfer coefficient for carbon dioxide, liquid phase. $[Lt^{-1}]$
A = cross-sectional area of water column. $[L^2]$
V = volume of water column. $[L^3]$

factors may at any one time limit growth. For this reason, the specific growth rate is given the model

$$\mu_p = \hat{\mu}_p \frac{I}{K_I + I} \frac{C_{CO_2}}{K_{CO_2} + C_{CO_2}} \frac{C_N}{K_N + C_N} \frac{C_P}{K_P + C_P} \qquad (4.6.5)$$

where I = light intensity

K_I, K_{CO_2}, K_N, K_P = saturation constants

Equation (4.6.5) holds for any particular depth below the water surface. To obtain the average specific growth rate over the depth of the water column being modeled, Eq. (4.6.5) can be written as

$$\mu_p = \hat{\mu}_p \sum_{k=1}^{n} \frac{\delta}{z_T} \frac{I_k}{K_I + I_k} \frac{C_{CO_2}}{K_{CO_2} + C_{CO_2}} \frac{C_N}{K_N + C_N} \frac{C_P}{K_P + C_P} \qquad (4.6.6)$$

where z_T = total depth of water column

n = number of depth increments

k = index number of depth increment

$\delta = z_T/n$ = length of each depth increment

Consumer System

The rate of change in concentration of consumer biomass can be expressed as

$$\frac{dX_c}{dt} = R(8) - R(4) - R(7) \qquad (4.6.7)$$

The grazing term $R(8)$ in reality is a growth term:

$$R(8) = \mu_c X_c = \hat{\mu}_c \frac{X_p}{K_{X_p} + X_p} X_c \qquad (4.6.8)$$

Detritus System

The detritus system is composed of particulate matter resulting from the death of individuals of the phytoplankton and consumer groups. Although not included in the present model, fecal material eliminated by the consumer group constitutes an additional source of detritus. Some of the detritus settles from the epilimnion, while the remainder is decomposed by bacteria and fungi. The rate of change in the concentration of detritus can be expressed as

$$\frac{dC_D}{dt} = R(7) + R(7) - R(5) - R(10) \qquad (4.6.9)$$

Nitrogen, Phosphorus, and Organic Carbon Systems

The rates of change in concentration of nitrogen, phosphorus, and organic carbon systems are given by

$$\frac{dC_N}{dt} = R(5) + R(6) + R(5) + R(1) - R(2) \quad (4.6.10)$$

$$\frac{dC_P}{dt} = R(5) + R(6) + R(5) + R(1) - R(2) \quad (4.6.11)$$

$$\frac{dC_{ON}}{dt} = R(1) - R(5) \quad (4.6.12)$$

$$\frac{dC_{OP}}{dt} = R(1) - R(5) \quad (4.6.13)$$

$$\frac{dC_{OC}}{dt} = R(1) - R(5) \quad (4.6.14)$$

Carbon Dioxide System

The carbon dioxide system has several sources and one important sink. The sink, of course, is the photosynthetic process involved in primary productivity. The rate of change in concentration of free carbon dioxide ($CO_2 + H_2CO_3$) in solution can be expressed as

$$\frac{dC_{CO_2}}{dt} = R(5) + R(3) + R(4) + R(4) + R(5) + R(1) + R(9) - R(2) \quad (4.6.15)$$

The carbonate system, consisting of normal carbonates and bicarbonates, provides an internal capacitance in which carbon dioxide is stored or withdrawn more or less instantaneously with the accumulation or reduction of carbon dioxide in solution. For modeling purposes the rate at which free carbon dioxide is withdrawn from (or stored in) the carbonate system can be expressed by

$$R(9)_{t+1} = (\alpha_1 C_T + \alpha_2 C_T)_t - (\alpha_1 C_T + \alpha_2 C_T)_{t+1} \quad (4.6.16)$$

where α_1, α_2 = distribution coefficients for bicarbonate and normal carbonate ions, respectively

C_T = total concentration of inorganic carbon

The distribution coefficients can be computed using Eqs. (2.3.5) and (2.3.6). The total concentration of inorganic carbon can be computed using

$$C_T = \frac{C_{CO_2}}{1 - \alpha_1 - \alpha_2} \quad (4.6.17)$$

It is to be noted that the distribution coefficients are a function of pH. The hydrogen-ion concentration in the system can be computed with Eq. (2.3.10).

Light Attenuation

The subsystems in aquatic ecosystems are interrelated by the interchange of carbon, nitrogen, phosphorus, and other materials. To function, the ecosystem has to be energized by an input of energy. Such energy is furnished by solar radiation. The solar energy flux reaching the water surface under a clear sky at a given time can be modeled by

$$I_t = I_n \sin 2\pi \frac{t - t_{sr}}{t_{ss} - t_{sr}} \qquad t_{sr} < t < t_{ss}$$
$$I_t = 0 \qquad t_{sr} > t > t_{ss} \qquad (4.6.18)$$

where I_t = solar energy flux at water surface at time t under clear sky. $[HL^{-2}t^{-1}]$

I_n = solar energy flux reaching upper atmosphere at noontime. $[HL^{-2}t^{-1}]$

t_{sr}, t_{ss} = time of sunrise and sunset, respectively. $[t]$

Adjusting for degree of cloud cover, one has

$$I_0 = (1 - 0.65F)I_t \qquad (4.6.19)$$

where I_0 = solar energy flux at water surface adjusted for cloudiness. $[HL^{-2}t^{-1}]$

F = degree of cloud cover (decimal)

The light energy reaching depth z_k, I_k, can be estimated using Eq. (3.6.2). The value of I_k is then used as an input to the solution of Eq. (4.6.6).

An information flow diagram for the ecological system shown in Fig. 4.6.1 is presented in Fig. 4.6.2. The external variables in the system are so labeled. The internal variables are found to the right of the subsystem boxes. The parameters of the system are not shown, but they would have to be furnished in a simulation of the system. The information flow diagram in Fig. 4.6.2 is useful in the construction of block diagrams in simulation either of the individual subsystem or the system as a whole.

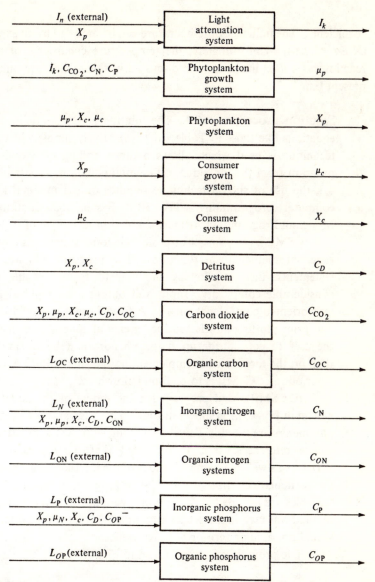

FIGURE 4.6.2
Information flow diagram for ecological system shown in Fig. 4.6.1.

It must be emphasized that temperature is a key external variable in any aquatic ecosystem. All the biologic, chemical, and physical processes involved in such systems are temperature-dependent. The effect of temperature is exerted on the individual-reaction-rate and specific-growth-rate terms as well as on the equilibrium constants, saturation-concentration terms, and transfer coefficients. The influence of temperature can be modeled by an expression such as Eq. (2.2.18). Unfortunately, the temperature coefficients of many of the reaction rates used in the models discussed in this section are unknown at the present time.

Solution procedure 4.6.1 The solution of sets of nonlinear differential equations like those in Table 4.6.2 is relatively simple when a simulation technique is used which allows a direct coding of a digital computer program from the individual differential equations. As indicated in Sec. 4.3, the use of such a technique requires only a minimal knowledge of computer programming. The IBM System/360 Continuous System Modeling Program (CSMP) is a simulation technique of this type.

CSMP includes a set of basic functional statements covering a wide range of mathematical functions. The program also accepts Fortran statements, thereby allowing the user to readily handle problems of considerable complexity. Input and output are simplified by means of user-oriented control statements. A fixed format is provided for printing and print plotting at selected increments of time. These features permit the user to concentrate upon the phenomenon being simulated rather than on the mechanics of implementing the simulation. A more detailed description of CSMP is to be found elsewhere.[1]

A CSMP-coded program for the model of the aquatic ecological system in Fig. 4.6.1 is shown in Fig. 4.6.3. The values of the variables and parameters are listed in Table 4.6.4 along with their simulation names and source references. It is to be noted that the values of the external variables as well as the initial values of the internal variables depend upon the particular situation. In the CSMP program in Fig. 4.6.3, the external variables are included in the parameter input list. An external integration method was used in the program so that the values of the internal variables would not assume zero or negative values.

The program output is shown in Fig. 4.6.4. ////

Solution procedure 4.6.2 Steady-state situations rarely, if ever, occur in nature. However, in the analysis of environmental system behavior it is often useful to estimate the steady-state values of internal variables that are expected in response to a set of time-invariant values of the external variables of the system. Such estimates at least provide insight into the general magnitude of values to be expected.

---

[1] IBM System/360 Continuous System Modeling Program, *User's Manual* H20-0367-3, 1969.

```
****CONTINUOUS SYSTEM MODELING PROGRAM****

    ***PROBLEM INPUT STATEMENTS***

*                C SMP
TITLE            ECOLOGIC MODEL
*       CARBONATE EQUILIBRIA
MACRO            ALP1,ALP2,PH=CARB(K1,K2,KW,CZ,CO2)
                 CO2M=0.001*CO2
                 A=CZ
                 B=-(K1*CO2M)-KW
                 C=-2.0*K1*K2*CO2M
                 SROOT=SQRT(B**2-4.0*A*C)
                 HI=(-B+SROOT)/(2.0*A)
                 ALP1=1.0/(1.0+(HI/K1)+(K2/HI))
                 ALP2=1.0/(1.0+(HI/K2)+(HI**2/(K1*K2)))
                 PH=ALOG10(1.0/HI)
ENDMAC
*       DIURNAL VARIATION OF SOLAR RADIATION
                 FREQ=6.2831853/24.0
                 F2=SINE(0.0,FREQ,0.0)
                 F1=COMPAR(F2,0.0)
                 AIT=F1*F2*AIN
                 AIO=(1.0-0.65*F**2)*AIT
*       LIGHT ATTENUATION EFFECT ON GROWTH OF PHYTOPLANKTON
PROCEDURE  MUP=ALI(KN,KP,KC,KI,MUHP,ZT,N,ALPHA,BETA,CN,CP,CO2,AIO,XP)
                 PROD=(CN/(KN+CN))*(CP/(KP+CP))*(CO2/(KC+CO2))
                 SUM=0.0
                 XDELT=ZT/N
                 ZK=XDELT
                 K=0.0
     10          K=K+1.0
                 AIK=AIO*EXP(-(ALPHA+BETA*XP)*ZK)
                 SUM=SUM+(XDELT/ZT)*(AIK/(KI+AIK))*PROD
                 ZK=ZK+XDELT
                 IF(K.LT.N)GO TO 10
                 MUP=MUHP*SUM
ENDPRO
*       PHYTOPLANKTON SYSTEM
                 XPDOT=RP-RRP-RMP-RG
                 RP=MUP*XP
                 RRP=KRP*XP
                 RMP=KMP*XP
                 RG=MUC*XC/YCP
                 XP=INTGRL(IXP,XPDOT)
*       CONSUMER SYSTEM
                 XCDOT=RC-RRC-RMC
                 RC=MUC*XC
                 RRC=KRC*XC
                 RMC=KMC*XC
                 MUC=MUHC*(XP/(KXP+XP))
                 XC=INTGRL(IXC,XCDOT)
*       DETRITUS SYSTEM
                 CDDOT=RMP+RMC-RDD-RSD
                 RDD=KDD*CD
                 RSD=KSD*CD
                 CD=INTGRL(ICD,CDDOT)
*       CARBON DIOXIDE SYSTEM
                 CO2DOT=ROC+RR1+RR2+RDD+RT+RCB-RP
                 ROC=KDOC*COC
                 RR1=(1.0-YCP)*MUC*XC/YCP
                 RR2=KRP*XP+KRC*XC
```

FIGURE 4.6.3
CSMP-coded program for model of aquatic ecological system in Fig. 4.6.1.

(continued overleaf)

```
            RT=KL*(AREA/VOL)*(CO2S-CO2)
            ALP1,ALP2,PH=CARB(K1,K2,KW,CZ,CO2)
            CT=CO2/(1.0-ALP1-ALP2)
            CTC=ALP1*CT-ALP2*CT
            RCB=LCTC-CTC
            CO2=INTGRL(ICO2,CO2DOT)
*           INORGANIC NITROGEN SYSTEM
            CNDOT=RON+RCN+RDN+LN-RPN
            RON=KDON*CON
            RCN=KEN*XC
            RDN=KDDN*CD
            RPN=GAMPN*MUP*XP
            CN=INTGRL(ICN,CNDOT)
*           INORGANIC PHOSPHORUS SYSTEM
            CPDOT=ROP+RCP+RDP+LP-RPP
            ROP=KDOP*COP
            RCP=KEP*XC
            RDP=KDDP*CD
            RPP=GAMPP*MUP*XP
            CP=INTGRL(ICP,CPDOT)
*           ORGANIC CARBON SYSTEM
            COCDOT=LOC-ROC
            COC=INTGRL(ICOC,COCDOT)
*           ORGANIC NITROGEN SYSTEM
            CONDOT=LON-RON
            CON=INTGRL(ICON,CONDOT)
*           ORGANIC PHOSPHORUS SYSTEM
            COPDOT=LOP-ROP
            COP=INTGRL(ICOP,COPDOT)
NOSORT
            LCTC=CTC
SORT
METHOD      SIMP
TIMER       DELT=0.1,PRDEL=24.0,FINTIM=480.0
PRINT       XP,XC,CD,CN,CP
INCON       IXP=1.0E-3,IXC=1.0E-6,ICD=1.0E-5,...
            ICO2=3.0E-3,ICN=3.0E-2,ICP=3.0E-3,...
            ICOC=3.0E-3,ICON=1.0E-4,ICOP=1.0E-5,...
            LCTC=0.0
PARAM       LOC=3.0E-4,LON=3.0E-5,LOP=3.0E-6,...
            LN=3.0E-5,LP=3.0E-6,AIN=1.0,...
            ZT=5.0E+2,N=10.0,ALPHA=2.98E-3,BETA=8.2E-6,...
            KI=3.0E-2,KC=1.36E-3,KN=1.43E-2,KP=1.61E-3,...
            MUHP=8.33E-2,MUHC=8.33E-2,KXP=8.33E-1,...
            KRP=1.25E-3,KMP=1.25E-3,...
            KRC=1.25E-3,KMC=2.08E-3,...
            KDD=4.17E-4,KSD=1.25E-3,...
            KDOC=4.17E-3,KDON=4.17E-3,...
            KEN=1.25E-3,KDDN=4.17E-3,...
            KDOP=4.17E-3,KEP=1.25E-4,...
            KDDP=4.17E-3,YCP=0.4,...
            KL=2.0,CO2S=3.0E-3,...
            AREA=1.0,VOL=500.0,...
            K1=4.45E-7,K2=4.69E-11,KW=1.01E-14,...
            GAMPN=0.172,GAMPP=7.75E-3,...
            CZ=2.0E-3,F=0.65
END
STOP
```

FIGURE 4.6.3 (continued)

ECOLOGICAL SYSTEMS 127

Table 4.6.4 VALUES OF VARIABLES AND PARAMETERS USED IN THE SIMULATION EXAMPLE IN SOLUTION PROCEDURE 4.6.1

Internal variables	X_p	XP	0.555	mmol of C/l	a
	X_c	XC	1.5×10^{-3}	mmol of C/l	a
	C_D	CD	0.2	mmol of C/l	a
	C_{CO_2}	CO2	3.0×10^{-3}	mmol of C/l	a
	C_N	CN	3.0×10^{-3}	mmol of N/l	a
	C_{ON}	CON	3.0×10^{-4}	mmol of N/l	a
	C_P	CP	3.0×10^{-3}	mmol of P/l	a
	C_{OP}	COP	3.0×10^{-5}	mmol of P/l	a
	C_{OC}	COC	3.0×10^{-1}	mmol of C/l	a
External variables	L_{OC}	LOC	3.0×10^{-3}	(mmol of C/l)/h	a
	L_N	LN	3.0×10^{-4}	(mmol of N/l)/h	a
	L_{ON}	LON	3.0×10^{-4}	(mmol of N/l)/h	a
	L_P	LP	3.0×10^{-5}	(mmol of P/l)/h	a
	L_{OP}	LOP	3.0×10^{-5}	(mmol of P/l)/h	a
	I_n	AIN	1.0	langley/min = (cal/cm²)/min	a
Multiple terms	R_p	RP		(mmol of C/l)/h	
	R_{rp}	RRP		(mmol of C/l)/h	
	R_{mp}	RMP		(mmol of C/l)/h	
	R_g	RG		(mmol of C/l)/h	
	R_c	RC		(mmol of C/l)/h	
	R_{rc}	RRC		(mmol of C/l)/h	
	R_{mc}	RMC		(mmol of C/l)/h	
	R_{dD}	RDD		(mmol of C/l)/h	
	R_{sD}	RSD		(mmol of C/l)/h	
	R_{OC}	ROC		(mmol of C/l)/h	
	R_{r1}	RR1		(mmol of C/l)/h	
	R_{r2}	RR2		(mmol of C/l)/h	
	R_T	RT		(mmol of C/l)/h	
	R_{cb}	RCB		(mmol of C/l)/h	
	R_{ON}	RON		(mmol of N/l)/h	
	R_{cN}	RCN		(mmol of N/l)h	
	R_{DN}	RDN		(mmol of N/l)/h	
	R_{pN}	RPN		(mmol of N/l)/h	
	R_{OP}	ROP		(mmol of P/l)/h	
	R_{cP}	RCP		(mmol of P/l)/h	
	R_{DP}	RDP		(mmol of P/l)/h	
	R_{pP}	RPP		(mmol of P/l)/h	
	μ_p	MUP		t^{-1}	
	μ_c	MUC		t^{-1}	
	C_T	CT		mol of C/l	
	α_1	ALP1			
	α_2	ALP2			
	$[H^+]$	HI		mol of H⁺/l	
	$[H_2CO_3]$	MCO2		mol of C/l	
	I_t	AIT		langley/min	
	I_o	AIO		langley/min	
	C_{Tc}	CTC		mol of C/l	

(*continued overleaf*)

Table 4.6.4 VALUES OF VARIABLES AND PARAMETERS USED IN THE SIMULATION EXAMPLE IN SOLUTION PROCEDURE 4.6.1 (continued)

Type of term	Term	Simulation name	Value	Units	References
Parameters	$\hat{\mu}_p$	MUHP	8.33×10^{-2}	h^{-1}	b
	$\hat{\mu}_c$	MUHC	8.33×10^{-3}	h^{-1}	a
	K_{xp}	KXP	0.834	mmol of C/l	a
	K_C	KC	1.36×10^{-3}	mmol of C/l	a
	K_N	KN	1.43×10^{-2}	mmol of N/l	c
	K_P	KP	1.61×10^{-3}	mmol of P/l	a
	K_I	KI	3.0×10^{-2}	langley/min	a
	z_T	ZT	500	cm	a
	n	N	10		a
	k_{rp}	KRP	1.25×10^{-3}	h^{-1}	d
	k_{mp}	KMP	1.25×10^{-3}	h^{-1}	a
	k_{rc}	KRC	1.25×10^{-3}	h^{-1}	a
	k_{mc}	KMC	2.08×10^{-3}	h^{-1}	a
	k_{dD}	KDD	4.17×10^{-4}	h^{-1}	e
	k_{sD}	KSD	1.25×10^{-3}	h^{-1}	d
	k_{dOC}	KDOC	4.17×10^{-3}	h^{-1}	a
	k_{dON}	KDON	4.17×10^{-3}	h^{-1}	a
	k_{eN}	KEN	1.25×10^{-3}	h^{-1}	a
	k_{dDN}	KDDN	4.17×10^{-4}	h^{-1}	e
	k_{dOP}	KDOP	4.17×10^{-3}	h^{-1}	a
	k_{eP}	KEP	1.25×10^{-4}	h^{-1}	a
	k_{dDP}	KDDP	4.17×10^{-4}	h^{-1}	e
	$Y_{c/p}$	YCP	0.4		a
	K_L	KL	2	cm/h	f
	C_s	CO2S	3.0×10^{-3}	mmol of C/l	a
	A	AREA	1.0	cm^2	a
	V	VOL	500	cm^3	a
	K_1	K1	4.45×10^{-7}	m/l	g
	K_2	K2	4.69×10^{-11}	m/l	g
	K_w	KW	1.01×10^{-14}	(m/l)^2	g
	γ_{pN}	GAMPN	0.172		a
	γ_{pP}	GAMPP	7.75×10^{-3}		a
	C_z	CZ	2.0×10^{-3}	m/l	a
	α	ALPHA	2.95×10^{-3}	cm^{-1}	h
	β	BETA	8.20×10^{-6}	(mmol of C/l)$^{-1}$ cm^{-1}	a

[a] Estimated.
[b] F. Shelef et al., Algal Reactor for Life Support Systems, *J. Sanit. Eng. Div.*, ASCE, vol. 96, no. SA1, proc. pap. 7105, pp. 91–110, 1970.
[c] R. W. Eppley et al., Half-saturation Constants for Uptake of Nitrogen and Ammonium by Marine Phytoplankton, *Limnol. Oceanogr.*, vol. 14, no. 4, pp. 920–921, November 1969.
[d] J. D. H. Strickland, Measuring the Production of Marine Phytoplankton, *Bull.* 122, The Fish. Res. Board of Canada, Ottawa, p. 172, 1960 (reprint 1966).
[e] M. M. Varma and F. DiGiano, Kinetics of Oxygen Uptake by Dead Algae, *J. WPCF*, pp. 613–626, April 1968.
[f] J. Kanwisher, Effect of Wind on CO Exchange across the Sea Surface, *J. Geophys. Res.*, vol. 68, no. 13, pp. 3921–3927, July 1963.
[g] L. G. Rich, "Unit Processes of Sanitary Engineering," Wiley, New York, 1963.
[h] H. S. Azad and J. A. Borchardt, A Method for Predicting the Light Intensity on Algal Growth and Phosphate Assimilation, *J. Water Pollut. Control Fed.*, R392–418, November 1969.

ECOLOGIC MODEL   SIMP   1

TIME	XP	XC	CD	CN	CP
0.0	1.0000E-03	1.0000E-06	1.0000E-05	3.0000E-02	3.0000E-03
2.4000E 01	1.3035E-03	9.2589E-07	4.6247E-05	3.0702E-02	3.0762E-03
4.8000E 01	1.7078E-03	8.5811E-07	9.2374E-05	3.1451E-02	3.1620E-03
7.2000E 01	2.2500E-03	7.9626E-07	1.5182E-04	3.2234E-02	3.2576E-03
9.6000E 01	2.9831E-03	7.4010E-07	2.2936E-04	3.3037E-02	3.3634E-03
1.2000E 02	3.9815E-03	6.8944E-07	3.3160E-04	3.3840E-02	3.4805E-03
1.4400E 02	5.3536E-03	6.4423E-07	4.6791E-04	3.4622E-02	3.6102E-03
1.6800E 02	7.2495E-03	6.0453E-07	6.5147E-04	3.5349E-02	3.7548E-03
1.9200E 02	9.8824E-03	5.7059E-07	9.0077E-04	3.5981E-02	3.9175E-03
2.1600E 02	1.3545E-02	5.4292E-07	1.2417E-03	3.6462E-02	4.1027E-03
2.4000E 02	1.8626E-02	5.2233E-07	1.7102E-03	3.6718E-02	4.3178E-03
2.6400E 02	2.5560E-02	5.1011E-07	2.3534E-03	3.6669E-02	4.5733E-03
2.8800E 02	3.4917E-02	5.0818E-07	3.2324E-03	3.6211E-02	4.8838E-03
3.1200E 02	4.7535E-02	5.1967E-07	4.4284E-03	3.5193E-02	5.2688E-03
3.3600E 02	6.4648E-02	5.4995E-07	6.0535E-03	3.3395E-02	5.7533E-03
3.6000E 02	8.7605E-02	6.0828E-07	8.2549E-03	3.0573E-02	6.3743E-03
3.8400E 02	1.1853E-01	7.1189E-07	1.1236E-02	2.6351E-02	7.1791E-03
4.0800E 02	1.5740E-01	8.9100E-07	1.5204E-02	2.0737E-02	8.2523E-03
4.3200E 02	2.0151E-01	1.1986E-06	2.0303E-02	1.4233E-02	9.7187E-03
4.5600E 02	2.4177E-01	1.7197E-06	2.6450E-02	8.5309E-03	1.1761E-02
4.8000E 02	2.6856E-01	2.5788E-06	3.3266E-02	5.4902E-03	1.4547E-02

FIGURE 4.6.4
Output of CSMP program in Fig. 4.6.3.

At steady state, the differential terms at the left of the equals sign of the equations in Table 4.6.2 are equal to 0. Let it be assumed that inorganic nitrogen is the sole limiting nutrient and that the steady-state values of C_N and X_p are sufficiently small so that

$$\hat{\mu}_p \frac{C_N}{K_N + C_N} \approx \frac{\hat{\mu}_p}{K_N} C_N \quad (4.6.20)$$

and

$$\hat{\mu}_c \frac{X_p}{K_{X_p} + X_p} \approx \frac{\hat{\mu}_c}{K_{X_p}} X_p \quad (4.6.21)$$

Then, for the steady-state situation the set of equations in Table 4.6.2 can be written as

1 $\dfrac{\hat{\mu}_p}{K_N} C_N - k_{rp} - k_{mp} - \dfrac{\hat{\mu}_c X_c}{K_{X_p} Y_{c/p}} = 0$

2 $\dfrac{\hat{\mu}_c}{K_{X_p}} X_p - k_{rc} - k_{mc} = 0$

3 $k_{mp} X_p + k_{mc} X_c - k_{aD} C_D - k_{sD} C_D = 0$

4 $k_{aOC} C_{OC} + \dfrac{1 - Y_{c/p}}{Y_{c/p}} \dfrac{\hat{\mu}_c}{K_{X_p}} X_p X_c + k_{rp} X_p + k_{rc} X_c$

   $+ k_{aD} C_D + K_L \dfrac{A}{V} C_s - K_L \dfrac{A}{V} C_{CO_2} + f(C_{CO_2}) - \dfrac{\hat{\mu}_p}{K_N} C_N X_p = 0$

5 $k_{dON}C_{ON} + k_{eN}X_c + k_{dDN}C_D + L_N - \gamma_{pN}\dfrac{\hat{\mu}_p}{K_N}C_N X_p = 0$

6 $k_{dOP}C_{OP} + k_{eP}X_c + k_{dDP}C_D + L_P - \gamma_{pP}\dfrac{\hat{\mu}_p}{K_N}C_N X_p = 0$

7 $L_{OC} - k_{dOC}C_{OC} = 0$

8 $L_{ON} - k_{dON}C_{ON} = 0$

9 $L_{OP} - k_{dOP}C_{OP} = 0$

The problem for which a solution is sought is to find the values of the set of internal variables

$$\{X_p, X_c, C_D, C_{CO_2}, C_N, C_P, C_{OC}, C_{ON}, C_{OP}\}$$

for given values of the set of external variables

$$\{I_n, L_N, L_P, L_{OC}, L_{ON}, L_{OP}\}$$

and the models' parameters. By assuming that C_N is the limiting growth factor, I_n can be ignored.

If the set of steady-state equations above were linear, their solution for values of the internal variables could be facilitated through the use of matrix methods designed for the solution of simultaneous equations. However, examination of eqs. 4 to 6 reveals terms consisting of the products of some of the internal variables. For such nonlinear equations, other solution techniques must be used.

The solution procedure to be considered here involves the decomposition of the steady-state equation set into individual equations, or small groups of equations, so that they may be solved in a given order of precedence. Although for the set of equations to be solved here a less formalized procedure (even visual examination) may be sufficient, the decomposition of larger systems usually requires methods of the type to be considered.

The first step in the solution procedure involves the construction of an *occurrence matrix*. Such a matrix is constructed by labeling the rows to correspond with the equation numbers and the columns to correspond with the internal variables. If a particular internal variable is present in a particular equation, its presence is indicated by assigning the value of 1 to the matrix element located in the corresponding row and column. If the variable is absent in that equation, then the element is assigned the value of 0. For the set of equations being considered here, the occurrence matrix takes the form of that shown in Fig. 4.6.5a.

Equation No.	X_p	X_c	C_D	C_{CO_2}	C_N	C_P	C_{OC}	C_{ON}	C_{OP}
1	0	1	0	0	1	0	0	0	0
2	1	0	0	0	0	0	0	0	0
3	1	1	1	0	0	0	0	0	0
4	1	1	1	1	1	0	1	0	0
5	1	0	1	0	1	0	0	1	0
6	1	0	1	0	1	0	0	0	1
7	0	0	0	0	0	0	1	0	0
8	0	0	0	0	0	0	0	1	0
9	0	0	0	0	0	0	0	0	1

(a)

	X_c	X_p	C_D	C_{CO_2}	C_N	C_{OC}	C_{ON}	C_{OP}
1	1	0	0	0	1	0	0	0
2	0	1	0	0	0	0	0	0
3	1	1	1	0	0	0	0	0
4	1	1	1	1	1	1	0	0
5	0	1	1	0	1	0	1	0
7	0	0	0	0	0	1	0	0
8	0	0	0	0	0	0	1	0
9	0	0	0	0	0	0	0	1

(b)

FIGURE 4.6.5
Occurrence matrices. (a) Initial matrix; (b) revised matrix.

There are nine internal variables and nine equations. Since no equation contains the term C_P, the column so labeled can be deleted as well as the row corresponding to the steady-state equation developed from the weight balance for inorganic phosphorus, eq. 6. Eight variables remain as well as eight independent equations.

The next step is to identify an *admissible output set* of elements. An admissible output set is a set of matrix elements selected from all the nonzero elements in the occurrence matrix so that a member of the set appears once and only once in each row and, at the same time, once and only once in each column. If all elements in the main diagonal have the value of unity, these elements constitute an admissible output set. For this reason, it is convenient to arrange the rows and columns in the occurrence matrix so that all elements in the main diagonal have the value

	1	2	3	4	5	7	8	9
1	0	0	1	1	0	0	0	0
2	0	0	1	1	1	0	0	0
3	0	0	0	1	1	0	0	0
4	0	0	0	0	0	0	0	0
5	1	0	0	1	0	0	0	0
7	0	0	0	1	0	0	0	0
8	0	0	0	0	1	0	0	0
9	0	0	0	0	0	0	0	0

(a)

$$R = \begin{vmatrix} 0 & 1 & 0 \\ 0 & 0 & 1 \\ 1 & 0 & 0 \end{vmatrix} \quad R^2 = \begin{vmatrix} 0 & 0 & 1 \\ 1 & 0 & 0 \\ 0 & 1 & 0 \end{vmatrix} \quad R^3 = \begin{vmatrix} 0 & 1 & 0 \\ 0 & 0 & 1 \\ 1 & 0 & 0 \end{vmatrix}$$

(b)

FIGURE 4.6.6
Relation matrices. (*a*) Initial matrix; (*b*) matrix reduced to the first, second, and third powers.

of 1 and then use these elements as members of the admissible output set. Toward this end, the matrix in Fig. 4.6.5*a* is rearranged to give the revised occurrence matrix in Fig. 4.6.5*b*. It is to be noted that besides deleting the column labeled C_p and the row for eq. 6, the columns labeled X_p and X_c have been rearranged so that all elements in the main diagonal have nonzero values. The revised occurrence matrix is then used to construct a relation matrix.

A *relation matrix* is a matrix the elements of which indicate the connections existing between the equations of the set. The connections are, in fact, the internal variables of the system. For present purposes, consider the equations in the occurrence matrix which corresponds to the nonzero values in the main diagonal as being source equations from which the values of the variables identified at the top of the columns flow to the other equations in the column where nonzero entries occur. The relation matrix is constructed so that the rows are labeled to correspond with the source equations and the columns labeled to indicate the equations to which the flow of the variables can occur. If a particular source equation is linked to a column equation, the relation-matrix element $a_{ij} = 1$. If not, $a_{ij} = 0$. The subscripts indicate the row and column numbers, respectively.

Consider Fig. 4.6.6*a*. Shown here is a relation matrix developed from the occurrence matrix in Fig. 4.6.5*b*. From the pattern of nonzero values in the first row of the matrix, it is seen that the internal variable X_c links the source eq. 1 to eqs. 3 and 4, whereas X_p links source eq. 2 to eqs. 3 to 5. It is to be noted that for present purposes the elements $a_{ii} = 0$.

In examining the relation matrix, it is to be noted that columns 2 and 7 to 9 have all 0 elements. The significance here is that eqs. 2 and 7 to 9 have no ancestors; hence they can be solved first. In fact, an examination of these equations reveals that for given values of the external variables L_{OC}, L_{ON}, and L_{OP} the values of X_p, C_{OC}, C_{ON}, and C_{OP} can be obtained directly from individual solutions of the equations. After one has determined which equations are to be solved first, the columns and rows for these equations are deleted from the matrix.

Further examination of the matrix will reveal that row 4 has all 0 entries, indicating that eq. 4 has no descendants. Consequently, eq. 4 must be solved last. This having been determined, row 4 and column 4 are deleted from the matrix.

The matrix in Fig. 4.6.6a is a one-step matrix, in that the linkages represented are achieved in a one-step sequence. If the one-step matrix is squared using Boolean algebra, a two-step matrix is obtained—a matrix in which the nonzero elements identify the linkages of equations that are achieved in two steps.

Boolean algebra uses the properties

$$a + b = \max(a,b)$$

and

$$a \cdot b = \min(a,b)$$

For example, matrix multiplication is carried out as follows:

$$\begin{matrix} \text{Row} & \text{Column} \end{matrix}$$

$$\begin{bmatrix} 1 & 0 & 1 & 1 \end{bmatrix} \begin{bmatrix} 1 \\ 0 \\ 1 \\ 0 \end{bmatrix} = 1 + 0 + 1 + 0 = 1$$

When the matrix in Fig. 4.6.6a is taken to the second and third powers by Boolean algebra, the matrices in Fig. 4.6.6b are obtained. It is to be noted that no columns or rows appear with all 0 entries. In fact, a repeat pattern is observed in matrices R and R^3. This indicates that eqs. 1, 3, and 5 must be solved simultaneously in the solution procedure.

In summary, the solution procedure is as follows:

Step 1 Equations 2 and 7 to 9 are solved separately.
Step 2 Equations 1, 3, and 5 are solved simultaneously.
Step 3 Equation 4 is solved.

A more detailed discussion of such decomposition procedures is to be found elsewhere.[1] Estimated and experimentally determined values of system parameters are presented in Table 4.6.4. ////

EXERCISES

4.1.1 Water flows through an open tank. Flow from the tank is controlled by an orifice with a free discharge located at the bottom of the tank. Flow into the tank is a time-varying function independent of the discharge. Formulate the system's model in terms of the dynamic behavior of the depth of water in the tank. Prepare an information flow diagram for the model.

4.1.2 Water flows into and out of an open tank through two fixed valves located at the bottom of the tank. Flow through the tank is controlled by the pressure on the upstream side of the inlet valve and the pressure on the downstream side of the outlet valve. Formulate the system's model in terms of the dynamic behavior of the water depth in the tank. Prepare an information flow diagram for the model.

4.1.3 Solve Exercise 4.1.2 for a situation in which the tank is enclosed and airtight.

4.1.4 The influent to the system described in Exercise 4.1.1 contains a material in solution which decays in the tank in a first-order reaction. Formulate the model for the complete system. Prepare an information flow diagram for the model.

4.2.1 A system consists of a tank through which an aqueous solution flows at a rate of 10 m³/h. The tank contents occupy a volume of 5 m³ and are completely mixed. While in the tank, the concentration of the material in solution is reduced by a first-order reaction with a rate constant of 2 h^{-1}. If the concentration of the material in the influent is 200 mg/l, what is the steady-state concentration in the effluent?

4.2.2 For the system described in Exercise 4.2.1, the steady-state conditions prevail for $t < 0$. At $t = 0$, a step increase of 200 mg/l is made in the concentration of the influent. Plot a curve giving the relationship between the concentration of the material in the effluent of the tank and time, for $-1 \text{ h} < t < +2 \text{ h}$.

4.2.3 For the system in Exercise 4.2.2 and the conditions described for $t \geq 0$, plot the ratio of the concentration of the material in the effluent to the initial concentration as a function of time. Prepare similar curves using rate constants equal to 0.5 and 5 h^{-1}. Compare the three curves. What would be the effect on such curves of increasing the retention time?

4.2.4 For the system described in Exercise 4.2.1, the steady-state conditions prevail for $t < 0$. For $t \geq 0$, the influent concentration is increased to $200 - 100t$ mg/l. Plot a curve giving the relationship between the concentration of the material in the effluent of the tank and time, for $-1 \text{ h} < t < +2 \text{ h}$.

[1] D. M. Himmelblau and K. B. Bischoff, "Process Analysis and Simulation," Wiley, New York, 1968.

4.2.5 Lake Erie has a volume of 113 mi^3 and an average flow-through rate of 201,000 ft^3/s. If the present chloride concentration is 3 mg/l, what will be the chloride concentration in the lake 1 year later if the lake receives a step increase of 25 × 10^6 lb/d of salt?

4.3.1 Prepare a block diagram to be used for coding the model formulated in Exercise 4.1.1.

4.3.2 Prepare a block diagram to be used for coding the model formulated in Exercise 4.1.2.

4.3.3 Prepare a block diagram to be used for coding the model formulated in Exercise 4.1.3.

4.3.4 Prepare a block diagram to be used for coding the model formulated in Exercise 4.1.4.

4.4.1 Are analytic solutions to Eqs. (4.4.3) and (4.4.4) convenient? Why? What effect will the assumption that $S_1 \gg K_s$ have upon the convenience of an analytic solution?

4.4.2 Assuming $S_1 \gg K_s$, convert Eq. (4.4.3) to an expression for batch growth kinetics. For an initial concentration of microorganisms equal to 5 mg/l and for values of $\hat{\mu}$ and k_d equal to 0.2 and 2.0 h^{-1}, respectively, solve the integrated form of the expression for the concentration at $t = 2$ h.

4.4.3 For an influent substrate concentration of 1,000 mg/l, what will be the minimum retention time possible for maintaining a steady-state concentration of microorganisms if the maximum specific growth rate and saturation constant have values of 0.2 h^{-1} and 50 mg/l, respectively?

4.4.4 Equation (4.4.3) is an expression for a continuous-flow completely mixed system when no microorganisms are recycled. Develop a similar model for situations in which microorganisms are recycled, and prepare a block diagram to be used for coding a program for simulation on a digital computer.

4.4.5 From the steady-state data below determine the biologic coefficients $\hat{\mu}$, K_s, k_d, and Y.

Dilution rate, h^{-1}	Feed substrate concentration, mg/l	Substrate concentration in discharge, mg/l	Cell mass concentration in discharge, mg/l
0.059	1,000	5	422
0.091	1,000	8	429
0.124	1,000	13	431
0.177	1,000	20	428
0.241	1,000	30	421
0.302	1,000	37	433
0.358	1,000	43	420
0.425	1,000	58	420
0.485	1,000	75	413
0.546	1,000	97	433
0.610	1,000	112	426
0.662	1,000	161	434

4.4.6 Assuming that $K_d \ll \hat{\mu}$, estimate that retention time for which the system in Exercise 4.4.5 will yield the greatest rate of substrate removal.

4.5.1 What are the general types of pesticides in use today? How are they introduced to the environment? What happens to pesticides entering the aquatic environment?

4.5.2 For a given trophic level in an ecosystem, what influence does an increase in death rate have upon (a) the time required for the trophic level to respond to a disturbance in the ecosystem and (b) the average life-span of members of the trophic level?

4.5.3 How is the concentration of a pesticide in a trophic level affected by (a) decreasing the average life-span of its members, (b) increasing the total biomass in the level, and (c) increasing the net retention of pesticide in the trophic level?

4.5.4 What factors influence the net retention of a pesticide in a trophic level?

4.6.1 In what ways can eutrophication impair the quality of water?

4.6.2 Explain the difference between "standing crop" and "primary productivity." How is each measured?

4.6.3 What major factors influence primary productivity?

4.6.4 In what forms can algae derive their macronutrient requirements?

4.6.5 How are the macronutrients for algae introduced to surface waters?

4.6.6 How is eutrophication of lakes affected by thermal stratification?

4.6.7 A particular lake has the characteristics of a continuous-flow completely mixed regime. Pollution control activities were implemented to impede eutrophication through the removal of phosphorus in wastes discharged to the lake. Over a period of years, the phosphorus levels in the lake dropped significantly. Why can it be concluded that phosphorus was not the limiting nutrient in eutrophication?

4.6.8 Describe the interactions that take place in an aquatic ecosystem between living organisms and their nutrients.

4.6.9 Construct a model of the dissolved-oxygen subsystem that could be included in the model of the aquatic ecosystem depicted in Fig. 4.6.1.

4.6.10 Discuss the possible sources and sinks of carbon dioxide in an aquatic ecosystem. How is pH influenced by the level of carbon dioxide in water?

4.6.11 Show why a steady-state pH level is possible during the day when carbon dioxide is growth-limiting but not when nitrogen or phosphorus is limiting.

4.6.12 Using the ecological model coded in CSMP in Fig. 4.6.3 with the values of the variables and parameters listed in Table 4.6.4, make a 2-week computer simulation run of the aquatic ecosystem in Fig. 4.6.1.

4.6.13 A new community is to be located on the shore of an oligotrophic lake. The thermocline during the summer and early fall is found to be poised at a depth of 10 m, and the volume of water above the thermocline has been determined to be 200,000 m^3. If the community is to have a population of 5,000, estimate the steady-state values of the phytoplankton concentration and the detritus flux to the hypolimnion if the liquid waste from the community is discharged into the lake following secondary treatment. Use the values of the parameters listed in Table 4.6.4.

4.6.14 An environmental system composed of 10 subsystems has the steady-state models given below. The equations are independent but nonlinear.

(a) Make a flow sketch of the system. (Assume that positive coefficients indicate flow to the subsystem.)

(b) Which are the internal variables?

(c) Decompose the system into a calculation sequence for a solution of the equations.

Subsystem	Steady-state model
1	$0 = a_1 X_1 - a_2 X_2 - a_3 X_3$
2	$0 = a_4 X_3 + a_5 X_5 - a_6 X_4$
3	$0 = a_7 X_2 + a_8 X_4 - a_9 X_5 - a_{10} X_6$
4	$0 = a_{11} X_7 - a_{12} X_8 - a_{13} X_{11}$
5	$0 = a_{14} X_8 - a_{15} X_9$
6	$0 = a_{16} X_6 + a_{17} X_9 - a_{18} X_{10} + a_{19} X_{15}$
7	$0 = a_{20} X_{10} + a_{21} X_{11} - a_{22} X_{12} - a_{23} X_{13}$
8	$0 = a_{24} X_{13} - a_{25} X_{14} - a_{26} X_{15}$
9	$0 = a_{27} X_{14} - a_{28} X_{17}$
10	$0 = a_{29} X_{12} - a_{30} X_{16}$

5
NATURAL TRANSPORT SYSTEMS

5.1 BASIC MODELS

The earth's fluids—water and air—provide the media in which most natural transport takes place. Either medium can be modeled as a continuum or as a system of finite volumes. In situations where the fundamental characteristics of the medium are relatively homogeneous from one point to another, the *continuous model* can be used. In those situations where such characteristics vary significantly, the *finite-volume model* provides a more convenient representation. The continuous model can be developed either for a one-, two-, or three-dimensional mode. However, for present purposes only the one-dimensional model will be considered.

Continuous Model

The one-dimensional continuous model as applied to the transport of a material across a cross-sectional area normal to the direction of fluid flow can be expressed as a generalized form of Eq. (1.1.10):

$$\frac{\partial c}{\partial t} = \frac{1}{A(x,t)} \frac{\partial}{\partial x} \left[E(x,t) A(x,t) \frac{\partial c}{\partial x} \right] - \frac{1}{A(x,t)} \frac{\partial}{\partial x} [Q(x,t)c] - S(c,x,t) \qquad (5.1.1)$$

where c = concentration of transported material. $[FL^{-3}]$

A = cross-sectional area. $[L^2]$

E = dispersion coefficient. $[L^2 t^{-1}]$

Q = rate of volume transport of fluid. $[L^3 t^{-1}]$

S = sources and sinks of transported material. $[FL^{-3} t^{-1}]$

Equation (5.1.1) is a fundamental expression that can be applied to the transport of any material, conservative or nonconservative.

The first group of terms to the right of the equals sign in Eq. (5.1.1) represents the dispersion component of the flux across the cross-sectional area. It is to be noted that both the dispersion-coefficient and the cross-sectional-area terms are identified as functions both of space and time.

The second group of terms, consisting of the volume transport rate and concentration, represents the advective component of the flux. The volume transport rate also can be a function both of space and time.

The source and sink term S can be a function of space, time, and concentration. The nature of the component sources and sinks will depend upon the particular system to which Eq. (5.1.1) is being applied. The sources and sinks of the dissolved-oxygen system are discussed in Sec. 5.2.

Finite-volume Model

Consider Fig. 5.1.1. Represented therein is an environmental medium segmented in a two-dimensional lattice defining a system of discrete fluid volumes. As materials are transported through the medium, they pass from one volume to another. Transport between adjacent volumes is indicated by the arrows located at the boundaries between volumes. Each volume is considered to be homogeneous along the third dimension.

For a particular material being transported, a generalized expression can be given for a weight balance struck across any volume.[1]

$$V_i \frac{dc_i}{dt} = \sum_j (G_{ji} + D_{ji}) - S_i \qquad (5.1.2)$$

---

[1] R. V. Thomann, "The Use of Systems Analysis to Describe the Time Variation of Dissolved Oxygen in a Tidal Stream," thesis presented in partial fulfillment of the requirements for the degree of doctor of philosophy, New York University, 1962.

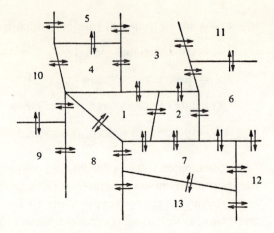

FIGURE 5.1.1
Environmental medium segmented in a two-dimensional lattice defining a system of discrete fluid volumes.

where dc_i/dt = time rate of change in concentration of material in volume i

V_i = volume of fluid in volume i

G_{ji} = transport of material from volume j to volume i by advection

D_{ji} = transport of material from volume j to volume i by dispersion

S_i = sources and sinks of material in volume i

The advective term may be expressed as

$$G_{ji} = Q_{ji}[\delta_{ji} c_j + (1 - \delta_{ji}) c_i] \qquad (5.1.3)$$

where Q_{ji} = volume transport of fluid from volume j to volume i. $[L^3 t^{-1}]$

δ_{ji} = net advection factor

The advection factor δ_{ji} is a proportionality constant used to represent the concentration of the material at the boundary in terms of the average concentrations within adjacent volumes. Normally the value of this factor ranges from 0.5 to 1.0, depending upon the general nature of the environmental system being considered.

The dispersion term can take the form of

$$D_{ji} = E'_{ji}(c_j - c_i) \qquad (5.1.4)$$

where E'_{ji} = mixing coefficient. $[L^3 t^{-1}]$

The mixing coefficient is analogous to the coefficient of dispersion. However, it is to be noted that the relationship expressed by Eq. (5.1.4) is linear whereas the dispersion term in the continuous model reflects a second-order phenomenon.

Substitution of Eqs. (5.1.3) and (5.1.4) into Eq. (5.1.2) yields

$$V_i \frac{dc_i}{dt} = \sum_j \{Q_{ji}[\delta_{ji}c_j + (1 - \delta_{ji})c_i] + E'_{ji}(c_j - c_i)\} - S_i \quad (5.1.5)$$

Equation (5.1.5) is the un-steady-state form of the finite-volume equation. The steady-state form is obtained by equating the differential term dc_i/dt to 0.

Let it be assumed that S_i contains no terms which are dependent upon the concentration c_i and consists only of a single term representing an introduction of the material from a source exterior to the medium f_i. For the steady state, then, Eq. (5.1.5) can be rearranged to give

$$\left\{\sum_j [E'_{ji} - (1 - \delta_{ji})Q_{ji}]\right\} c_i - \sum_j [(\delta_{ji}Q_{ji} + E'_{ji})c_j] = f_i \quad (5.1.6)$$

For $i = 1$, Eq. (5.1.6) can be written in the form of

$$a_{11}c_1 + a_{21}c_2 + a_{31}c_3 + \cdots + a_{n1}c_n = f_1 \quad (5.1.7)$$

where

$$a_{11} = \sum_j^n [E'_{j1} - (1 - \delta_{j1})Q_{j1}] \quad \text{for } j \neq 1 \quad (5.1.8)$$

$$a_{j1} = -(\delta_{j1}Q_{j1} + E'_{j1}) \quad \text{for } j = 2, 3, \ldots, n \quad (5.1.9)$$

Since $Q_{ji} = -Q_{ij}$, $E'_{ji} = E'_{ij}$, and $\delta_{ji} = \delta_{ij}$, Eq. (5.1.7) can be modified to obtain

$$a_{11}c_1 + a_{12}c_2 + a_{13}c_3 + \cdots + a_{1n}c_n = f_1 \quad (5.1.10)$$

where

$$a_{11} = \sum_j^n [E'_{1j} + (1 - \delta_{1j})Q_{1j}] \quad \text{for } j \neq 1 \quad (5.1.11)$$

$$a_{1j} = \delta_{1j}Q_{1j} - E'_{1j} \quad \text{for } j = 2, 3, \ldots, n \quad (5.1.12)$$

Note should be made of the fact that the above expressions have been developed with the convention that the volume transport rate Q is positive in those instances where the direction of flow is to section i and negative where the direction of flow is from section i.

The same procedure can be followed for volumes $i = 2, 3, \ldots, n$, thereby generating the family of linear equations with constant coefficients:

$$\begin{aligned} a_{11}c_1 + a_{12}c_2 + \cdots + a_{1n}c_n &= f_1 \\ a_{21}c_1 + a_{22}c_2 + \cdots + a_{2n}c_n &= f_2 \\ \cdots\cdots\cdots\cdots\cdots\cdots\cdots\cdots\cdots\cdots & \\ a_{n1}c_1 + a_{n2}c_2 + \cdots + a_{nn}c_n &= f_n \end{aligned} \quad (5.1.13)$$

Equations (5.1.13) provide the steady-state model of the system represented in Fig. 5.1.1. Since each volume of this system seldom shares a common boundary with more than four other volumes, many of the coefficients a_{ij} will be 0.

Such a system of equations can be expressed in matrix form:

$$\begin{bmatrix} a_{11} & a_{12} & \cdots & a_{1n} \\ a_{21} & a_{22} & \cdots & a_{2n} \\ \multicolumn{4}{c}{\dotfill} \\ a_{n1} & a_{n2} & \cdots & a_{nn} \end{bmatrix} \begin{bmatrix} c_1 \\ c_2 \\ \vdots \\ c_n \end{bmatrix} = \begin{bmatrix} f_1 \\ f_2 \\ \vdots \\ f_n \end{bmatrix} \quad (5.1.14)$$

In matrix notation Eq. (5.1.14) becomes

$$\mathbf{AC} = \mathbf{F} \quad (5.1.15)$$

and the solution for the vector of the unknown concentrations is given by

$$\mathbf{C} = \mathbf{A}^{-1}\mathbf{F} \quad (5.1.16)$$

where $\mathbf{A}^{-1}$ = inverse of matrix $\mathbf{A}$, called the *unit-loading matrix*

Once having solved for the unit-loading matrix, the term can be multiplied by any vector $\mathbf{F}$ to determine the corresponding concentration vector $\mathbf{C}$. Furthermore, because of the nature of the unit-loading matrix it is possible to determine the portion of the concentration in any volume resulting from a source in any other volume. For example, ϕ_{ij}, the element in the ith row and jth column of the unit-loading matrix $\mathbf{A}^{-1}$, has a numerical value equal to the concentration in volume i resulting from a unit-value source in volume j. By multiplying the element ϕ_{ij} by the actual value of the source term in volume j, that portion of the concentration in volume i resulting from the source term can be obtained. In mathematical notation, then, the total concentration in volume i is given by

$$c_i = \sum_j \phi_{ij} f_j \quad (5.1.17)$$

The element ϕ_{ij} is sometimes referred to as the *steady-state transfer function*.

5.2 DISSOLVED-OXYGEN SYSTEM

The dissolved oxygen present in surface waters is often used as an index of water quality. Such use stems from the fact that the biologic systems present in these waters are quite sensitive to dissolved-oxygen concentrations, which at most will seldom exceed 10 mg/l. At least 4.0 mg/l are required to maintain a balance of desirable species in the water. At concentrations less than 1.0 mg/l, conditions become anaerobic resulting in odors of hydrogen sulfide and organic sulfides, coloration of the water, and the destruction of desirable fish species and many other aquatic organisms.

The basic transport models discussed in Sec. 5.1 are applicable to the dissolved-oxygen system as well as to other systems of conservative and nonconservative materials in solution. However, the sink and source terms in each system are unique to each particular system. The sink and source terms for dissolved oxygen (DO) are as follows.

Reaeration

Water undersaturated with respect to dissolved oxygen undergoes atmospheric reaeration. The time rate of change in dissolved-oxygen concentration as the result of reaeration is given by Eq. (1.2.7):

$$\frac{dc}{dt} = k_2(c^* - c) \qquad (5.2.1)$$

where c^* = concentration of dissolved oxygen when solution is in equilibrium with air. $[FL^{-3}]$

k_2 = reaeration coefficient. $[t^{-1}]$

The reaeration coefficient has been found to be a function of turbulence:[1]

$$k_2 = \frac{(D_m U)^{1/2}}{H^{3/2}} \qquad (5.2.2)$$

where D_m = molecular diffusion coefficient for oxygen in water. $[L^2 t^{-1}]$

U = average velocity of flow. $[Lt^{-1}]$

H = average depth. $[L]$

When Eq. (5.2.2) is applied to the estimation of the reaeration coefficient in an estuarine situation, U refers to the mean tidal velocity over a complete cycle, and H to the average depth over a cycle.

The molecular diffusion coefficient is temperature-dependent. Expressed in units of square centimeters per second it is[2]

$$D_m = 2.037 \times 10^{-5}(1.037)^{T-20} \qquad (5.2.3)$$

where T = temperature, °C

---

[1] D. J. O'Connor and W. E. Dobbins, Mechanism of Reaeration in Natural Streams, *Trans. ASCE*, vol. 123, pp. 641–666, 1958.
[2] W. E. Dobbins, BOD and Oxygen Relationships in Streams, *J. Sanit. Eng. Div., ASCE*, vol. 90, no. SA3, pp. 53–78, June 1964.

The temperature dependence of the reaeration coefficient has been expressed as

$$k_{2(T)} = k_{2(20)}(1.025)^{T-20} \qquad (5.2.4)$$

The saturation value of dissolved oxygen c^* can be estimated using the function

$$c^* = 14.652 - 41.0222 \times 10^{-2}T + 79.9 \times 10^{-4}T^2 - 77.77 \times 10^{-6}T^3 \qquad (5.2.5)$$

The reaeration coefficients for nontidal waters will range in value from 0.2 to 10 d^{-1}.

Biochemical Oxygen Demand

BOD can be the most significant oxygen sink in streams and estuaries. In considering BOD, distinction should be made between removal of BOD through biologic activity and removal by physical factors. When a portion of the BOD exerting material is in particulate form, sedimentation may occur, especially when flow velocities are sufficiently low. Furthermore, when the ratio of bottom area to flow volume is relatively large, bottom slimes may result in the removal of a significant portion of BOD exerting materials through adsorption. Effectively, BOD removed through sedimentation and/or adsorption is no longer available to exert a demand on the dissolved oxygen in the water from which it was removed.

The time rate change in dissolved-oxygen concentration caused by the removal of BOD through biologic activity can be expressed as a first-order reaction similar to Eq. (3.4.2):

$$\frac{dc}{dt} = -k_1 L \qquad (5.2.6)$$

where $L =$ first-stage BOD. $[FL^{-3}]$

$k_1 =$ deoxygenation constant. $[t^{-1}]$

The time rate of change in the concentration of BOD, however, is given by

$$\frac{dL}{dt} = -k_1 L - k_3 L \qquad (5.2.7)$$

where $k_3 =$ rate constant for BOD removal by sedimentation and/or adsorption. $[t^{-1}]$

As was discussed in Sec. 3.4, biochemical oxygen demand consists of two components. One component consists of the demand exerted by the carbon fraction of organic materials, whereas the other consists of the demand exerted by the nitrogen fraction and ammonia. Since nitrifying bacteria have growth rates considerably less than the rates of those bacteria responsible for the demand exerted by the carbon fraction, the latter often occurs before the demand exerted by the nitrogen fraction. As a result, two distinct stages may occur—the first caused by the oxidation of the carbon present in the organics and the second by the nitrogenous materials.

In those cases where a waste is discharged into a stream after a large percentage of the carbon has been removed by treatment, two distinct stages of BOD exertion may not be observed. The oxidation of the nitrogenous materials can be quite important in the dissolved-oxygen economy of the stream.

Ammonia exists in the dissolved state. Consequently, sedimentation and adsorption do not play a large role in the removal of nitrogenous BOD.

The total rate at which BOD is exerted may be expressed as the sum of the rates at which carbonaceous BOD and the nitrogenous BOD are exerted:

$$k_1 L = k_{1OC} L_{OC} + k_{1ON} L_{ON} \quad (5.2.8)$$

where k_{1OC}, k_{1ON} = rate constants for BOD exertion by carbonaceous and nitrogenous materials, respectively. $[t^{-1}]$

L_{OC}, L_{ON} = concentration of carbonaceous and nitrogenous material, respectively, in terms of their BOD. $[FL^{-3}]$

Bottom Deposits and Runoff

Organic material once it settles out along the bottom of a stream, undergoes decomposition, both anaerobic and benthic. As the result of such decomposition, organic materials in the form of organic acids and reduced gases are released to the water above. These materials then become part of the BOD of the flowing stream.

The rate at which BOD is added to a stream from bottom deposits is generally expressed in a term that includes also the rate at which BOD is added to the stream by surface runoff:

$$\frac{dL}{dt} = L_a \quad (5.2.9)$$

where L_a = rate of addition of BOD by local runoff and/or by resuspension of organics from bottom sludge deposits. $[FL^{-3} t^{-1}]$

Photosynthesis

The photosynthetic source of dissolved oxygen depends upon many factors, such as sunlight, temperature, concentration of phytoplankton, and nutrients. If the rate of photosynthesis is assumed to vary with the sunlight intensity during the day and to be 0 at night, then this source can be represented by the function

$$P(t) = P_m \sin \frac{2\pi(t - t_{sr})}{t_{ss} - t_{sr}} \qquad t_{sr} < t < t_{ss}$$
$$P = 0 \qquad t_{sr} > t > t_{ss} \qquad (5.2.10)$$

where P_m = maximum rate of photosynthetic oxygen production. $[FL^{-3}t^{-1}]$

t_{sr}, t_{ss} = time of sunrise and sunset, respectively

Equation (5.2.10) describes a half-sine wave coinciding with the daylight hours.

Respiration

The respiration of both microscopic and macroscopic aquatic plants may be a significant sink for dissolved oxygen. Respiration is continuous and independent of the diurnal photosynthetic pattern. Hence, respiration may be expressed as a constant rate term R.

In many cases, however, it is convenient to combine photosynthesis, respiration, and other sources and sinks not explicitly included in the weight balance for dissolved oxygen in a constant rate term:

$$\frac{dc}{dt} = -S_R \qquad (5.2.11)$$

where S_R = rate at which dissolved oxygen changes as a result of photosynthesis, respiration, and other sources and sinks not explicitly included in weight balance for oxygen. $[FL^{-3}t^{-1}]$

5.3 STREAMS

For present purposes a stream can be defined as a body of flowing water in which the velocity is the only significant component of the flux through a cross section normal to the direction of flow and in which longitudinal dispersion may be neglected without serious error. Consequently, the time rate change in concentration of a substance being transported in a stream can be described mathematically with a modification of Eq. (5.1.1):

$$\frac{\partial c}{\partial t} = -\frac{1}{A(x,t)} \frac{\partial [Q(x,t)c]}{\partial x} - S(c,x,t) \qquad (5.3.1)$$

Both flow and cross-sectional area may vary with distance and time. The application of Eq. (5.3.1) to these cases is discussed elsewhere.[1]

Equation (5.3.1) can be used to describe the transport of conservative and nonconservative materials in streams. For many applications of the equation to the transport of dissolved oxygen, both volumetric flow rate and cross-sectional area can be considered to remain constant. For such cases, the equation can be written as

$$\frac{\partial c}{\partial t} = -\frac{Q}{A}\frac{\partial c}{\partial x} + k_2(c^* - c) - k_1 L(x,t) - S_R(x,t) \qquad (5.3.2)$$

where c = dissolved-oxygen concentration. $[FL^{-3}]$

t = time. $[t]$

Q = volumetric rate of flow. $[L^3 t^{-1}]$

A = cross-sectional area of stream. $[L^2]$

x = distance. $[L]$

k_2 = reaeration coefficient. $[t^{-1}]$

c^* = dissolved-oxygen saturation concentration. $[FL^{-3}]$

k_1 = deoxygenation constant. $[t^{-1}]$

L = first-stage BOD. $[FL^{-3}]$

S_R = rate at which dissolved-oxygen concentration changes as result of remaining sources and sinks in stream. $[FL^{-3}t^{-1}]$

In Eq. (5.3.2), the term S_R includes the combined effects of photosynthesis and aigal respiration. The sign convention is such that when S_R is positive the rate of removal is greater than the rate of addition.

The dissolved-oxygen concentration is often expressed in terms of the dissolved-oxygen saturation deficit D:

$$D = c^* - c \qquad (5.3.3)$$

Through substitution, Eq. (5.3.2) becomes

$$\frac{\partial D}{\partial t} = -\frac{Q}{A}\frac{\partial D}{\partial x} - k_2 D + k_1 L(x,t) + S_R(x,t) \qquad (5.3.4)$$

The assumption that the spatial variation of flow Q and area A are negligible applies most correctly to reaches of streams with contributory drainage basins varying in size from 500 to 5,000 km².

---

[1] D. J. O'Connor, The Temporal and Spatial Distribution of Dissolved Oxygen in Streams, *Water Resour. Res.*, vol. 3, pp. 65–79, 1967.

Flow in most streams drops to low, relatively stable values during late summer and early fall, a time of year when temperatures are the highest. Consequently, stream conditions are most critical at such times, and steady-state flow conditions are frequently assumed in the determination of the spatial distribution of dissolved-oxygen concentrations. The time required to reach steady state in a stream where dispersion is insignificant is simply the time of travel from the point of pollution to the location under consideration.

At steady state, $\partial D/\partial t = 0$, and Eq. (5.3.4) becomes

$$U\frac{dD}{dx} = -k_2 D + k_1 L(x) + S_R(x) \quad (5.3.5)$$

where $U = Q/A$

By applying Eq. (5.3.2) to the transport of the organic material expressed in terms of its biochemical oxygen demand, one has

$$\frac{\partial L}{\partial t} = -\frac{Q}{A}\frac{\partial L}{\partial x} - k_1 L(x,t) - k_3 L(x,t) + L_a(x,t) \quad (5.3.6)$$

where k_3 = rate constant for BOD removal through sedimentation and/or adsorption. $[t^{-1}]$

L_a = rate of addition of BOD by local runoff or by resuspension of organics from bottom sludge deposits. $[FL^{-3}t^{-1}]$

At steady state, $\partial L/\partial t = 0$, and Eq. (5.3.6) becomes

$$U\frac{dL}{dx} = -(k_1 + k_3)L(x) + L_a(x) \quad (5.3.7)$$

By assuming that U, k_1, k_3, and L_a remain constant for a given reach of the stream, Eq. (5.3.7) can be integrated over the length of the reach to yield

$$L(x) = L_0 F_1 + \frac{L_a}{k_1 + k_3}(1 - F_1) \quad (5.3.8)$$

where

$$F_1 = \exp\left[-(k_1 + k_3)\frac{x}{U}\right] \quad (5.3.9)$$

L_0 = first-stage BOD at $x = 0$. $[FL^{-3}]$

Equation (5.3.8) is the steady-state equation for the spatial distribution of BOD over the length of the stream reach under consideration. By substituting this expression into Eq. (5.3.5), one has

$$\frac{dD}{dx} = -D\frac{k_2}{U} + \left[L_0 F_1 + \frac{L_a}{k_1 + k_3}(1 - F_1)\right]\frac{k_1}{U} + \frac{S_R}{U} \quad (5.3.10)$$

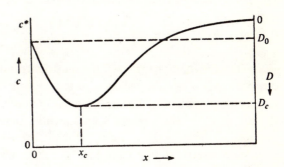

FIGURE 5.3.1
Oxygen-sag curve.

By assuming that k_1, k_2, k_3, U, L_a, and S_R remain constant over the reach of stream being considered, Eq. (5.3.10) can be integrated to give[1]

$$D(x) = D_0 F_2 + \frac{k_1}{k_2 - (k_1 + k_3)}\left(L_0 - \frac{L_a}{k_1 + k_3}\right)(F_1 - F_2)$$
$$+ \left[\frac{S_R}{k_2} + \frac{k_1 L_a}{k_2(k_1 + k_3)}\right](1 - F_2) \quad (5.3.11)$$

where $\quad F_2 = \exp\left(-k_2 \dfrac{x}{U}\right) \quad$ (5.3.12)

D_0 = dissolved-oxygen saturation deficit at $x = 0$. $[FL^{-3}]$

A typical plot of Eq. (5.3.11) yields the curve shown in Fig. 5.3.1. Such a plot is commonly known as an *oxygen-sag curve*. Oxygen-sag curves provide a visual representation of the spatial distribution of the dissolved-oxygen concentration in a polluted stream. Of particular interest is the low point along the curve located at a distance x_c from the origin at $x = 0$. The deficit at this point, called the *critical deficit* D_c, is the maximum deficit that will occur under the given conditions of loading and stream phenomena.

By differentiating Eq. (5.3.11), equating the derivative to 0, and solving for x, an expression is obtained for the critical distance x_c:

$$x_c = \frac{U}{k_2 - (k_1 + k_3)} \ln\left[\frac{k_2}{k_1 + k_3} + \frac{k_2 - (k_1 + k_3)}{(k_1 + k_3)L_0 - L_a}\right.$$
$$\left. \times \left(\frac{L_a}{k_1 + k_3} - \frac{k_2 D_0 - S_R}{k_1}\right)\right] \quad (5.3.13)$$

---

[1] W. E. Dobbins, BOD and Oxygen Relationships in Streams, *J. Sanit. Eng. Div., ASCE*, vol. 90, no. SA3, pp. 53–78, 1964.

The solution of Eq. (5.3.13) requires a knowledge of the BOD at $x = 0$, L_0. The value of L_0 can be calculated from a rearrangement of Eq. (5.3.11):

$$L_0 = \frac{D - [S_R/k_2 + k_1 L_a/k_2(k_1 + k_3)](1 - F_2) - D_0 F_2}{\{k_1/[k_2 - (k_1 + k_3)]\}(F_1 + F_2)} + \frac{L_a}{k_1 + k_3} \quad (5.3.14)$$

The cross-sectional area, flow, and reaeration coefficient may vary considerably over a relatively short distance. Furthermore, tributaries, dams, and waste-water inputs may result in significant discontinuities in these and other characteristics of the stream. Consequently, it is often desirable to segment streams in order to achieve a more realistic correspondence between the stream and its mathematical model. In such cases, the stream is divided so that within each segment the geometric and hydraulic characteristics are reasonably uniform. Segmentation is arranged so that points of discontinuity coincide with boundaries between segments.

The continuous model can also be applied to a segmented stream. Equations (5.3.8) and (5.3.11) can be solved for points along the first reach upstream. At the boundary between the first and second reaches, the necessary changes are made in the parameters, and the BOD and dissolved-oxygen deficit computed at the downstream end of the first segment become input variables at the upper end of the second segment. This procedure is continued throughout the length of the stream.

Solution procedure 5.3.1 The terms k_1, k_3, and L_a are independent of each other. The term k_1 is a specific reaction rate for a biologic process, whereas k_3 and L_a are measures of two independent physical processes. In many cases, k_3 and L_a are negligible. The term k_1 should be determined in the laboratory using samples from the river reach under consideration. For waters polluted by domestic sewage, k_1 will normally have a value in the neighborhood of 0.25 d^{-1}.

The values of k_3 and L_a for any particular situation can be determined from stream field data. The procedure for determining these terms will be illustrated by three cases, in each of which it is assumed that k_1 has been determined from a laboratory analysis.[1]

CASE I The BOD L decreases along the reach, and L is less than is predicted by

$$L = L_0 e^{-k_1(x/U)} \quad (5.3.15)$$

---

[1] Ibid.

This indicates that BOD is decreasing faster than the rate of removal due to oxidation. Therefore, it can be concluded that k_3 is positive and that the reduction in BOD due to sedimentation and/or adsorption is greater than any addition of BOD to the flowing load from the benthic deposit or other sources. In such cases, assume that $L_a = 0$, and compute the effective value of k_3 from Eqs. (5.3.8) and (5.3.9) using stream data.

CASE II The BOD L decreases along the reach, but L is greater than is predicted by Eq. (5.3.15). This indicates that the rate of addition of BOD along the reach exceeds the rate of removal by sedimentation and/or adsorption. In such cases, assume that $k_3 = 0$, and compute the value of L_a from Eqs. (5.3.8) and (5.3.9) using stream data.

CASE III The BOD L remains constant or increases along the reach. Such cases as this are handled in the same manner as Case II was handled.

The procedure outlined in the three cases above treats any net reduction in BOD in excess of that predicted by Eq. (5.3.15) as being proportional to the concentration present L, whereas a net increase is treated as being due to a uniform addition along the reach. Actually, both removal and addition can be occurring simultaneously in a grossly polluted stream that contains sludge deposits.

As was pointed out previously, the term S_R can be positive or negative, depending on the relative magnitudes of photosynthesis, algal respiration, and the oxygen demand of any aerobic layer at the top of a benthic deposit. Compared with photosynthesis and respiration, the oxygen demand exerted by benthic deposits is considered to be large. More than likely, the principal effect of benthic deposits is to contribute BOD to the flowing load, thereby exerting its demand in this manner.

If k_1, k_3, and L_a are known (or can be calculated), then only S_R and k_2 remain unknown for the solution of Eq. (5.3.11). Either one can be calculated from Eq. (5.3.11) providing the other can be established by an independent procedure. Inasmuch as the value of the reaeration coefficient k_2 depends upon a physical process, it is more readily subjected to theoretical estimation. Values of k_2 can be estimated using Eq. (5.2.2).

Table 5.3.1 contains the physical, hydraulic, and biologic characteristics for stream A. This information is used in the solutions of Exercises 5.3.1, 5.3.2, 5.3.4, 5.3.6, and 5.4.9.

Solution procedure 5.3.2 A problem often confronted in stream pollution control is that of determining the BOD concentration in the stream which results in some maximum acceptable dissolved-oxygen saturation deficit. The solution of such problems is facilitated through the use of a procedure developed by Loucks.[1] This procedure involves an iterative process, the solutions of which quickly converge to the BOD concentration corresponding to the maximum acceptable deficit. The procedure is as follows: With Eq. (5.3.14), the value of the BOD at $x = 0$, L_0, is determined using the maximum allowable deficit D_{max} for the value of D and the total length of stream reach x_{max} for the value of x. The value of L_0 thus obtained may or may not result in a greater deficit at some distance less or greater than x_{max}. Using this value of L_0, the critical distance x_c is determined with Eq. (5.3.13). If $x_c \geq x_{max}$, then the maximum deficit within the reach occurs at the end of the reach, and L_0 is the maximum initial BOD concentration that can exist without violating the minimum dissolved-oxygen requirement. See Fig. 5.3.2a. If, however, $x_c < x_{max}$, then L_0 will result in a greater deficit than that allowed in the reach. In such a case, L_0 is decreased so that at the critical distance x_c the minimum dissolved-oxygen concentration equals the minimum allowable one. This is accomplished through an iterative process in which successive values of $L_0(i)$ are determined with Eq. (5.3.14) using values of $x_c(i-1)$ which, in turn, were determined from values of $L_0(i-1)$. See Fig. 5.3.2b. The process is carried out until the difference between successive values of x_c is less than an acceptable minimum. Because $x_c(i) \lesseqgtr x_c(i-1)$, the procedure includes a logical step which takes cognizance of the following

[1] D. P. Loucks, "A Probabilistic Analysis of Wastewater Treatment Systems," Ph.D. thesis, Cornell University, Ithaca, N.Y., 1965.

Table 5.3.1 PHYSICAL, HYDRAULIC, AND BIOLOGIC CHARACTERISTICS OF STREAM A

Parameters	Stream reaches				
	1	2	3	4	5
Length, km	8.0	3.2	6.4	8.0	4.8
Average cross section, m^2	19.5	19.5	37.2	26.0	37.2
Average river flow, m^3/s	3.80	4.94	4.94	5.78	5.78
Average river depth, m	0.64	0.80	0.80	0.52	0.58
k_1, d^{-1}	0.95	0.95	0.95	0.95	0.95
k_3, d^{-1}	0.20	0.0	0.0	0.0	0.0
L_a, mg/(l)(d)	1.00	1.00	1.00	1.00	1.00
S_R, mg/(l)(d)	4.00	4.00	1.00	−2.00	−2.00
Water temperature, °C	28.0	28.0	28.0	28.0	28.0

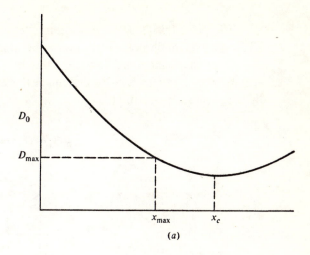

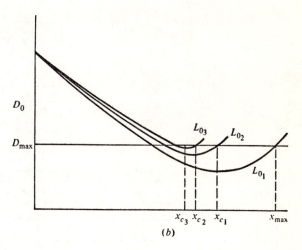

FIGURE 5.3.2
Definition sketch for Loucks' solution procedure.

relationships:

$$\text{If} \quad L_a > \frac{k_1 + k_3}{k_2}(k_2 D_0 - S_R) \quad \text{then } x_c(i) > x_x(i-1)$$

$$\text{If} \quad L_a < \frac{k_1 + k_3}{k_2}(k_2 D_0 - S_R) \quad \text{then } x_c(i) < x_c(i-1)$$

$$\text{If} \quad L_a = \frac{k_1 + k_3}{k_2}(k_2 D_0 - S_R) \quad \text{then } x_c(i) = x_c(i-1)$$

The procedure is detailed for computer computation in Algorithm 5.3.1. An explanation of algorithms can be found in the Appendix.

Once the maximum allowable value of L_0 has been computed, a weight balance can be used to determine the maximum allowable BOD discharge into the stream at $x = 0$:

$$L_0(Q_s + Q_w) = L_s Q_s + L_w Q_w \quad (5.3.16)$$

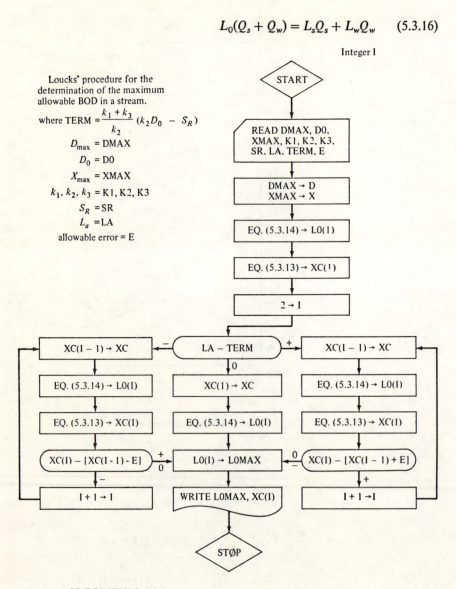

ALGORITHM 5.3.1

and
$$L_w = \frac{L_0(Q_s + Q_w) - L_s Q_s}{Q_w} \quad (5.3.17)$$

where L_w = BOD of waste discharge. $[FL^{-3}]$

Q_w = volumetric flow rate of waste discharge. $[L^3 t^{-1}]$

L_s = BOD of stream water above point of waste discharge. $[FL^{-3}]$

Q_s = volumetric rate of flow of stream above point of discharge. $[L^3 t^{-1}]$ ////

5.4 ESTUARIES

An *estuary* is defined as a semienclosed coastal body of water which is subject to tidal action and in which the sea water is measurably diluted by fresh water. Therefore, the lower reaches of streams flowing into marine waters are classified as estuaries. Some estuaries are stratified in the sense that the salt water does not mix significantly with the fresh water in a vertical direction. In such estuaries, two prisms are formed—one of which is salt water (*tidal prism*) and the other being fresh water (*advective prism*). As a general rule, stratified estuaries are relatively deep.

Most estuaries along the eastern coastal plain of the United States are considered to be nonstratified. The one-dimensional transport models can be applied to those estuaries of this type which are assumed to have uniform lateral characteristics. Only the nonstratified estuary will be considered here.

Water movement in an estuary varies significantly from that in a stream. In those reaches of a river that are subjected to tides, the motion of the water is caused not only by flow due to gravity but also by the rise and fall of tides, density currents, and wind effects. Between ebb tide and flood tide, water movement will be upstream. As flood tide gives way to ebb tide, both fresh-water flow and receding tide contribute to the movement toward the sea. Pollutants discharged into such an estuary are mixed with the water and are gradually diminished in concentration as they are transported back and forth over many tidal cycles. Ultimately, the pollutants are translated to the open sea.

Not only does the velocity of flow vary over a tidal cycle, but the cross section varies as well. Figure 5.4.1 illustrates the variation in stage height over one tidal cycle.

The finite-volume model is widely used to describe the transport of both conservative and nonconservative materials. At the steady state, it can be shown that a system of linear equations can be developed for dissolved-oxygen con-

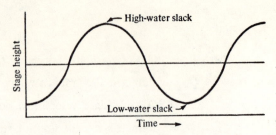

FIGURE 5.4.1
Variation of stage in time due to tides.

centrations similar to the system given by Eq. (5.1.14). The coefficients of the equations are determined by the relationships

$$a_{ij} = \delta_{ij}Q_{ij} - E'_{ij} \quad \text{for } j \neq i \tag{5.4.1}$$

$$a_{ii} = \sum_{j}^{n} [E'_{ij} + (1 - \delta_{ij})Q_{ij}] + V_i k_{2i} \tag{5.4.2}$$

$$f_i = V_i k_{2i} c^* - V_i k_{1i} L_i - S_{Ri} \tag{5.4.3}$$

where Q_{ij} = net flow from volume i to volume j. $[L^3 t^{-1}]$

δ_{ij} = net advection factor

E'_{ij} = tidal mixing coefficient. $[L^3 t^{-1}]$

V_i = midtide value of volume i. $[L^3]$

k_{2i} = reaeration coefficient at air-water interface of volume i. $[t^{-1}]$

c^* = saturation concentration of dissolved oxygen. $[FL^{-3}]$

k_{1i} = deoxygenation constant for volume i. $[t^{-1}]$

L_i = concentration of first-stage BOD in volume i. $[FL^{-3}]$

S_{Ri} = rate of change of dissolved-oxygen concentration as a result of remaining sources and sinks in volume i. $[FL^{-3}t^{-1}]$

Certain coefficients must be modified for those volumes where water boundary conditions exist. For volumes where flow is entering the system (where Q_{ik} is negative),

$$a_{ii} = \sum_{j}^{n} [E'_{ij} + (1 - \delta_{ij})Q_{ij}] + V_i k_{2i} + \delta_{ik}Q_{ik} + E'_{ik} \tag{5.4.4}$$

and

$$f_i = V_i k_{2i} c^* - V_i k_{1i} L_i - S_{Ri} + [E'_{ik} - (1 - \delta_{ik})Q_{ik}]c_{ik} \tag{5.4.5}$$

where subscript k = region exterior to boundary of system

Where flow is leaving the system (where Q_{ik} is positive),

$$a_{ii} = \sum_{j}^{n} [E'_{ij} + (1 - \delta_{ij})Q_{ij}] + V_i k_{2i} + \delta_{ik}Q_{ik} + E'_{ik} \tag{5.4.6}$$

and
$$f_i = V_i k_{2i} c^* - V_i k_{1i} L_i - S_{Ri} + (E'_{ij} - \delta_{ik} Q_{ik}) c_{ik} \quad (5.4.7)$$

A similar system of linear equations can be developed for BOD concentrations L_i. For these equations, the coefficients are

$$b_{ij} = \delta_{ij} Q_{ij} - E'_{ij} \quad \text{for } j \neq i \quad (5.4.8)$$

$$b_{ii} = \sum_j^n [E'_{ij} + (1 - \delta_{ij}) Q_{ij}] + V_i (k_{1i} + k_{3i}) \quad (5.4.9)$$

$$f_i = W_i \quad (5.4.10)$$

where k_{3i} = rate constant for BOD removal through sedimentation and/or adsorption. $[t^{-1}]$

W_i = input of first-stage BOD to volume i. $[FL^{-3} t^{-1}]$

For those volumes where flow is entering the system (where Q_{ik} is negative),

$$b_{ii} = \sum_j^n [E'_{ij} + (1 - \delta_{ij}) Q_{ij}] + V_i (k_{1i} + k_{3i}) + \delta_{ik} Q_{ik} + E'_{ik} \quad (5.4.11)$$

and
$$f_i = W_i + [E'_{ik} - (1 - \delta_{ik}) Q_{ik}] L_{ik} \quad (5.4.12)$$

Where flow is leaving the system (where Q_{ik} is positive),

$$b_{ii} = \sum_j^n [E'_{ij} + (1 - \delta_{ij}) Q_{ij}] + V_i (k_{1i} + k_{3i}) + \delta_{ik} Q_{ik} + E'_{ik} \quad (5.4.13)$$

and
$$f_i = W_i + (E'_{ik} - \delta_{ik} Q_{ik}) L_{ik} \quad (5.4.14)$$

The tidal mixing coefficient E' includes the mixing effect of the tidal phenomenon as well as the effects of turbulence, velocity gradients, and density currents. It has been related to the coefficient of dispersion E by[1]

$$E'_{ij} = \frac{E_{ij} A_{ij}}{0.5(x_i + x_j)} \quad (5.4.15)$$

where A_{ij} = midtide cross-sectional area of boundary between volumes i and j

x_i, x_j = lengths of segments i and j

As was discussed in Sec. 5.1, the advection factor δ_{ij} is a proportionality constant used to represent the concentration of material at the boundary in terms of the average concentrations within adjacent volumes. In volumes dominated by tidal action it is generally assumed that $\delta_{ij} = 0.5$. In nontidal streams, the quantity of material transported downstream at a boundary is related only to the material upstream, and the downstream concentration has no effect. In such cases, $\delta_{ij} = 1.0$.

---

[1] R. V. Thomann, "The Use of Systems Analysis to Describe the Time Variation of Dissolved Oxygen in a Tidal Stream," Ph.D. thesis, New York University, 1963.

Solution procedure 5.4.1 The estuarine reaches of streams discharging into the sea can be represented by a one-dimensional form of the finite-volume model. Such reaches are segmented so that there are no significant gradients within a single segment. With the exception of the first and last segments, each segment has a common boundary with only two other segments. The first and last segments are each adjacent to a single other segment and a water boundary of the system.

The equation for each segment contains only three coefficients—$a_{i,i-1}$, $a_{i,i}$, and $a_{i,i+1}$, except the first equation, in which the coefficients $a_{i,i}$ and $a_{i,i+1}$ appear, and the last equation, in which $a_{i,i-1}$ and $a_{i,i}$ appear. The coefficient matrix, therefore, consists of nonzero values along the main diagonal and the two adjacent diagonals only.

Physical data and waste loadings for two one-dimensional estuarine systems are presented in Tables 5.4.1 and 5.4.2. These data are used in the solutions of Exercises 5.4.3 to 5.4.5. A solution procedure for the finite-volume model is given in Algorithm 5.4.1. In this algorithm, δ_{ij} is assumed to be equal to 0.5. ////

Solution procedure 5.4.2 As was stated in Sec. 5.1, the set of equations defining the steady-state finite-volume model can be represented in matrix form. For dissolved oxygen

$$\mathbf{C} = \mathbf{A}^{-1}[Vk_1 L] + \mathbf{A}^{-1}\mathbf{S} \qquad (5.4.16)$$

where $\mathbf{C}$ = dissolved-oxygen-concentration vector ($n \times 1$ matrix)

$\mathbf{A}^{-1}$ = inverse of coefficient matrix with elements a_{ij} ($n \times n$ matrix)

$[Vk_1 L]$ = vector of terms given by product $V_i k_{1i} L_i$ ($n \times 1$ matrix)

$\mathbf{S}$ = vector of constants given by known terms in expressions for forcing functions f_i ($n \times 1$ matrix)

Equation (5.4.16) can be modified to give

$$\mathbf{C} = \mathbf{A}^{-1}[\overline{Vk_1}]\mathbf{L} + \mathbf{A}^{-1}\mathbf{S} \qquad (5.4.17)$$

where $[\overline{Vk_1}]$ = diagonal matrix with term Vk_1 in main diagonal ($n \times n$ matrix)

$\mathbf{L}$ = vector of BOD terms L_i ($m \times 1$ matrix)

A similar matrix expression can be developed for the BOD system:

$$\mathbf{L} = \mathbf{B}^{-1}\mathbf{W} - \mathbf{B}^{-1}\mathbf{R} \qquad (5.4.18)$$

Table 5.4.1 PHYSICAL DATA AND WASTE LOADINGS FOR A ONE-DIMENSIONAL ESTUARINE SYSTEM, NUMBER 1*†

Section i	Length l_i, ft	Volume V_i, ft$^3 \times 10^6$	Average tidal velocity U_i, ft/h	Average depth H_i, ft	Area $A_{i,i+1}$, ft$^2 \times 10^3$	Dispersion coefficient $E_{i,i+1}$, mi^2/d	Average flow Q_i, ft^3/s	Total BOD source W_i, lb/d	Total DO sink S_{Ri}, lb/d
0	21,000	242	1,500	18	6.5	4.0	2,500	7,799	22,280
1	20,000	364	1,500	16	15.8	4.0	2,500	8,957	8,622
2	20,000	460	1,500	19	21.4	4.0	2,500	5,997	4,140
3	20,000	532	1,500	18	24.6	4.0	2,500	6,648	2,700
4	20,000	636	2,000	18	28.5	4.0	2,500	4,347	4,800
5	20,000	756	2,000	17	34.1	4.0	2,500	7,308	5,040
6	20,000	455	2,000	19	41.4	4.0	2,700	2,105	890
7	10,000	504	2,000	20	49.6	4.0	2,700	6,409	2,125
8	10,000	533	2,000	18	51.2	5.0	2,700	12,496	2,250
9	10,000	582	2,000	22	55.4	5.0	2,700	143,695	2,250
10	10,000	630	2,500	23	60.9	5.0	2,700	22,259	5,760
11	10,000	655	2,500	31	65.0	5.0	2,700	5,585	1,350
12	10,000	694	2,500	36	66.0	5.0	2,700	85,215	3,240
13	10,000	805	2,500	24	72.7	5.0	2,700	197,836	3,960
14	10,000	1,860	2,500	23	88.3	5.0	2,700	79,389	14,700
15	20,000	2,030	3,000	21	98.0	5.0	3,400	187,481	6,750
16	20,000	2,184	3,000	19	104.9	5.0	3,400	67,623	11,475
17	20,000	2,396	3,000	21	113.4	5.0	3,400	30,723	7,200
18	20,000	2,692	3,000	21	126.3	5.0	3,400	67,848	16,200
19	20,000	2,932	3,000	21	142.8	5.0	3,400	1,071	15,750
20	20,000	1,512	3,500	21	150.4	5.0	3,700	111,048	6,930
21	10,000	1,574	3,500	24	151.9	5.0	3,700	117,049	6,000
22	10,000	1,698	3,500	21	162.9	5.0	3,700	2,176	13,050
23	10,000	1,792	3,500	21	176.8	7.0	3,700	791	11,000
24	10,000	1,850	3,500	19	181.7	7.0	3,700	855	9,300
25	10,000	1,924	4,000	16	188.4	7.0	3,700	3,916	12,000
26	10,000	2,054	4,000	17	196.4	7.0	3,700	992	15,000
27	10,000	2,248	4,000	17	214.5	7.0	3,700	10,808	15,000
28	10,000	4,896	4,000	21	235.0	7.0	3,700	5,011	0
29	20,000	5,620	4,000	22	254.5	7.0	3,700	5,509	0
30	20,000				307.4				

* After G. D. Pence et al., Time-varying Dissolved Oxygen Model, tables 1 and 2, *J. Sanit. Eng. Div., ASCE*, vol. 94, no. SA2, pp. 381–402, April 1968.
† Temperature in all segments = 25°C.

Table 5.4.2 PHYSICAL DATA AND WASTE LOADINGS FOR A ONE-DIMENSIONAL ESTUARINE SYSTEM, NUMBER 2*†

Section i	Length x_i, ft	Volume V_i, ft$^3 \times 10^6$	Area $A_{i,i+1}$, ft$^2 \times 10^3$	Dispersion coefficient $E_{i,i+1}$, mi^2/d	Average flow Q_i, ft^3/s	Reaeration coefficient k_{2i}, d^{-1}	Total BOD source W_i, lb/d	Total DO sink S_{Ri}, lb/d
0	14,203	250	4.60	0.000	500	0.120	0	0
1	14,203	240	30.00	0.050	500	0.140	0	−700
2	11,035	330	24.30	0.100	500	0.220	64	−10,300
3	9,240	400	28.95	0.150	502	0.330	0	−24,800
4	6,916	450	34.26	0.200	502	0.270	1,239	−25,300
5	9,873	520	32.70	0.250	528	0.400	4,572	−37,200
6	11,140	760	47.64	0.350	864	0.400	0	−60,400
7	13,569	660	49.86	0.450	903	0.730	0	−38,000
8	11,563	790	43.62	0.550	903	0.500	75	−41,800
9	12,883	650	58.98	0.650	903	0.540	147	−28,400
10	10,876	920	66.48	0.750	907	0.440	23	−30,300
11	11,880	2,520	84.90	0.900	910	0.390	0	−68,200
12	20,856	1,700	110.67	1.200	913	0.400	0	−32,100
13	15,523	2,500	129.75	1.500	913	0.300	0	−52,200
14	15,470	3,300	171.09	1.950	913	0.410	0	−78,100
15	23,284	4,450	116.73	2.800	913	0.380	0	−81,300
16	22,492	5,200	221.52	3.900	913	0.360	0	−94,600
17	23,337	5,350	220.20	5.300	913	0.360	0	−111,200
18	24,129	6,200	242.52	6.000	913	0.360	0	−59,900
19	28,300	5,250	193.92	6.000	913	0.360	0	−86,600
20	27,456	6,500	216.96	6.000	913	0.360	0	−100,200
21	30,043		222.60	6.000	913			

* Taken from R. E. Bunce and L. J. Hetling, A Steady-state Segmented Estuary Model, *Tech. Pap.* 11, FWPCA, Dep. of Interior, 1970.
† Temperature at all segments = 20°C.

NATURAL TRANSPORT SYSTEMS

Solution procedure for a finite volume model applied to the determination of the dissolved oxygen concentrations in a one-dimensional estuary

where $Q = Q$
$E = E$
$E' = EP$
$V = V$
$A = A$
$x = XL$
$U = UT$
$H = H$
$S_R = SR$
$k_1 = K1$
$k_2 = K2$
$k_3 = K3$
$W = W$
$T = T$
$c^* = CS$

and a_{ij} for DO = AC(I, J)

a_{ij} for BOD = AL(I,J)

f_i for DO = FC(I)

f_i for BOD = FL(I)

Q_{ik} = QIN, QØUT

E'_{ik} = EPIN, EPØUT

c_{ik} = CIN, CØUT

L_{ik} = LIN, LØUT

SINQ = Subroutine for the solution of a set of algebraic simultaneous equations

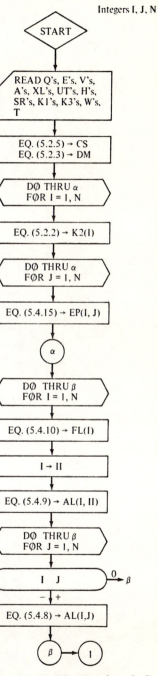

Integers I, J, N

ALGORITHM 5.4.1

(continued overleaf)

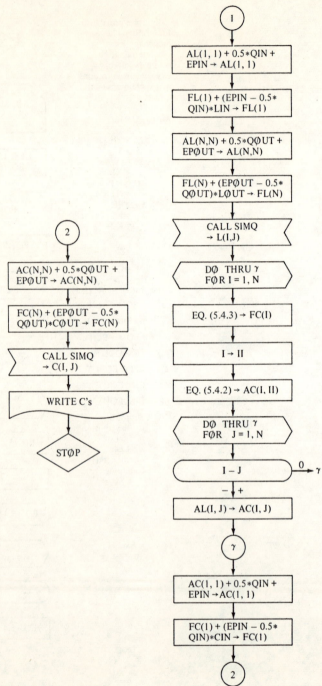

ALGORITHM 5.4.1 (*continued*)

where $\mathbf{B}^{-1}$ = inverse of coefficient matrix with elements b_{ij} ($n \times n$ matrix)

$\mathbf{W}$ = vector of loading terms W_i ($n \times 1$ matrix)

$\mathbf{R}$ = vector of constants given by known terms in expressions for forcing functions f_i ($n \times 1$ matrix)

Substitution of Eq. (5.4.18) into Eq. (4.4.17) yields

$$\mathbf{C} = \mathbf{A}^{-1}[\overline{Vk_1}]\mathbf{B}^{-1}\mathbf{W} + \mathbf{A}^{-1}[\overline{Vk_1}]\mathbf{B}^{-1}\mathbf{R} + \mathbf{A}^{-1}\mathbf{S} \qquad (5.4.19)$$

Through matrix multiplication

$$\mathbf{C} = \mathbf{X}\mathbf{W} + \mathbf{Y} \qquad (5.4.20)$$

where $\mathbf{X}$ = steady-state transfer-function matrix, also called a *unit-loading matrix* ($n \times n$ matrix)

$\mathbf{Y}$ = vector of constants ($n \times 1$ matrix)

Equation (5.4.20) can be used for the direct matrix solution for the dissolved-oxygen concentrations in the system $\mathbf{C}$ for a given set of loading conditions $\mathbf{W}$.

Using the elements of the unit-loading matrix it is possible to determine that portion of the dissolved-oxygen concentration in any volume resulting from a unit BOD loading in any other volume. For example, ϕ_{ij}, the element in the ith row and jth column of the unit-loading matrix $\mathbf{X}$, has a numerical value equal to the dissolved-oxygen concentration in volume i resulting from a unit-value BOD source in volume j. By multiplying the element ϕ_{ij} by the actual value of the BOD-source term in volume j, that portion of the dissolved-oxygen concentration in volume i resulting from the source term can be obtained. In mathematical notation, then, the total concentration of the dissolved oxygen in volume i is given by

$$c_i = \sum_j \phi_{ij} W_j \qquad (5.4.21)$$

where c_i = dissolved-oxygen concentration in volume i

W_j = BOD loading in volume j ////

Solution procedure 5.4.3 Harbors and bays may require two-dimensional finite-volume models to adequately describe transport behavior. Such is the case when these bodies of water are large and irregular in shape and/or when they are segmented naturally by island structures.

The procedure outlined in Algorithm 5.4.1 can be used for harbors and bays with little modification. Water boundary conditions may exist at more than two segments in the system. Also, fresh-water flow entering

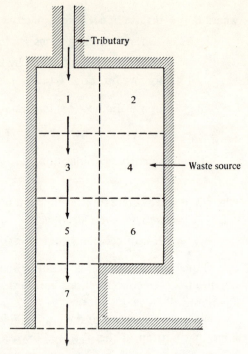

FIGURE 5.4.2
Two-dimensional estuarine system for Exercise 5.4.7.

the system is assumed to exit to the ocean along major navigational channels via the most direct route to the ocean.

The estimation of the tidal mixing coefficients offers considerable difficulty. In many cases, these estimates are based on experience.

Figure 5.4.2 is a sketch of a two-dimensional estuarine system for Exercise 5.4.7. Table 5.4.3 lists the physical data and the waste loadings discharged into the system.   ////

Solution procedure 5.4.4 The finite-volume model can be used to describe transport in a nontidal stream. In such cases the stream is segmented in reaches shorter than it would be for the application of the continuous model. No reach should be long enough to permit significant gradients to exist in the reach.

In applying the finite-volume model to a nontidal stream, the assumption can be made that E'_{ij} equals 0 and δ_{ij} has the value of unity. In effect, it is assumed that the water volume in each segment is completely mixed.   ////

Table 5.4.3 PHYSICAL DATA AND WASTE LOADINGS FOR THE TWO-DIMENSIONAL ESTUARINE SYSTEM SHOWN IN FIG. 5.4.2*

Volume parameters

Section i	Volume characteristic length x_i, ft	Volume V_i, ft$^3 \times 10^6$	Reaeration coefficient k_{2i}, d^{-1}	Total BOD source W_i, lb/d	Total DO sink S_{Ri}, lb/d
1	20,000	250	0.120	0	−10,000
2	20,000	200	0.330	0	−5,000
3	20,000	750	0.400	0	−10,000
4	20,000	650	0.400	100,000	−10,000
5	20,000	1,500	0.360	0	−15,000
6	20,000	1,200	0.400	0	−15,000
7	20,000	2,500	0.400	0	−15,000

Interface parameters

Interface number	Area A, ft$^2 \times 10^3$	Dispersion coefficient E, mi^2/d	Average flow Q, ft^3/s
In–1	3	4.0	1,000
1–2	15	4.0	0
1–3	25	4.0	1,000
3–4	24	4.0	0
3–5	30	5.0	1,000
5–6	28	5.0	0
5–7	35	6.0	1,000
7–out	40	6.0	1,000
2–4	20	4.0	0
4–6	26	5.0	0

* Temperature in all segments = 25°C.

5.5 TRANSPORT IN THE AIR ENVIRONMENT

Models of the general nature of those discussed in the preceding sections have been used to describe the transport of materials in the atmosphere. Such models include transport by convection and by turbulent diffusion, as well as the production of materials and their dissipation through chemical reaction. Convection (called *advection* in the water environment) in these models includes transport by thermal phenomena and winds. However, because of the complexity and variability of atmospheric systems the application to these systems of transport models of the type used in water transport has been limited. Instead, steady-state models have been used which describe transport by a diffusing plume convected by means of winds. The concentrations of materials in the plume are assumed to be distributed in a gaussian profile. One such model has the form of

$$c(x,y,z=0) = \frac{Q}{\pi \sigma_y \sigma_z U} \exp\left[-\left(\frac{y^2}{2\sigma_y^2} - \frac{h^2}{2\sigma_z^2}\right)\right] \quad (5.5.1)$$

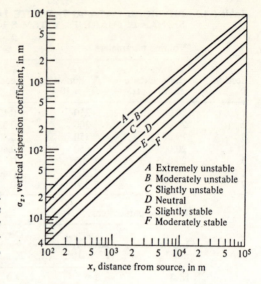

FIGURE 5.5.1
Horizontal dispersion coefficient as a function of distance from source. (*Taken from F. A. Gifford, Jr., Use of Routine Meteorological Observations for Estimating Atmospheric Dispersion, Nucl. Saf., vol. 2, p. 48, 1961.*)

where
- c = concentration of transported material. $[FL^3]$
- Q = emission rate at source. $[Ft^{-1}]$
- U = average wind speed. $[Lt^{-1}]$
- x, y, z = rectangular coordinates with x downwind, y crosswind, and z vertical (origin at source and ground level). $[L]$
- h = height of source above ground. $[L]$
- σ_y, σ_z = standard deviation of plume profile in crosswind and vertical directions. $[L]$

The terms σ_y and σ_z, which are generally referred to as dispersion coefficients, are dependent on downwind distance from the source and atmospheric stability. Values of these terms for different meteorological conditions can be estimated from the curves found in Figs. 5.5.1 and 5.5.2. The categories of meteorological conditions are correlated with surface wind speed and other meteorological factors in Table 5.5.1.

Expressions such as Eq. (5.5.1) are applicable to relatively short time periods and to average conditions. Typically, separate calculations are made for 1- or 2-h segments over a 24-h period, and then a 24-h average is determined from all the separate calculations. Since fallout (sedimentation) is not accounted for in Eq. (5.5.1), application of the equation to the transport of particulates is limited to sizes on the order of 20 μm or less.

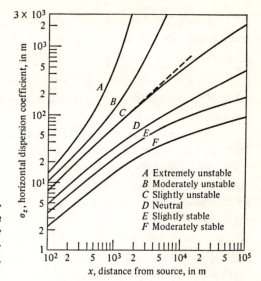

FIGURE 5.5.2
Vertical dispersion coefficient as a function of distance from source. (*Taken from F. A. Gifford, Jr., Use of Routine Meteorological Observations for Estimating Atmospheric Dispersion, Nucl. Saf., vol. 2, p. 48, 1961.*)

Table 5.5.1 METEOROLOGICAL CATEGORIES*†

Surface wind speed, m/s	Daytime insolation			Nighttime conditions	
	Strong	Moderate	Slight	Thin overcast or $\geq \frac{4}{8}$ cloudiness‡	$\leq \frac{3}{8}$ cloudiness
<2	A	A–B	B		
2	A–B	B	C	E	F
4	B	B–C	C	D	E
6	C	C–D	D	D	D
>6	C	D	D	D	D

* Taken from F. A. Gifford, Jr., Use of Routine Meteorological Observations for Estimating Atmospheric Dispersion, *Nucl. Saf.*, vol. 2, no. 48, 1961.
† *A*: extremely unstable conditions; *B*: moderately unstable conditions; *C*: slightly unstable conditions; *D*: neutral conditions (applicable to heavy overcast, day or night); *E*: slightly stable conditions; *F*: moderately stable conditions.
‡ The degree of cloudiness is defined as that fraction of the sky above the local apparent horizon which is covered by clouds. [Manual of Surface Observations (WBAN), *Circ. N*(7th ed.), paragraph 1210, U.S. Government Printing Office, Washington, July 1960.]

EXERCISES

5.1.1 Discuss the characteristic differences between the continuous and finite-volume models. How do the dispersion terms differ? What advantages does one model have over the other?

5.1.2 What is the most fundamental difference between models such as the continuous and finite-volume models and the aquatic ecosystem model discussed in Sec. 4.6?

5.1.3 What types of variables are included in the unit-loading matrix? Will the unit-loading matrix vary with external loading conditions? With other boundary conditions?

5.1.4 What is the steady-state transfer function, and what is its significance?

5.1.5 What is the significance of the advection factor?

5.2.1 If a reach of stream has a reaeration coefficient of 0.2 d^{-1} at 20°C, estimate its value at 15°C.

5.2.2 The average flow velocity through a reach of stream is 0.5 m/s, the average depth is 1 m, and the temperature is 15°C. Estimate the rate of oxygen transfer [in mg/(l)(d)] from the atmosphere to the water in the reach.

5.2.3 On what factors does the occurrence of a distinct, nitrogenous BOD stage in streams depend? What is the total oxygen demand exerted by 2 mg/l of ammonia?

5.2.4 In what ways do organic materials which have become part of a stream's bottom deposits exert a demand on the oxygen resources of the stream water?

5.2.5 Relate the phenomena depicted for the aquatic ecosystem in Fig. 4.6.1 with that which occurs in a flowing stream.

5.3.1 One hundred kg of 5-d 20°C BOD (BOD_5) are discharged each day into the upstream end of the first reach of stream A, the characteristics of which are listed in Table 5.3.1. Assuming that $D_0 = 0$, construct the oxygen-sag curve from the point of discharge to the point where the stream is 90 percent recovered.

5.3.2 For the situation described in Exercise 5.3.1, calculate the distance of the minimum point along the oxygen-sag curve from the point of waste discharge. Check this distance with the distance as given by the curve constructed in Exercise 5.3.1.

5.3.3 During a low flow period, a reach of stream 16 km in length has the following characteristics:

$Q = 60$ m³/s $L_s = 0.0$ mg/l
$U = 0.3$ m/s $D_0 = 0.0$ mg/l
$k_2 = 0.40$ d^{-1} $L_a = 1$ mg/(l)(d)
Temperature = 25°C $S_R = 0.0$ mg/(l)(d)

If a minimum of 5 mg/l of dissolved oxygen is to be maintained in the reach, how many kilograms of BOD can be discharged into the stream per day at the upper end of the reach if $k_1 = 0.25$ d^{-1} and $k_3 = 0.10$ d^{-1}?

5.3.4 If a minimum of 5 mg/l of dissolved oxygen is to be maintained in stream A, the characteristics of which are given in Table 5.3.1, how many kilograms of BOD_5 can be discharged per day into the upstream end of reach number 4? Assume $D_0 = 0$.

5.3.5 The BOD_5 of river water at the upper end of a reach is 62 mg/l. However, by the time the water flows to the lower end of the reach, a 0.345-d flow downstream, the BOD_5 is reduced to 45 mg/l. Assuming $k_1 = 0.23$ d^{-1}, estimate the values of L_a and k_3.

5.3.6 Stream A is segmented into five reaches, each of which has relatively uniform characteristics. These characteristics are listed in Table 5.3.1. Two waste discharges are introduced into the stream, the first being located at the upper end of reach number 1 and the second at the upper end of reach number 3. Let it be assumed that the characteristics of the stream water flowing into the upper end of reach number 1 are known. Prepare a computer program for the dissolved-oxygen profile down the total length of the five reaches using the continuous model. Use the program to explore the effects of various combinations of BOD loadings at points of waste discharge.

5.4.1 Discuss the physical characteristics and the transport mechanisms of estuarine systems.

5.4.2 What variables are included in the elements of the unit-loading matrix computed for the DO-BOD system in estuarine situations?

5.4.3 Given the data in Table 5.4.1 and assuming that $L_{ik} = 0$, $c_{ik} = 4$ mg/l, and $\delta_{ij} = 0.5$, determine the steady-state dissolved-oxygen concentration in all volumes. Use any method of solution for simultaneous algebraic equations.

5.4.4 Determine the DO-BOD unit-loading matrix for the system given in Table 5.4.1. Using the unit-loading matrix, solve for the dissolved-oxygen-concentration vector. Assume $L_{ik} = 0$, $c_{ik} = 4$ mg/l, and $\delta_{ij} = 0.5$.

5.4.5 Given the data in Table 5.4.2 determine the steady-state dissolved-oxygen concentrations in all volumes. Assume $L_{ik} = 0$, $c_{ik} = 4$ mg/l, and $\delta_{ij} = 0.5$. Use any method of solution for simultaneous algebraic equations.

5.4.6 Determine the DO-BOD unit-loading matrix for the system given in Table 5.4.2. Using the unit-loading matrix, solve for the dissolved-oxygen-concentration vector. Assume $L_{ik} = 0$, $c_{ik} = 4$ mg/l, and $\delta_{ij} = 0.5$.

5.4.7 A two-dimensional estuarine system is shown in Fig. 5.4.2. The physical data and waste loadings for the estuary are listed in Table 5.4.3. Assuming $L_{ik} = 0$, $c_{ik} = 4$ mg/l, and $\delta_{ij} = 0.5$, determine the dissolved-oxygen concentration in all segments. Use any method of solution for simultaneous algebraic equations.

5.4.8 Determine the DO-BOD unit-loading matrix for the system given in Table 5.4.3. Using the unit-loading matrix, solve for the dissolved-oxygen-concentration vector. Assume $L_{ik} = 0$, $c_{ik} = 4$ mg/l, and $\delta_{ij} = 0.5$.

5.4.9 Assume $L_{ik} = 0$ and $c_{ik} = 4$ mg/l, determine the DO-BOD unit-loading matrix for stream A, the characteristics of which are listed in Table 5.3.1.

5.5.1 Why have transport models of the type used in water systems found limited application in atmospheric systems?

5.5.2 On what factors do the values of the dispersion coefficients used in plume models depend? How are these values usually determined when use is made of the plume model?

5.5.3 Make a sketch defining the spatial significance of the parameters found in the plume model.

6
PLANNING FACTORS

6.1 WATER-QUALITY CRITERIA AND STANDARDS

The distinction between *water-quality criteria* and *water-quality standards* is subtle but important. Water-quality criteria can be defined as statements concerning the limiting levels of concentration or intensity of key quality parameters established by *the intended use* of the water. On the other hand, water-quality standards can be defined as regulations as to the limiting levels of concentration or intensity of key quality parameters established by *regulatory authority* for the purpose of protecting or preparing a water resource for one or more uses. A correlation of water-quality criteria with water-quality standards is illustrated by the bar diagram in Fig. 6.1.1. Shown there are criteria for three quality parameters established by three beneficial uses of the water. The standards for quality parameters 1 and 3 are established at the minimum values of criteria for those parameters. The standard for quality parameter 2 is established at the maximum value. Dissolved oxygen is an example of the type of quality parameter for which a standard is established to meet the maximum of the minimum values.

There are many beneficial uses made of water, but those that involve

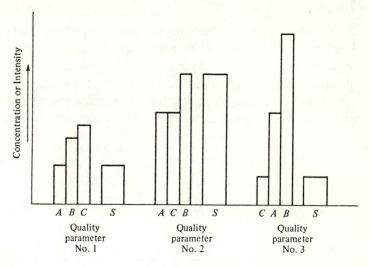

FIGURE 6.1.1
Correlation of water-quality criteria for beneficial uses *A*, *B*, and *C* with water-quality standards *S*.

quality criteria are grouped as follows:

1 Public water supplies
2 Recreation and aesthetics
3 Fish, other aquatic life, and wildlife
4 Agriculture
5 Industrial water supplies

Detailed and complete criteria for each of these uses have been published elsewhere.[1] For public water supplies, the criteria are established so that the water *after* treatment by conventional treatment systems will meet the limits of the U.S. Public Health Service Drinking Water Standards listed in Table 6.1.1. The standards contained therein define the physical, chemical, and microbiological requirements of a water ready for human consumption. Two types of limits are used in these standards: *recommended* and *rejection*. Materials having recommended limits when present in concentrations above the limit are either objectionable to an appreciable number of people or exceed the levels required by good water-quality control practice. Materials having rejection limits may cause adverse effects on health when present in concentrations above the limit.

[1] National Technical Advisory Committee, "Water Quality Criteria," FWPCA, Washington, 1968.

In establishing standards for the protection of any facet of the environment, the regulatory authority must select for each parameter a standard's objective, or goal. Such objective will fall at some level on the standard's objective spectrum.[1] (See Fig. 6.1.2.) Normally, the decision as to level depends upon health considerations, technological and economic resources, and political pressures. These factors determine public policy, which in turn determines

[1] F. M. Stead, *Proc., Joint Semin. Water Qual. Res.*, Calif. State Dep. Public Health Bur. Sanit. Eng., June 1965.

Table 6.1.1 U.S. PUBLIC HEALTH SERVICE DRINKING WATER STANDARDS*

Substance	Recommended limit	Rejection limit
Physical parameters:		
Color, units	15	
Odor, TON (threshold odor number)	3	
Turbidity, units	5	
Microbiological parameters, number/100 ml:		
Coliform organisms		1
Toxic chemicals, mg/l:		
Arsenic	0.01	0.05
Barium		1.0
Cadmium		0.01
Chromium (+6 valence)		0.05
Cyanide	0.01	0.02
Fluoride	1.3	2.6
Lead		0.05
Nitrate-N	10	
Phenols	0.001	
Selenium		0.01
Inorganic chemicals, mg/l:		
Chloride	250	
Copper	1	
Iron	0.3	
Manganese	0.05	
Silver		0.05
Sulfate	250	
Total dissolved solids (TDS)	500	
Zinc	5	
Organic parameters, mg/l:		
Carbon chloroform extract (CCE)	0.2	
Methylene blue active substances (MBAS)	0.5	

* U.S. Dep. of HEW, Public Health Service Drinking Water Standards, Revised 1962, *U.S. Public Health Serv., Publ.* 956, 1962.

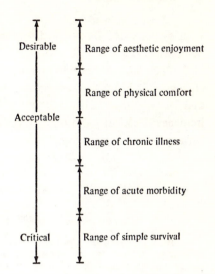

FIGURE 6.1.2
Standard's objective spectrum.

standards. In the past, standards for public water supplies have risen up the spectrum from the lowest range to some level in the top two ranges.

Standards to be effective must be quantitative. Unfortunately, water-quality criteria and, hence, water-quality standards have been established for the most part on bases other than precise experimental determination. Standards have been established on the basis of current practice, technological attainability, economic factors, experimental investigations on laboratory animals, and even guesses. Some standards may have their origins in decisions forced by public clamor.

Water-quality management is assisted by the application of stream and effluent standards. Stream standards involve a system of stream classification based on the intended use of the water. Any one stream may be zoned into several reaches, each having its own classification. For each classification, threshold values for key quality parameters are specified. Most stream standards include specifications as to fecal coliform, pH, dissolved oxygen, phenolic compounds, and temperature. Fecal coliform organisms, while nonpathogenic themselves, serve as indicators of fecal pollution. Their presence indicates that pathogenic organisms of fecal origin *may* be present in the water, thus constituting a health hazard to humans who use the water as a source of water supply. All surface waters must be disinfected prior to domestic use. However, disinfection processes are not infallible, and, statistically, surface waters with low fecal coliform counts are safer to use as raw water sources for drinking purposes than are waters with high counts. Pathogenic organisms that may be found in surface waters polluted by fecal discharges include those bacteria and viruses that cause intestinal disorders. Infectious hepatitis has been shown to be water-borne.

Phenolic compounds in water cause tastes and odors when the water is chlorinated. Many of these compounds have also been found to be toxic to man and aquatic life. Dissolved oxygen, pH, and temperature are factors that influence drastically the aquatic ecosystem. Typical standards for nonmarine surface waters are listed in Table 6.1.2.

Table 6.1.2 TYPICAL STANDARDS FOR NONMARINE SURFACE WATERS

Temperature standards

All waters of lakes and reservoirs shall not exceed a monthly average temperature of 90°F at any time after adequate mixing of heated and normal waters as a result of heated liquids, nor shall the monthly average water temperature after passing through an adequate zone for mixing be more than 3°F greater than that of water unaffected by the heated discharge. The size of the mixing zone will be determined on an individual project basis and will be based on normal engineering considerations, and the area affected shall be kept to a minimum. The mixing zone shall not prevent free passage of fish or cause fish casualty.

Compliance with temperature standards in streams below impoundments (includes those which are used for cooling purposes) shall be based on measurements in the receiving waters below the impoundment.

Quality standards for class AA waters

Class AA water is suitable for use for domestic and food processing purposes with disinfection and pH adjustments as the only treatment required. It is suitable also for trout survival where so specified and for uses requiring water of lesser quality.

Items	Specifications
1 Sewage, treated waste, thermal discharges, or other waste effluents	None.
2 Dissolved oxygen	Not less than 6 mg/l with a daily average of 7 mg/l.
3 Toxic wastes, deleterious substances, colored or other wastes	None in amounts to exceed limitations set forth in the latest edition of U.S. Public Health Service Drinking Water Standards.
4 Fecal coliform	Not to exceed 20/100 ml as a monthly arithmetic average.

Quality standards for class A waters

Class A water is suitable for use as swimming water. It is suitable also for other uses requiring waters of lesser quality.

Items	Specifications
1 Fecal coliform	Not to exceed a geometric mean of 200/100 ml, nor shall more than 10% of the total samples during any 30-d period exceed 400/100 ml.

Table 6.1.2 TYPICAL STANDARDS FOR NONMARINE SURFACE WATERS (*continued*)

Items	Specifications
2 Phenolic compounds	Not greater than 1 μg/l unless caused by natural conditions.
3 pH	Range between 6.0 and 8.0, except that swamp waters may range from 5.0 to 8.0.
4 Dissolved oxygen	Not less than 5 mg/l, except that swamp waters may have an average of 4 mg/l.

Quality standards for class B waters

Class B water is suitable for domestic supply after complete treatment in accordance with requirements of the state board of health. It is suitable also for propagation of fish, industrial and agricultural uses, and other uses requiring water of lesser quality.

Items	Specifications
1 Fecal coliform	Not to exceed a log mean of 1,000/100 ml based on five consecutive samples during any 30-d period, nor to exceed 2,000/100 ml in more than 20% of the samples examined during such period (not applicable during or following periods of rainfall).
2 pH	Range between 6.0 and 8.5, except that swamp waters may range from 5.0 to 8.5.
3 Dissolved oxygen	Daily average not less than 5 mg/l with a low of 4 mg/l, except that swamp waters may have an average of 4 mg/l.
4 Phenolic compounds	Not greater than 1 μg/l unless caused by natural conditions.

Quality standards for class C* waters

Class C water is suitable for fish survival,† industrial and agricultural uses, and other uses requiring water of lesser quality.

Items	Specifications
1 pH	Range between 6.0 and 8.5, except that swamp waters may range between 5.0 and 8.5.
2 Dissolved oxygen	Not less than 3 mg/l, except that swamp waters may have a low of 2.5 mg/l.
3 Fecal coliform	Not to exceed a log mean of 1,000/100 ml based on five consecutive samples during any 30-d period, nor to exceed 2,000/100 ml in more than 20% of the samples examined during such period (not applicable during or immediately following periods of rainfall).

* To apply only to streams receiving waste prior to May 4, 1950, and not to be applied to streams with a 7-day once-in-10-years-occurrence flow of more than 22.5 mg/d; nor shall this classification be assigned to interstate streams.

† "Fish survival" as used in this standard means the continued existence of individual fish normally indigenous to water of this type.

Effluent standards are used either to restrict the amount of pollutants discharged into a surface water or to specify the degree of treatment required before discharge. Although effluent standards are easier to enforce than the stream standards, they do not take into account the natural purification capacity of the surface water. Most regulations in force at the present time are a combination of stream and effluent standards. They specify standards similar to those in Table 6.1.2 as well as a minimum level of treatment for discharges.

6.2 AIR POLLUTION AND ITS CONTROL

There are three modes of man's exposure to air pollution. These are (1) personal exposure resulting from social habits such as smoking, (2) exposure under occupational conditions, and (3) exposure of large or small segments of the general population through no action of their own, either resulting from man-made or natural sources of environmental pollution. Personal exposure can be controlled only by the individual himself, and occupational exposure can be altered by controlling the source of exposure or the movements of the individuals involved. The control of exposure to environmental pollution is the subject of this section.

Air-pollution effects are manifested in two types of human health hazards: *episodes* and *chronic diseases*. During episodes atmospheric levels of pollutants build up to a point where relatively large numbers of individuals with particular susceptibilities to air pollution become ill or succumb and die. Such episodes are generally dramatic but short-term.

Chronic effects are not as dramatic, but the long-term effects are just as deadly. Most of the chronic ailments occur in the respiratory system. *Emphysema*, which is the fastest-growing cause of death in the United States, is a progressive breakdown of air sacs in the lungs usually brought on by chronic infection or irritation of the bronchial tubes. The disease progressively diminishes the ability of the lungs to transfer oxygen to and carbon dioxide from the bloodstream. Deaths from emphysema are twice as high in the city as in the country. *Bronchial asthma*, *pneumonia*, and even the *common cold* are conditions that appear to be aggravated by air pollution. Air pollution is also suspect as one of the causes of lung cancer.

Air pollution also causes considerable damage to vegetation as well as severe economic losses through corrosion. Aesthetic costs are no doubt significant.

Many types of materials pollute the air. However, the principal pollutants fall into one of five classes: particulate matter, sulfur oxides, carbon monoxide, nitrogen oxides, and hydrocarbons.

Particulate Matter[1]

Particulates—solid and occasionally liquid substances—in the air may be large enough that they settle rapidly to the ground or small enough to remain suspended in the air until removed by such natural phenomena as rain and wind or by people who breathe them into their lungs. The major portion of the particulates results from fuel combustion, incineration of waste materials, and industrial-process losses. Particulates decrease visibility as well as provide a substrate which catalyzes many atmospheric reactions.

Sulfur Oxides

The sulfur oxides result primarily from the burning of high-sulfur fuels in stationary sources. When such fuel burns, the sulfur burns too, producing sulfur dioxide gas and, to a much smaller extent, sulfur trioxide, which in the atmosphere converts immediately to sulfuric acid. Sulfur dioxide gas alone can irritate the upper respiratory tract, and absorbed on particulate matter it can be carried deep into the lung where it can injure delicate tissue. Sulfuric acid in the right particle size can also penetrate deep into the lung and damage tissue. There is considerable evidence that sulfur oxides aggravate existing respiratory disease in humans and contribute to its development. High levels of sulfur pollutants, as well as other contaminants, have been observed during the more dramatic air-pollution episodes.

Carbon Monoxide

Carbon monoxide is a poison. When the gas enters the bloodstream, it replaces the oxygen needed to carry on the body's metabolism. At low concentrations it brings on headaches and slowing down of physical and mental activity; at higher concentrations it kills quickly. Motor-vehicle exhaust accounts for most of the carbon monoxide emissions. Inside an automobile operating in traffic the concentration of carbon monoxide may reach high enough levels to affect the driver, thereby creating a safety hazard. At approximately 100 parts per million, most people experience dizziness, headache, lassitude, and other symptoms of poison-

---

[1] Material in this and succeeding paragraphs was extracted from U.S. Dep. of HEW, U.S. Public Health Ser., "The Effects of Air Pollution," 1966.

ing. Concentrations higher than this occasionally occur in garages, in tunnels, or behind automobiles.

Nitrogen Oxides

Most of the nitrogen oxides result from stationary combustion sources as well as from motor-vehicle exhausts. At the temperatures commonly reached when fuel combusts, nitrogen in the air combines with oxygen to form nitric oxide, which is relatively harmless but which usually converts in the atmosphere to the more dangerous nitrogen dioxide. The latter is usually considered as one of the photochemical oxidants responsible for smog. Little information is available on the health effects of the nitrogen oxides at levels commonly found in polluted air. However, low concentrations are suspected to increase the susceptibility of animals to infection.

Hydrocarbons

There are many sources of hydrocarbon pollution, but by far the major source is the incomplete burning of gasoline in automobiles. The hydrocarbons constitute a large group of organic chemicals, most of which are directly harmful only in very high concentrations. However, a number of hydrocarbons at low concentrations will photochemically react in the atmosphere with the nitrogen oxides to produce photochemical smog. The reactions which result in smog formation are very complex. However, a reduction in hydrocarbons is thought to result directly in a reduction in smog. Other oxidants believed to play a role in photochemical smog formation include ozone and peroxyacyl nitrates (PAN). Discomfort and a variety of pulmonary effects are attributed to smog.

Meteorology

In an atmospheric situation where the air is well-mixed, the air temperature decreases about 1°C for each 100 m in altitude. This decrease, which results from changes in pressure, is referred to as the *adiabatic lapse rate*. At times, the lapse rate will be greater or less than adiabatic. When the lapse rate is greater than adiabatic, the air is said to be *superadiabatic*; and a situation exists in which the atmosphere is unstable, and there is considerable vertical mixing of the air. Air pollutants discharged during a superadiabatic situation are transported vertically and dispersed rapidly.

When the lapse rate is less than the adiabatic lapse rate, the atmosphere is more stable, and dispersion is hindered. If the lapse rate becomes negative, i.e., temperature increases with altitude, an *inversion* is created. Possible lapse rates are indicated in Fig. 6.2.1.

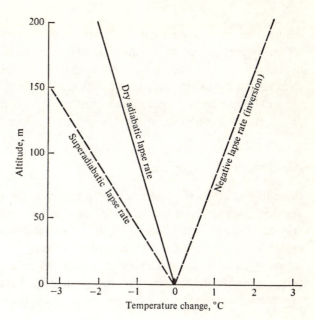

FIGURE 6.2.1
Atmospheric air-temperature lapse rates.

An inversion occurs when a warm air mass moves over the cooler air below. The warm air mass cannot move through the heavier cool air; hence, it must move above it. The warm air from the ground will rise until it hits the still warmer air above. When such a situation occurs, the effect is as if a lid were clamped over the city. Air pollutants become trapped by the inversion, and if the wind dies away, the concentrations of these pollutants build up to high levels.

Temperature inversions can result from the normal diurnal cooling cycle. After sunset, the ground cools off quickly, cooling also the air immediately above the ground. The air at still higher altitudes remains warm, thus establishing an inversion. The next day sunlight destroys the inversion.

Control

Particulate matter can be removed at the source by a variety of unit operations. These operations include settling chambers, filters, centrifugal collectors, electrostatic precipitators, and wet collectors, sometimes referred to as scrubbers. The unit operation chosen for a particular emission will depend on the size distribution of the particulates, the specific gravity, electrostatic charge, and the chemical nature of the material out of which the particulates are composed.

Gaseous emissions from stationary sources can be controlled by passing the gas stream through a gas-liquid absorber, such as a packed tower, a tray tower, a spray tower, or a venturi scrubber. Certain emissions of a gaseous nature lend themselves to the conversion to nonpollutants by combustion. Combustion operations may be accomplished by flame or by catalysts.

Pollution from vehicular exhausts can be controlled by devices which reduce the emission of hydrocarbons and carbon monoxide.

Air-Quality Management

Basic to any air-quality management program is the need for an emission inventory. Such an inventory provides information useful in designing air surveillance systems and is necessary in the establishment of standards and the evaluation of progress made toward air-management goals. The emission inventory involves the systematic collection and collation of detailed data concerning the air-pollution emissions in a given area. The inventory should contain as much information as possible concerning the types of sources as well as their

Table 6.2.1 CLASSIFICATION OF SOURCE TYPES FOR PURPOSE OF PREPARING AN AIR-POLLUTION EMISSIONS INVENTORY

1 Fuel combustion
   a. Residential
   b. Commercial and institutional
   c. Industrial
   d. Steam-electric power plants

2 Industrial-process losses
   a. Chemical process industries
   b. Food and agricultural industries
   c. Metallurgical industries
   d. Mineral products industries
   e. Petroleum refining and petrochemical operations
   f. Wood processing
   g. Petroleum storage

3 Solid-waste disposal
   a. Incineration
   b. Open burning

4 Transportation
   a. Motor vehicles
   b. Off-highway fuel usage
   c. Aircraft
   d. Railroads
   e. Vessels
   f. Gasoline-handling evaporative losses

5 Miscellaneous

Table 6.2.2 TYPICAL AMBIENT AIR-QUALITY STANDARDS*

Pollutant	Measuring interval	$\mu g/m^3$ †‡
Sulfur dioxide	1 h	790
	24 h	260
	annual	45
Suspended particulates	24 h	250
	annual G.M.§	60
Carbon monoxide	1 h	25×10^3
	8 h	10×10^3
Photochemical oxidant	1 h	100
Nonmethane hydrocarbons	3 h	130
Gaseous fluorides (as HF)	30 d	0.3 $\mu g/(cm^2)(month)$
Oxides of nitrogen	Annual	100

* Taken from "Air Pollution Control Regulation and Standards," adopted by the South Carolina Pollution Control Authority, January 1972.
† Arithmetic average except in case of suspended particulates.
‡ At 25°C and 760 mm Hg.
§ Geometric mean.

contributions in terms of types and rates of pollutant discharge. Emissions are generally reported in terms of weight of pollutant per unit time. Table 6.2.1 shows a classification of source types for use in making an emission inventory.

Air-quality standards are of two general types: The first type establishes the limit on source discharge. The prohibition of open burning or a limit placed on smoke emission from a stack are examples of this type of standard. The second type of standard establishes limits on ambient air quality. The latter depends upon the contribution made to air quality by the community as a whole and not just upon a single source. An example of ambient air-quality standards is shown in Table 6.2.2.

6.3 RADIOLOGICAL HEALTH

Radiological health can be defined as the science and practice of protecting man and his environment from unacceptable exposure to radiation. The exposure of mankind to a variety of low-level man-made radiation sources is increasing.

Such sources include x-rays, radioactive materials, and electronic devices in the home, the office, and industry. The increase both in the number of sources and in the number of people exposed is raising concern over the hazard to human health.

Radiation can be defined as energy transmitted through space. There are two types of energy: electromagnetic and particulate. The electromagnetic radiation of traditional concern is gamma radiation and x-rays. As may be seen in the electromagnetic spectrum in Fig. 3.6.2, this radiation has wavelengths of 10^{-4} to 10^3 Å. Gamma radiation and x-rays have the property of inducing ionization in matter. Particulate radiation takes the form of alpha particles (helium nuclei with a mass of 4 and a positive charge of 2), beta particles (free electrons), protons, and neutrons.

In living organisms, radiation damages the complex molecules within the cell and interferes with the cell's chemical machinery to a point where, in extreme cases, the organism dies. The structure of genes is particularly vulnerable to radiation, and a cell that is not killed may be damaged to an extent that it cannot duplicate itself. Sometimes the radiation is so weak that the damaged cell can still divide, but the cell has been so changed that it and its descendants no longer can function normally. In humans, such an effect may cause skin cancer, leukemia, and other diseases.

Radiation may cause mutations in the sex cells, and since these cells are passed on from parent to child, succeeding generations are affected. In fact where sex cells are involved, the relatively mild effect of the mutations is considered more serious than the drastic one of nondivision. For this reason, it is generally maintained that there is no "safe" level of radiation as far as genetic effects are concerned.

Most radiation damage to biologic tissue has been associated with ionizing radiation. It appears that nonionizing radiation has less impact on the general population than ionizing radiation.

Dosimetry

The quantity of a radioactive isotope is commonly expressed in terms of the rate at which the nuclear disintegrations occur. One *curie* is equal to 3.7×10^{10} disintegrations per second. To be complete, the description of the radiation must also include the type of radiation (such as alpha, beta, and gamma) and the energy of each type of radiation, usually expressed in terms of electron volts.

The interaction of radiation with matter is called *dosimetry*. Alpha, beta, and gamma radiation interact with matter producing ion pairs—a positively charged atom and a free electron. The *roentgen* unit is defined as the quantity of x-ray radiation or gamma radiation producing 1 esu of electricity in 1 cm³ of dry air, an equivalent to the formation of 2.08×10^9 ion pairs. The basic physical unit now used to describe a radiation dose is the *rad*. This unit is defined as the amount of any radiation producing an absorption of 100 ergs/g of any material. The unit that is most frequently used in expressing biologic dose is the *rem* (roentgen equivalent man). The rem is equal in value to the product

$$\text{Rem} = \text{RBE} \times \text{rad}$$

where RBE = relative biologic effectiveness of different forms of radiation

Table 6.3.1 lists the RBE for different forms.

The total health risk associated with different sources of radioactivity such as nuclear power plants and nuclear-fuel reprocessing plants has been expressed in terms of *total man-rems* of exposure. The number of man-rems associated with a radiation source is determined by adding the exposures of all individuals of a population to estimate the cumulative exposure received by the population. For example, 10,000 people of a community each receiving 2 mrem/year of radiation would have had a total exposure of 20 man-rems.

Sources of Radiation

The earth's population is subjected to two types of radiation: natural radiation and man-made radiation. (The radiation exposure of the general population in 1970 is shown in Table 6.3.2.) Two general sources contribute to the earth's natural background of radiation. Cosmic radiation continually bombards the earth's atmosphere. Some of the radiations are absorbed by the atmosphere; the remainder penetrates to the earth's surface. The dose received by a person at any point on earth from cosmic sources varies with altitude and latitude. The other major source of natural radiation is the radioactive isotopes found in mineral deposits and water.

Table 6.3.1 RELATIVE BIOLOGIC EFFECTIVENESS

Type of radiation	RBE
Beta	1
Gamma	1
X-rays	1
Alpha	10
Neutrons	5

Medical uses of radiation represent by far the largest single source of man-made radiation. This source accounts for over 90 percent of the total man-made radiation. Atmospheric nuclear tests currently contribute approximately 5 percent. The isotopes of greatest concern in fallout are strontium 90 (Sr^{90}) with a half-life of 28 years, and cesium 137 (Cs^{137}), with a half-life of 30 years. Their half-lives are such that the isotopes will persist in the environment for many years. Another isotope, tritium H^3, with a half-life of 12.26 years, is also produced in significant quantities by weapons testing.

Reactors in nuclear power plants discharge effluents containing small amounts of radioactivity. Krypton 85 (Kr^{85}), with a half-life of 10.76 years, and tritium are released in such effluents as well as from the effluents released from nuclear-fuel reprocessing plants. However, the effluents from the latter contain those isotopes in greater quantities. Exposure to persons living near a typical nuclear power reactor has been found to range from 1 to 10 mrem/year. The amount of krypton in the atmosphere is increasing, and by the year 2000 it is estimated that the yearly dose to the world's population will be about 0.02 mrem. However, there is no evidence now that krypton will concentrate in food chains. Tritium is extremely difficult to remove from liquid discharges. Here again, there is no evidence that the isotope will concentrate in food chains. Furthermore, the anticipated population dose from tritium by the year 2000 is no more than 0.002 mrem/year.

Nuclear power reactors will not detonate in the manner of nuclear weapons. For those reactors currently in commercial use, the most serious hazard faced is the possibility that the core could melt from excessive heating and spill large quantities of radioactive materials into the environment. An elaborate set of design and operating features is employed to make the possibility remote.

Radiation is reduced only by time. Consequently, the major objective in disposing of radioactive wastes is to concentrate and store. No concentration

Table 6.3.2 RADIATION EXPOSURE IN 1970*

Source	Average per capita dose, mrem/year
Natural background	130
Medical diagnostic x-ray	90
Weapons-test fallout	5.1
Nuclear power	less than 0.01

* Taken from J. A. Lieberman, Ionizing-radiation Standards for Population, *Phys. Today*, November 1971.

process is 100 percent efficient, however, and some level of activity must be released to the environment, where it is subject either to dispersion and dilution or to concentration by physical or biologic processes. Decisions concerning levels of released activity must be based on considerations of short-term and long-term risks to man and his environment.

Two categories of radioactive wastes are produced by the nuclear power industry: low-level and high-level. Low-level wastes are gaseous, liquid, and solid. The radioactivity in liquid discharges is reduced through a variety of methods including precipitation and ion exchange. Low concentrations of radioactivity are permitted to be released to the atmosphere and surface waters. Low-level solid wastes are packaged and shipped to AEC-owned or AEC-licensed burial sites for storage.

High-level wastes result from nuclear-fuel-element reprocessing activities. These wastes, which originally are in liquid form, are concentrated and stored in tanks under controlled conditions at the reprocessing plant. It is anticipated that in the future the stored liquids will be converted into solid forms and stored at some isolated place.

Standards

Two general types of standards are employed in radiological health: external standards and internal standards. External standards are based on maximum exposure to external surfaces of the human body, whereas internal standards are based on maximum exposure suffered by the internal tissues of the body as a result of breathing air and drinking water contaminated with radioactivity.

Atomic Energy Commission external standards applying to restricted areas in an AEC plant for persons employed at the plant are as follows:

Whole-body exposure	1.25 rd/calendar quarter
Hands and feet exposure	18.75 rd/calendar quarter
Skin	7.25 rd/calendar quarter

A whole-body exposure up to 3 rd/calendar quarter is allowed provided the accumulated dose to the whole body is less than 5 (age in years—18) and a full record of exposure is maintained. Stated another way, the employee should not work with ionizing radiation until he is 18 years old and then should not be exposed to more than an average of 5 rd/year. Minors are restricted to a maximum of 10 percent of the tabulated values. For unrestricted areas the AEC standards limit whole-body exposure to any individual to no more than 0.5 rd/year with a maximum 7-d dose of 0.10 rd.

AEC external standards are expressed in terms of *maximum permissible concentrations* in air $(MPC)_a$ and water $(MPC)_w$ for employees of AEC plants. These standards are as follows:

Isotope	$(MPC)_a$, μci/ml	$(MPC)_w$, μci/ml
Cs^{137}	1×10^{-8}	4×10^{-4}
Sr^{90}	3×10^{-10}	4×10^{-6}
H^3	5×10^{-6}	1×10^{-1}

U.S. Public Health Service Drinking Water Standards relative to radiation protection are as follows:

Isotope	Concentration, μμci/ml
Ra^{226}	3
Sr^{90}	10

If radium and strontium are absent, gross beta-ray concentration should not exceed 1,000 μμci/l.

Shielding

Electromagnetic radiations such as beta and gamma rays attenuate when passing through matter according to the Beer-Lambert law expressed by Eq. (3.6.2). A quantified estimate of shielding requirements for any particular level of radiation must take into account not only the type of radiation and the material out of which the shield is to be constructed but, also, the geometry of the relationship between the radiation source and the shield. It is important to know that alpha particles can be stopped easily by a sheet of paper or by the horny outer layer of human skin. Beta particles can be stopped by an inch of wood or by a thin sheet of aluminum foil. Gamma radiation can penetrate several inches of lead, depending upon its energy.

6.4 ENVIRONMENTAL IMPACT STATEMENTS

The Environmental Policy Act of 1969 requires for all proposed construction and land-use development an environmental impact statement which addresses itself not only to the need for the proposed project and the customary cost-benefit analysis but also to a detailed assessment of the probable effect of the project on the environment. The statement is required to be comprehensive and should be prepared and reviewed by interdisciplinary teams consisting of physical and biologic scientists, social scientists, and engineers. A primary purpose of the statement is to ensure that the impact of alternate actions is evaluated and considered in project planning.

The environmental impact statement results from a planning procedure which consists of an ordered sequence of steps. The steps are as follows:

Step 1 Identify the major objective of the project.
Step 2 Analyze the technological possibilities of achieving the objective.
Step 3 Identify one or more actions for achieving the stated objective.
Step 4 Evaluate the characteristics and conditions of the *existing* environment.
Step 5 Prepare alternate engineering plans and monetary cost-benefit analyses for each action identified in step 3.
Step 6 Evaluate the impact each plan in step 5 will have on the environment.
Step 7 Compare the various plans and their impacts.
Step 8 Make recommendations.

Steps 4 and 5 result in detailed reports—the engineering report and the report characterizing the existing environment. These are utilized in step 6.

An environmental impact statement should consist of four sections. These sections should contain the following information:[1]

1 Analysis of need should include information developed in steps 1 to 3 of the planning procedure. This section should be a justification which considers the full range of values to be derived, not simply the usual cost-benefit analysis.

2 Characterization of the existing environment should include information developed in step 4. This information should (*a*) give a detailed description of the existing environmental elements and factors, with special emphasis on those rare or unique aspects, both good and bad, that might not be common to other similar areas, (*b*) provide sufficient information to permit an objective evaluation of the environmental factors which could be affected by the proposed actions, and (*c*) include all the factors which together make up the ecosystem of the area.

3 Details of proposed actions should contain information developed in step 5. Such information should include discussion of possible alternative engineering methods to accomplish the proposed objective. This should be done in sufficient detail so that all actions that may have impact upon the environment can be checked.

4 Environmental impact assessment should include information developed

---

[1] U.S. Dep. of Interior, A Procedure for Evaluating Environmental Impact, *Geol. Surv. Circ.* 645, 1971.

in steps 6 to 8. This section should consist of three basic elements: (*a*) a listing of the effects on the environment which would be caused by the proposed project and an estimate of the *magnitude* of each, (*b*) an evaluation of the *importance* of each of these effects, and (*c*) the combining of *magnitude* and *importance* estimates in terms of a summary evaluation. The environmental impact assessment is prepared in the form of a *matrix* and a supporting *text*.

Matrix

The matrix is constructed as a two-coordinate system of boxes which can be used for marking purposes. The vertical axis lists the existing environmental characteristics and conditions, each of which identifies a row of the box matrix. The horizontal axis lists the actions which can cause an impact on the environment. Each of the actions identifies a column. A comprehensive list of the existing environmental characteristics and conditions is given in Table 6.4.1. Those items covering the spectrum of actions having possible impact on the environment are listed in Table 6.4.2. Although these lists include the items most likely to be involved in any particular situation, the coding and format are designed for easy inclusion of additional items. Only a few items in each list normally apply to a typical situation.

Use of the matrix for a particular situation involves first the marking with a diagonal slash those boxes which identify a significant interaction (impact). Then the relative *magnitude* of the interaction is rated with a number from 1 to 10 in the upper left-hand corner; 10 represents the greatest magnitude, and 1 the least. The relative *importance* of the interaction is rated in a like manner with the rating marked at the lower right-hand corner of the box. The *magnitude* of the interaction should be rated in terms of degree, extensiveness, or scale. For instance, a project may involve surface excavations that could result in considerable erosion. The *magnitude* of such an interaction might be rated with the number 7. However, if the streams draining the area generally carry a high sediment load anyway, then the *importance* of the erosion interaction may be rated 1 or 2. It is to be noted that the evaluation of the *magnitude* rating can be more objective than the evaluation of the *importance* rating. An interaction may be short-term, in which case the importance may not be as great as it would be otherwise.

Beneficial interactions can be identified with a plus sign before the relative importance rating.

After assigning ratings to all significant interactions, a reduced matrix may be constructed consisting only of those interactions. An example of a reduced matrix is shown in Fig. 6.4.1. It is to be noted that such a matrix yields

Table 6.4.1 EXISTING CHARACTERISTICS AND CONDITIONS OF THE ENVIRONMENT

1 Physical and chemical characteristics
   a. Earth
      (1) Mineral resources
      (2) Construction material
      (3) Soils
      (4) Land form
      (5) Force fields and background radiation
      (6) Unique physical features
   b. Water
      (1) Surface
      (2) Ocean
      (3) Underground
      (4) Quality
      (5) Temperature
      (6) Recharge
      (7) Snow, ice, and permafrost
   c. Atmosphere
      (1) Quality (gases, particulates)
      (2) Climate (micro, macro)
      (3) Temperature
   d. Processes
      (1) Floods
      (2) Erosion
      (3) Deposition (sedimentation, precipitation)
      (4) Solution
      (5) Sorption (ion exchange, complexing)
      (6) Compaction and settling
      (7) Stability (slides, slumps)
      (8) Stress-strain (earthquake)
      (9) Air movements
2 Biologic conditions
   a. Flora
      (1) Trees
      (2) Shrubs
      (3) Grass
      (4) Crops
      (5) Microflora
      (6) Aquatic plants
      (7) Endangered specie
      (8) Barriers
      (9) Corridors
   b. Fauna
      (1) Birds
      (2) Land animals including reptiles
      (3) Fish and shellfish
      (4) Benthic organisms
      (5) Insects
      (6) Microfauna
      (7) Endangered species
      (8) Barriers
      (9) Corridors

Table 6.4.1 EXISTING CHARACTERISTICS AND CONDITIONS OF THE ENVIRONMENT (*continued*)

3 Cultural factors
  a. Land use
     (1) Wilderness and open spaces
     (2) Wetlands
     (3) Forestry
     (4) Grazing
     (5) Agriculture
     (6) Residential
     (7) Commercial
     (8) Industrial
     (9) Mining and quarrying

  b. Recreation
     (1) Hunting
     (2) Fishing
     (3) Boating
     (4) Swimming
     (5) Camping and hiking
     (6) Picnicking
     (7) Resorts

  c. Aesthetics and human interest
     (1) Scenic views and vistas
     (2) Wilderness qualities
     (3) Open-space qualities
     (4) Landscape design
     (5) Unique physical features
     (6) Parks and reserves
     (7) Monuments
     (8) Rare and unique species or ecosystems
     (9) Historical or archaeological sites and objects
     (10) Presence of misfits

  d. Cultural status
     (1) Cultural patterns (life style)
     (2) Health and safety
     (3) Employment
     (4) Population density

  e. Man-made facilities and activities
     (1) Structures
     (2) Transportation network (movement, access)
     (3) Utility networks
     (4) Waste disposal
     (5) Barriers
     (6) Corridors

4 Ecological relationships
  a. Salinization of water resources
  b. Eutrophication
  c. Disease-insect vectors
  d. Food chains
  e. Salinization of surficial material
  f. Brush encroachment
  g. Other

Table 6.4.2 PROPOSED ACTIONS WHICH MAY CAUSE ENVIRONMENTAL IMPACT

1 Modification of regime
  a. Exotic flora or fauna introduction
  b. Biologic controls
  c. Modification of habitat
  d. Alteration of ground cover
  e. Alteration of groundwater hydrology
  f. Alteration of drainage
  g. River control and flow modification
  h. Canalization
  i. Irrigation
  j. Weather modification
  k. Burning
  l. Surface or paving
  m. Noise and vibration

2 Land transformation and construction
  a. Urbanization
  b. Industrial sites and buildings
  c. Airports
  d. Highways and bridges
  e. Roads and trails
  f. Railroads
  g. Cables and lifts
  h. Transmission lines, pipelines, and corridors
  i. Barriers including fencing
  j. Channel dredging and straightening
  k. Channel revetments
  l. Canals
  m. Dams and impoundments
  n. Piers, seawalls, marinas, and sea terminals
  o. Offshore structures
  p. Recreational structures
  q. Blasting and drilling
  r. Cut and fill
  s. Tunnels and underground structures

3 Resource extraction
  a. Blasting and drilling
  b. Surface excavation
  c. Subsurface excavation and retorting
  d. Well drilling and fluid removal
  e. Dredging
  f. Clear cutting and other lumbering
  g. Commercial fishing and hunting

4 Processing
  a. Farming
  b. Ranching and grazing
  c. Feed lots
  d. Dairying
  e. Energy generation
  f. Mineral processing
  g. Metallurgical industry
  h. Chemical industry
  i. Textile industry
  j. Automobile and aircraft
  k. Oil refining
  l. Food

Table 6.4.2 PROPOSED ACTIONS WHICH MAY CAUSE ENVIRONMENTAL IMPACT (*continued*)

    m. Lumbering
    n. Pulp and paper
    o. Product storage

5 Land alteration
    a. Erosion control and terracing
    b. Mine sealing and waste control
    c. Strip-mining rehabilitation
    d. Landscaping
    e. Harbor dredging
    f. Marsh fill and drainage

6 Resource renewal
    a. Reforestation
    b. Wildlife stocking and management
    c. Groundwater recharge
    d. Fertilization application
    e. Waste recycling

7 Changes in traffic
    a. Railway
    b. Automobile
    c. Trucking
    d. Shipping
    e. Aircraft
    f. River and canal traffic
    g. Pleasure boating
    h. Trails
    i. Cables and lifts
    j. Communication
    k. Pipeline

8 Waste emplacement and treatment
    a. Ocean dumping
    b. Landfill
    c. Emplacement of tailings, spoil, and overburden
    d. Underground storage
    e. Junk disposal
    f. Oil-well flooding
    g. Deep-well emplacement
    h. Cooling-water discharge
    i. Municipal waste discharge including spray irrigation
    j. Liquid effluent discharge
    k. Stabilization and oxidation ponds
    l. Septic tanks, commercial and domestic
    m. Stack and exhaust emission
    n. Spent lubricants

9 Chemical treatment
    a. Fertilization
    b. Chemical deicing of highways, etc.
    c. Chemical stabilization of soil
    d. Weed control
    e. Insect control (pesticides)

10 Accidents
    a. Explosions
    b. Spills and leaks
    c. Operational failure

		IIA.e	IIA.f	IIB.b	IIB.r	IID.e	IIH.h
		Alteration of ground cover	Alteration of drainage	Industrial sites and buildings	Cuts and fills	Energy generation	Cooling water discharge
IA.2d	Water quality				4/4		
IA.2e	Water temperature						5/2
IA.4b	Erosion		5/3		4/2		
IA.4c	Deposition				4/1		
IB.1a	Trees				4/2		
IC.1a	Wilderness and open spaces	3/2		4/4	4/4		
IC.3b	Wilderness qualities	3/2		4/4	4/4		
IC.3f	Parks and reserves	6/4		6/4			
IC.3h	Rare and unique ecosystem			4/10			
IC.5b	Transportation network					6/3	
IC.5c	Utility network					2/4	

FIGURE 6.4.1
Example of a reduced matrix.

at a glance all information considered in evaluating the environmental impact. Also, the coding allows easy reference to supporting discussion found in the text of the environmental impact assessment.

Text

The text of the environmental impact assessment consists of a discussion of the interactions identified in the matrix. The discussion should give the reasoning behind the assignment of the numerical ratings to the interactions. It should include also:

    1 The relationship between local short-term uses of man's environment and the maintenance and enhancement of long-term productivity

2 Any irreversible and irretrievable commitments of resources which will be involved if the proposed action is implemented

3 A discussion of any problems and objections raised by other federal, state, and local agencies and by private organizations and citizens in the review process along with information concerning the disposition of the issues involved

The environmental impact statement provides an adequate basis for the evaluation of the probable impact of a project on the environment. Not only does the statement provide a document for review by the appropriate state and federal reviewing agencies, but it provides, also, valid information for concerned citizens and citizen groups.

6.5 POPULATION-GROWTH MODELS

The estimation of future demands to be made on water, waste-water, and solid-waste collection and treatment systems is a key step in planning these systems. Furthermore, such demands relate directly to the populations to be served by the systems. Several methods are currently employed to forecast population growth. Most of the methods are derived from the population-growth curve shown in Fig. 6.5.1.

The population-growth curve is similar to the classic growth pattern exhibited by microorganisms. (See Fig. 3.3.2.) Theoretically, population initially grows unrestrained in a geometric pattern until a population is reached where factors—economic and/or geographic—begin to repress growth. From that population on, growth proceeds but at a diminishing rate of increase. Ultimately, a population is reached, the *saturation population*, where no further growth takes place. For modeling purposes the growth curve is segmented into three regions—a region of geometric increase, a region of arithmetic increase, and a region of decreasing rate of increase.

The model describing growth when the latter falls in the region of geometric increase is the expression for a first-order reaction:

$$\frac{dP}{dt} = k_G P \quad (6.5.1)$$

where P = population

k_G = constant

By integrating between the limits of t_1 and t_2, Eq. (6.5.1) becomes

$$P_2 = P_1 + e^{k_G(t_2 - t_1)} \quad (6.5.2)$$

PLANNING FACTORS 195

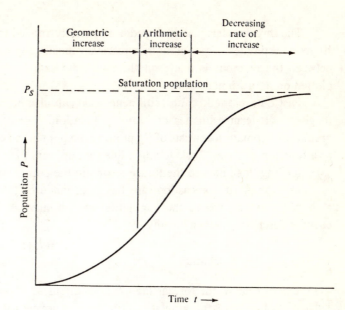

FIGURE 6.5.1
Population-growth curve.

In the region of arithmetic increase, the model is of zero order:

$$\frac{dP}{dt} = k_A \quad (6.5.3)$$

where k_A = constant

Through integration, the equation becomes

$$P_2 = P_1 + k_A(t_2 - t_1) \quad (6.5.4)$$

The region of decreasing rate of increase can be described by

$$\frac{dP}{dt} = k_D(P_S - P) \quad (6.5.5)$$

where P_S = saturation population

k_D = constant

Integration of Eq. (6.5.5) yields

$$P_2 = P_S - (P_S - P_1)e^{-k_D(t_2 - t_1)} \quad (6.5.6)$$

Use of Eq. (6.5.6) requires an estimate of the saturation population P_S.

The three models just discussed are generally applied to short-term (up to 10 years) estimates of population in relatively small geographic areas. Choice between the methods is made on the basis of visual inspection of past data plotted on an arithmetic scale.

Past data plotted on an arithmetic scale can also be extrapolated by eye to give short-term estimates of future population. Forecasting by means of graphic extrapolation consists of (1) plotting the population of past census years as a function of time, (2) sketching a line that appears to fit the past data, and (3) extending this line into the future to obtain the population for future years.

For long-term forecasting, use has been made of the *logistics curve*, an S-shaped curve similar to the population-growth curve in Fig. 6.5.1. Such a curve is described mathematically by

$$P = k_S(1 - e^{a-bt})^{-1} \qquad (6.5.7)$$

where $k_S, a, b =$ constants

Use of each of the forecasting methods discussed above requires a knowledge of past and present populations for the area concerned. Such information can be obtained from decennial census reports prepared by the Bureau of the Census and from a variety of local censuses. Records of births and deaths can be used to estimate intercensual populations.

The models discussed in this section are based primarily on an extension of existing trends. They do not take into account changing factors. Furthermore, such models are introspective in that no consideration is given explicitly to external influences on growth. The regional-growth model discussed in Sec. 6.6 ties forecasting with factors of an external nature that influence growth in a region.

Water demands, waste-water flows, and solid-waste generation can be correlated with future-population estimates using the general relationship

$$Q = CP^n \qquad (6.5.8)$$

where $Q =$ quantity
$n, C =$ constants

The exponent n has been found to have a value of approximately 1.3 for most applications of Eq. (6.5.8).

Solution procedure 6.5.1 The constants contained in the equations found in Sec. 6.5 can be estimated as follows:

Eq. (6.5.2):
$$k_G = \frac{\ln P_2 - \ln P_1}{t_2 - t_1} \qquad (6.5.9)$$

Eq. (6.5.4): $$k_A = \frac{P_2 - P_1}{t_2 - t_1} \qquad (6.5.10)$$

Eq. (6.5.6): $$k_D = \frac{\ln\left[(P_S - P_1)/(P_S - P_2)\right]}{t_2 - t_1} \qquad (6.5.11)$$

Eq. (6.5.7): $$k_S = \frac{2P_0 P_1 P_2 - P_1^2(P_0 - P_2)}{P_0 P_2 - P_1^2} \qquad (6.5.12)$$

$$a = \ln \frac{k_S - P_0}{P_0} \qquad (6.5.13)$$

$$b = \frac{1}{N}\left[\ln \frac{P_0(k_S - P_1)}{P_1(k_S - P_0)}\right] \qquad (6.5.14)$$

where P_0, P_1, P_2 = population at three different years, each equidistant from the other in succession

N = number of years between populations P_0 and P_1, and P_1 and P_2

The populations P_0, P_1, and P_2 are generally populations recorded in successive decennial censuses. ////

6.6 REGIONAL-GROWTH MODEL

Until recently, growth estimates made in connection with water needs, wastewater quantities, and solid-waste generation were confined to relatively small geographic areas constituting political subdivisions at the lowest levels. The scale of planning was governed largely by the immediate requirements of the local economy. Little attention was paid to economic interrelationships and those factors that appeared to have no direct influence on the objective of the planning studies.

Now, it is realized that in a modern industrial society there exists a myriad of interrelationships, primarily of an economic nature, which influence industrial growth. Growth or decline in any segment of the economic community of a region will have its impact on all other segments. Consequently, any studies relating to the future water needs or waste generation capacity of a region require a model which takes into account all direct as well as indirect factors involved. One such model can be developed from an *input-output analysis* of the economics of the region.

Input-output analysis is an econometric technique which focuses upon the interdependencies among the various industries or sectors of the economy and the relationships between these sectors and the final consumers. The technique derives its name from the method by which it classifies economic transactions. Each exchange of goods and services between sectors of the economy is recorded in double-entry fashion as both a sale of output and a purchase of input. The basic interindustry accounting system used in the technique is represented by the generalized matrix in Table 6.6.1.[1] Quadrant I contains the final-demand (or use) matrix, which essentially is the net output of the economy. Quadrant II contains a matrix the elements of which identify the interindustry transactions occurring within the region between industries that produce the commodities and those that are consumers. Quadrant III contains a row vector of the sums of the value added to the value of the raw materials in producing a commodity and the value of import purchases. Value added includes such items as salaries, wages, profits, taxes, etc. Quadrant IV contains a vector of the value added associated with the items included in the final-use categories listed in quadrant I.

The format followed in constructing the interindustry accounting matrix shown in Table 6.6.1 reflects the following mathematical relationships.

Given an economy divided into n different sectors (or industries),

$$X_i = x_{i1} + \cdots + x_{ij} + \cdots + x_{in} + Y_i \quad \text{for } i = 1, 2 \ldots, n \quad (6.6.1)$$

where X_i = total output of sector (or industry) i

x_{ij} = amount of commodity i required by sector (or industry) j

Y_i = total final use for commodity i

Equation (6.6.1) is a sum of the elements in the rows of quadrants I and II in Table 6.6.1. Also,

$$X_j = x_{1j} + \cdots + x_{ij} + \cdots + x_{nj} + V_j \quad \text{for } j = 1, 2, \ldots, n \quad (6.6.2)$$
$$= X_i \quad \text{for } i = j = 1, 2, \ldots, n$$

where X_j = total purchases made by sector (or industry) j

V_j = value added and purchases of imports by sector (or industry) j

Equation (6.6.2) is a sum of the elements in the columns of quadrants II and III. A working table containing the matrices in Table 6.6.1 is commonly referred to as a *transaction table*.

---

[1] For a more comprehensive discussion of the input-output analysis technique, the reader is referred to E. M. Lofting and P. H. McGauhey, "Economic Evaluation of Water, Part IV. An Input-Output and Linear Programming Analysis of California Water Requirements," Sanit. Eng. Res. Lab., University of Calif., August 1968.

Table 6.6.1 BASIC INTERINDUSTRY ACCOUNTING SYSTEM*

	Purchasing sectors							Gross output
	Intermediate use	Final use (net output)						
	Sector 1 ··· j ··· n	Investment	Consumption	Government	Exports	Total final use		Production
Producing sector 1	$x_{11} \cdots x_{1j} \cdots x_{1n}$ (quadrant II)	I_1	C_1	G_1 (quadrant I)	E_1	Y_1		X_1
2								
⋮								⋮
i	$x_{i1} \cdots x_{ij} \cdots x_{in}$	I_i	C_i	G_i	E_i	Y_i		X_i
⋮								⋮
n	$x_{n1} \cdots x_{nj} \cdots x_{nn}$	I_n	C_n	G_n	E_n	Y_n		X_n
Primary inputs (value added and imports)	V_1, V_j, V_n (quadrant III)	V_I	V_C	V_G (quadrant IV)	V_E			V
Total production	X_1, X_j, X_n	I	C	G	E	Y		X

* After E. M. Lofting and P. H. McGauhey, "Economic Evaluation of Water, Part IV. An Input-Output and Linear Programming Analysis of California Water Requirements," table I, Sanit. Eng. Res. Lab, University of Calif, August 1968.

The information contained in a transaction table can be used to construct a *table of technical coefficients*. The latter is a matrix of elements a_{ij}, where

$$a_{ij} = \frac{x_{ij}}{X_j} \quad \text{for } i, j = 1, 2, \ldots, n \quad (6.6.3)$$

where a_{ij} = technical coefficients

$X_j = X_i$ = total output of industry j

Implicit in input-output analysis is the assumption that elements of the technical-coefficient matrix are fixed and remain unchanged with changes in the technological fabric of the economy.

Through substitution, Eq. (6.6.1) becomes

$$X_i = a_{i1}X_1 + \cdots + a_{ij}X_j + \cdots + a_{in}X_n + Y_i \quad \text{for } i = 1, 2, \ldots, n \quad (6.6.4)$$

which in matrix form is

$$\mathbf{X} = \mathbf{AX} + \mathbf{Y} \quad (6.6.5)$$

or

$$(\mathbf{I} - \mathbf{A})\mathbf{X} = \mathbf{Y} \quad (6.6.6)$$

where $\mathbf{I}$ = identity matrix ($n \times n$ matrix whose main diagonal is composed of elements having value of unity and whose off-diagonal elements are 0s)

$(\mathbf{I} - \mathbf{A})$ = $n \times n$ matrix containing elements $1 - a_{ii}$ in main diagonal and elements $-a_{ij}$ in off-diagonal position

Through matrix inversion, Eq. (6.6.6) becomes

$$\mathbf{X} = (\mathbf{I} - \mathbf{A})^{-1}\mathbf{Y} \quad (6.6.7)$$

Equation (6.6.7) provides the general solution for an $n \times 1$ matrix of total output values for n sectors of the economy $\mathbf{X}$, given the $n \times 1$ matrix of final use (or demand) for the commodities of the n sectors $\mathbf{Y}$. The matrix $(\mathbf{I} - \mathbf{A})^{-1}$, which is derived from the table of technical coefficients, is often referred to as the *table of direct and indirect requirements*. The latter provides information concerning how much commodity must be produced to satisfy final demand. Sufficient commodity must be produced to satisfy the demand of the direct consumers, the demand by the industries in other sectors to produce final consumer products, the demand by industries in other sectors to produce the inputs for the final consumer products, and so on. Customarily, the table of direct and indirect requirements is prepared as the transpose of the matrix $(\mathbf{I} - \mathbf{A})^{-1}$. The interchange between rows and columns gives the convenience of reading the relative information along rows rather than down the columns.

The firm can be considered as the basic business unit of the economy, and all firms producing similar goods or services can be considered an industry. A sector can be thought of as a group of industries having similar characteristics. For example, a listing of such sectors might include an agricultural sector, a metal-industry sector, a utilities and utilities-services sector, etc. Often it is convenient to group together all final purchasers in a final-use sector. The number of sectors that is used to represent the economic spectrum in input-output analysis will be a compromise between the number required for a definitive representation on one hand and computational convenience on the other.

Water use and waste generation can be correlated with the level of economic activity. Such a correlation is made through the use of a water-use or waste coefficient which numerically gives the volume of water used or the quantity of waste generated per unit dollar volume of output by that sector.

A water-use or waste-generation matrix can be computed with

$$\mathbf{W} = {}_T(\mathbf{I} - \mathbf{A})^{-1}\mathbf{L} \qquad (6.6.8)$$

where $\mathbf{W} = n \times n$ water-use or waste-generation matrix

${}_T(\mathbf{I} - \mathbf{A})^{-1}$ = transpose of $(\mathbf{I} - \mathbf{A})^{-1}$

$\mathbf{L} = n \times n$ diagonal matrix with water-use or waste coefficients l_i appearing in main diagonal and with 0s in all off-diagonal locations

In order to predict future needs for water or future waste generation, the water-use or waste-generation matrix is postmultiplied by a vector of future final demands estimated for each sector:

$$\mathbf{X}_W = \mathbf{W}\mathbf{Y}_F \qquad (6.6.9)$$

where $\mathbf{X}_W = n \times 1$ matrix of future needs for water or future waste generation by each sector

$\mathbf{Y}_F = n \times 1$ matrix of future final demands

The vector of future final demands can be estimated on the basis of the growth of the *gross regional product* (GRP), which in turn may be correlated with the growth of the *gross national product* (GNP).

Solution procedure 6.6.1 An input-output economic analysis was applied to the nine-county San Francisco Bay Area.[1] Industries in the

---

[1] P. H. McGauhey et al., "Final Report, Economic Evaluation of Water," Sanit. Eng. Res. Lab., University of Calif., November 1969.

area were grouped into 14 sectors, the identifications of which are listed in Table 6.6.2. From survey data, the dollar values of all industry purchases and sales were determined and used to construct the interindustry transactions table shown in Table 6.6.3. The row and column numbers in Table 6.6.3 are identified in Table 6.6.2. With the use of Eq. (6.6.3) and

Table 6.6.2 SECTOR CLASSIFICATION, NINE BAY AREA COUNTIES*

Sector number and sector title

Rows

1. Agriculture, forestry, and fisheries
2. Mining
3. Construction
4. Food and kindred products
5. Paper and allied products
6. Chemicals and chemical products
7. Petroleum refining and related industries
8. Stone and clay products
9. Fabricated metal products
10. All other manufacturing
11. Transportation, communication, electric, gas, and sanitary services
12. Wholesale and retail trade
13. Finance, insurance, and real estate
14. Services and government enterprises
15. Interindustry inputs
16. Sectoral imports
17. Foreign imports
18. Total sectoral imports
19. Value added
20. Adjustment
21. Total gross outlay

Columns

1–14. Same as above (rows)
15. Intermediate outputs—total
16. Personal consumption expenditures
17. Gross private capital formation
18. Federal government expenditures
19. State and local government expenditures
20. Net exports
21. Total final demand (sum of row elements of columns 16–20)
22. Total gross outputs
23. Commodity imports

* Adapted from P. H. McGauhey et al., "Final Report, Economic Evaluation of Water," table III, Sanit. Eng. Res. Lab., University of Calif., November 1969.

Table 6.6.3 INTERINDUSTRY TRANSACTIONS TABLE (FLOW OF GOODS AND SERVICES BY INDUSTRY OF ORIGIN AND DESTINATION), NINE BAY AREA COUNTIES*†

Row	_____ Column _____							
	1	2	3	4	5	6	7	8
1	54.43	0.00	3.26	317.31	0.00	0.31	0.00	0.03
2	0.08	0.40	2.64	0.20	0.21	1.11	62.63	1.26
3	6.59	0.05	0.45	10.49	1.25	0.87	2.40	0.13
4	35.69	0.00	0.54	527.93	1.59	9.98	1.12	0.21
5	0.34	0.11	14.18	44.79	71.25	13.25	6.54	9.48
6	9.91	0.72	61.49	16.13	9.07	121.02	45.82	9.28
7	9.30	0.65	62.88	11.76	3.08	18.27	116.67	2.66
8	0.24	0.41	182.79	19.43	1.07	4.26	2.73	23.67
9	1.03	0.36	288.96	66.99	2.89	8.38	23.89	3.00
10	6.05	3.15	459.69	23.04	22.44	17.12	2.82	8.15
11	12.76	3.59	123.83	141.60	17.60	27.80	114.86	24.79
12	18.00	1.73	304.91	107.53	12.26	15.90	17.32	9.70
13	26.51	7.77	41.27	28.50	3.73	10.24	23.75	5.03
14	12.71	2.34	172.26	185.83	7.12	41.14	44.13	9.20
15	193.64	21.27	1,719.16	1,501.51	153.55	289.66	464.68	106.59
16	114.11	6.14	315.08	683.28	31.87	52.09	786.55	31.21
17	10.84	7.63	0.00	108.83	20.90	9.85	51.36	2.77
18	124.96	13.77	315.08	792.11	52.78	61.93	837.91	33.99
19	214.34	46.47	1,435.81	777.95	120.84	217.55	315.22	112.63
20	0.07	0.00	0.03	−0.02	0.00	−0.01	0.00	42.03
21	533.01	81.51	3,470.07	3,071.56	327.17	569.12	1,617.81	295.24

Row	_____ Column _____							
	9	10	11	12	13	14	15	16
1	0.00	25.75	0.49	1.83	34.39	12.14	449.96	58.77
2	0.03	4.71	5.90	0.02	0.51	0.79	80.48	0.74
3	0.44	8.98	89.31	28.75	270.57	111.51	531.79	0.00
4	0.01	2.82	4.24	22.24	2.81	124.54	733.71	1,725.23
5	4.73	74.60	2.26	22.93	5.69	31.05	301.19	25.32
6	5.61	98.20	3.50	7.54	5.18	42.55	436.01	117.77
7	3.21	14.41	72.87	27.49	21.87	20.88	386.00	266.42
8	3.85	31.22	1.05	6.63	0.86	8.16	286.36	8.88
9	36.90	124.02	7.10	6.42	0.83	8.37	579.13	14.75
10	204.04	1,424.80	72.37	50.47	35.18	531.29	2,860.60	1,294.10
11	17.60	153.62	297.67	121.57	77.89	425.20	1,560.39	748.36
12	23.25	181.37	54.67	57.33	61.04	132.48	997.50	2,300.30
13	8.67	70.18	91.60	245.27	523.61	260.22	1,346.35	2,192.00
14	16.93	165.45	261.93	340.27	225.18	427.13	1,911.63	1,560.61
15	325.28	2,380.12	964.97	938.77	1,265.61	2,136.31	12,461.11	10,313.24
16	68.75	579.37	96.40	29.61	87.86	200.88	3,083.19	516.76
17	3.67	135.02	51.26	1.08	3.77	61.50	468.47	134.00
18	72.42	714.38	147.65	30.69	91.63	262.37	3,551.66	650.76
19	263.71	2,046.65	1,718.94	2,571.89	2,785.64	2,516.16	15,143.81	76.00
20	−0.01	−32.38	0.02	−4.35	0.01	0.02	0.00	0.00
21	661.40	5,108.78	2,831.58	3,537.00	4,142.89	4,914.86	31,162.01	11,040.00

* From P. H. McGauhey et al., "Final Report, Economic Evaluation of Water," table IV, Sanit. Eng. Res. Lab., University of Calif., November 1969.
† All figures in millions of dollars.

(*continued overleaf*)

Table 6.6.3 INTERINDUSTRY TRANSACTIONS TABLE (FLOW OF GOODS AND SERVICES BY INDUSTRY OF ORIGIN AND DESTINATION), NINE BAY AREA COUNTIES*† (*continued*)

	Column						
Row	17	18	19	20	21	22	23
1	0.00	20.38	3.90	0.00	83.05	533.01	−1,066.97
2	0.00	0.26	0.04	0.00	1.03	81.51	−998.57
3	1,533.84	253.74	630.20	520.50	2,938.28	3,470.07	0.00
4	0.00	6.71	8.47	597.44	2,337.85	3,071.56	0.00
5	0.00	0.65	0.00	0.00	25.98	327.17	−92.61
6	0.00	10.15	5.20	0.00	133.12	569.13	−121.74
7	0.00	149.30	14.94	801.15	1,231.81	1,617.81	0.00
8	0.00	0.00	0.00	0.00	8.88	295.24	−100.84
9	61.59	4.92	1.01	0.00	82.27	661.40	−164.68
10	260.02	669.15	24.90	0.00	2,248.18	5,108.78	−1,385.85
11	0.38	49.52	44.71	428.22	1,271.19	2,831.58	0.00
12	0.39	34.48	7.19	197.14	2,539.50	3,537.00	0.00
13	0.19	4.45	17.08	582.82	2,796.54	4,142.89	0.00
14	0.00	764.01	81.23	597.38	3,003.23	4,914.86	0.00
15	1,856.41	1,967.72	838.87	3,724.66	18,700.90	31,162.01	−3,931.26
16	85.87	229.03	16.42	0.00	848.07	3,931.26	0.00
17	−0.00	143.00	−0.00	0.00	277.00	745.47	745.47
18	85.87	372.03	16.42	0.00	1,125.07	4,676.73	4,676.73
19	−0.00	866.20	949.50	0.00	1,891.70	17,035.51	0.00
20	0.00	0.00	0.00	0.00	0.00	0.00	0.00
21	1,942.27	3,205.95	1,804.79	3,724.66	21,717.67	52,879.68	0.00

* From P. H. McGauhey et al., "Final Report, Economic Evaluation of Water," table IV, Sanit. Eng. Res. Lab., University of Calif., November 1969.
† All figures in millions of dollars.

the appropriate values in Table 6.6.3, the technical-coefficient matrix **A** was computed. The latter is presented in Table 6.6.4.

The matrix $(I - A)^{-1}$ was computed from the technical-coefficient matrix. Its transpose $_T(I - A)^{-1}$ is presented in Table 6.6.5.

Table 6.6.6 lists typical values of fresh-water withdrawal coefficients for the different sectors selected for the analysis. These values were used to form the $n \times n$ diagonal matrix **L**. The matrix **L** was premultiplied by $_T(I - A)^{-1}$ to give the fresh-water withdrawal matrix **W** [Eq. (6.6.8)]. The elements of **W**, w_{ij}, give the total acre-feet of fresh water withdrawn directly and indirectly by sector j per 1 million dollars of deliveries to final demand by industry i.

Tables 6.6.7 and 6.6.8 give final-demand projections at 3 percent and 6 percent growth rates, respectively, for the 14 sectors. The values in

Table 6.6.4 INTERINDUSTRY TRANSACTIONS-TECHNICAL-COEFFICIENT MATRIX (DIRECT PURCHASES PER DOLLAR OF OUTPUT) 1963, NINE BAY AREA COUNTIES*

Column

Row	1	2	3	4	5	6	7	8	9	10	11	12	13	14
1	0.102118	0.000150	0.012364	0.066959	0.000638	0.018593	0.017448	0.000450	0.001932	0.011351	0.023940	0.033770	0.049736	0.023846
2	0.000000	0.004907	0.000613	0.000000	0.001350	0.008833	0.007974	0.005030	0.004417	0.038646	0.044044	0.021224	0.095326	0.028708
3	0.000939	0.000761	0.000130	0.000156	0.004086	0.017720	0.018121	0.052676	0.083272	0.132473	0.035685	0.087869	0.011893	0.049642
4	0.103306	0.000065	0.003415	0.171877	0.014582	0.005251	0.003829	0.006326	0.021810	0.007501	0.046100	0.035008	0.009279	0.060500
5	0.000000	0.000642	0.003821	0.004860	0.217777	0.027723	0.009414	0.003270	0.008833	0.068588	0.053795	0.037473	0.011401	0.021762
6	0.000545	0.001950	0.001529	0.017536	0.023281	0.212640	0.032102	0.003270	0.014724	0.030081	0.048846	0.027937	0.017992	0.072286
7	0.000000	0.038713	0.001483	0.000692	0.004043	0.028322	0.072116	0.001687	0.014767	0.001743	0.070997	0.010706	0.014680	0.027278
8	0.000102	0.004268	0.000440	0.000711	0.032109	0.031432	0.009010	0.080172	0.010161	0.027605	0.083966	0.032855	0.017037	0.031161
9	0.000000	0.000045	0.000665	0.000015	0.007151	0.008482	0.004853	0.005821	0.055791	0.308497	0.026610	0.035153	0.013109	0.025597
10	0.005040	0.000922	0.001758	0.000552	0.014602	0.019222	0.002821	0.006111	0.024276	0.278892	0.030070	0.035502	0.013737	0.032385
11	0.000173	0.002084	0.031541	0.001497	0.000798	0.001236	0.025735	0.000371	0.002507	0.025558	0.105125	0.019307	0.032349	0.092503
12	0.000517	0.000006	0.008128	0.006288	0.006483	0.002132	0.007772	0.001874	0.001815	0.014269	0.034371	0.016209	0.069344	0.096203
13	0.008301	0.000123	0.065309	0.000678	0.001373	0.001250	0.005279	0.000208	0.000200	0.008492	0.018801	0.014734	0.126388	0.544353
14	0.002470	0.000161	0.022688	0.025339	0.006318	0.008657	0.004248	0.001660	0.001703	0.108099	0.086513	0.026955	0.052946	0.086906

* From P. H. McGauhey et al., "Final Report, Economic Evaluation of Water," table V, Sanit. Eng. Res. Lab., University of Calif., November 1969.

Table 6.6.5 INTERINDUSTRY TRANSACTIONS (DIRECT AND INDIRECT REQUIREMENTS PER DOLLAR OF FINAL DEMAND) 1963, NINE BAY AREA COUNTIES*

Row	1	2	3	4	5	6	7	8	9	10	11	12	13	14
1	1.1257377	0.0014450	0.0227859	0.0940684	0.0059143	0.0310855	0.0259002	0.0034072	0.0088859	0.0411286	0.0511295	0.0510873	0.0769752	0.0572601
2	0.0020415	1.0057338	0.0122419	0.0027093	0.0051056	0.0154245	0.0128501	0.0071488	0.0086716	0.0738407	0.0647559	0.0315057	0.1202800	0.0543914
3	0.0038486	0.0025797	1.0090003	0.0055309	0.0165408	0.0356059	0.0268371	0.0609736	0.0980170	0.2526979	0.0772794	0.1118169	0.0398355	0.0951910
4	0.1418294	0.0009754	0.0144762	1.2235754	0.0268957	0.0169711	0.0129889	0.0105214	0.0326736	0.0564510	0.0885285	0.0602242	0.0385312	0.1088510
5	0.0027524	0.0020913	0.0125941	0.0115447	1.2847728	0.0515907	0.0195323	0.0129889	0.0189491	0.1497220	0.0982190	0.0627345	0.0339627	0.0615460
6	0.0059168	0.0048674	0.0121612	0.0325249	0.0432461	1.2790256	0.0498335	0.0072379	0.0128036	0.0262243	0.0968153	0.0997091	0.0521406	0.1298961
7	0.0011541	0.0425126	0.0085765	0.0039882	0.0087106	0.0419648	1.0834561	0.0128036	0.0036276	0.0199584	0.0302615	0.1001394	0.0335531	0.0541112
8	0.0017531	0.0058365	0.0092043	0.0051416	0.0490855	0.0492868	0.0176986	1.0892344	0.0166231	0.0734180	0.1223092	0.0210553	0.0379841	0.0667349
9	0.0037389	0.0012000	0.0081485	0.0039876	0.0213395	0.0262098	0.0113349	0.0110643	1.0735284	0.4783377	0.0629105	0.0622952	0.0366452	0.0663207
10	0.0092652	0.0020009	0.0095076	0.0055465	0.0300076	0.0386790	0.0095691	0.0110449	0.0392245	1.4253644	0.0659165	0.0607990	0.0371659	0.0725793
11	0.0023813	0.0039001	0.0432239	0.0068553	0.0050465	0.0080486	0.0344758	0.0040335	0.0100236	0.0749945	1.1416373	0.0350928	0.0564145	0.1305666
12	0.0035446	0.0007492	0.0194331	0.0123786	0.0115349	0.0073242	0.0124605	0.0040659	0.0061442	0.0503218	0.0594225	1.0277406	0.0935185	0.1251845
13	0.0118946	0.0008097	0.0793564	0.0050444	0.0050123	0.0070868	0.0107236	0.0055575	0.0093356	0.0482349	0.0414053	0.0309178	1.1570775	0.0844766
14	0.0092847	0.0012189	0.0362073	0.0367358	0.0153049	0.0202290	0.0121906	0.0061391	0.0121889	0.1912230	0.1270830	0.0484802	0.0828612	1.1324392

Column

* From P. H. McGauhey et al., "Final Report, Economic Evaluation of Water," table VI, Sanit. Eng. Res. Lab., University of Calif., November 1969.

Table 6.6.6 FRESH-WATER WITHDRAWAL COEFFICIENTS*

Sector	Acre-feet/ million dollars of output
1 Agriculture, forestry, and fisheries	4,345
2 Mining	950
3 Construction	13
4 Food and kindred products	16
5 Paper and allied products	92
6 Chemical and chemical products	26
7 Petroleum refining and related industries	16
8 Stone and clay products	26
9 Fabricated metal products	2
10 All other manufacturing	15
11 Transportation, communication, electric, gas, and sanitary services	160
12 Wholesale and retail trade	8
13 Finance, insurance, and real estate	13
14 Services and government enterprises	300

* Adapted from E. M. Lofting and P. H. McGauhey, "Economic Evaluation of Water, Part IV. An Input-Output and Linear Programming Analysis of California Water Requirements," p. 55, Sanit. Eng. Res. Lab., University of Calif., August 1968.

Table 6.6.7 FINAL-DEMAND PROJECTIONS—3 PERCENT GROWTH RATE*

Sector	1963	1968	1970	1975	1980	1990
1	83.05†	96.25	102.15	118.32	137.03	183.54
2	1.03	1.19	1.27	1.47	1.70	2.28
3	2,938.28	3,405.46	3,614.08	4,187.04	4,848.16	6,493.60
4	2,337.85	2,709.57	2,875.55	3,331.44	3,857.45	5,166.65
5	25.98	30.11	31.95	37.02	42.87	57.41
6	133.12	154.28	163.74	189.70	219.65	294.19
7	1,231.81	1,427.66	1,515.13	1,755.33	2,032.49	2,722.30
8	8.88	10.29	10.92	12.65	14.65	19.62
9	82.27	95.35	101.19	117.23	135.74	181.82
10	2,248.18	2,605.64	2,765.26	3,203.66	3,709.50	4,968.48
11	1,271.19	1,473.31	1,563.56	1,811.44	2,097.46	2,809.33
12	2,539.50	2,943.28	3,123.58	3,618.78	4,190.17	5,612.29
13	2,796.54	3,241.19	3,439.74	3,985.07	4,614.29	6,180.35
14	3,003.23	3,480.74	3,693.97	4,279.60	4,955.33	6,637.14

* From P. H. McGauhey et al., "Final Report, Economic Evaluation of Water," table VII, Sanit. Eng. Res. Lab., University of Calif., November 1969.
† Given in units of 10^6.

Table 6.6.8 FINAL-DEMAND PROJECTIONS—6 PERCENT GROWTH RATE*

Sector	Year					
	1963	1968	1970	1975	1980	1990
1	83.05†	111.29	124.57	167.76	224.23	402.79
2	1.03	1.38	1.54	2.08	2.78	4.99
3	2,938.28	3,937.29	4,407.42	5,935.32	7,933.35	14,250.66
4	2,337.85	3,132.72	3,506.77	4,722.46	6,312.19	11,338.57
5	25.98	34.81	38.97	52.48	70.14	126.00
6	133.12	178.38	199.68	268.90	359.42	645.63
7	1,231.81	1,650.62	1,847.71	2,488.25	3,325.88	5,974.28
8	8.88	11.90	13.32	17.94	23.98	43.06
9	82.27	110.24	123.40	166.18	222.13	399.00
10	2,248.18	3,012.56	3,372.27	4,541.32	6,070.09	10.903.67
11	1,271.19	1,703.39	1,906.78	2,567.80	3,432.21	6,165.27
12	2,539.50	3,402.93	3,809.25	5,129.79	6,856.65	12,316.57
13	2,796.54	3,747.36	4,194.81	5,649.00	7,550.65	13,563.22
14	3,003.23	4,024.33	4,504.84	6.066.52	8,108.72	14,565.66

* From P. H. McGauhey et al., "Final Report, Economic Evaluation of Water," table VIII, Sanit. Eng. Res. Lab., University of Calif., November 1969.
† Given in units of 10^6.

each column of these tables form elements for a future-final-demand vector Y_F. Using Eq. (6.6.9), a vector giving the future needs for fresh water X_W can be computed for each future-final-demand vector Y_F. ////

6.7 TIME-CAPACITY EXPANSION OF SYSTEMS

The amount of excess capacity to be provided within a system designed to serve a demand which increases with time is a problem of major concern to the engineer. On one hand, economics of scale favor increasing capacity in large increments, whereas, on the other, increasing the capacity in smaller increments allows savings in interest costs on capital investment. Often the engineer approaches the problem intuitively or by rule of thumb. A more rational method is available.[1]

The assumptions are made that the capacity of the system will equal or exceed the demand at all times and that the demand to be satisfied by the system

---

[1] T. M. Rachford, et al., Time-capacity Expansion of Waste Treatment Systems, *J. Sanit. Eng. Div.*, ASCE, vol. 95, no. SA6, pp. 1063–1077, December 1969.

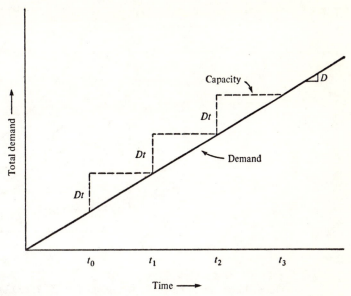

FIGURE 6.7.1
Time growth of demand and installed capacity. (*After T. M. Rachford et al., Time Capacity Expansion of Waste Treatment Systems, J. Sanit. Eng. Div., ASCE, vol. 95, no. SA6, pp. 1063–1077, fig. 1, 1969.*)

increases linearly with time. These assumptions are reflected in the time-capacity expansion relationship shown in Fig. 6.7.1. The straight line which is the demand curve has a slope equal to D, the rate at which the demand increases per year. The ordinate of each step represents the increment of capacity expansion required at points in time when the demand reaches the system's capacity and is equal to the product Dt, where t is the time interval in years between expansions.

The cost of a single expansion can be expressed as

$$C(Dt) = k(Dt)^a \quad (6.7.1)$$

where C = cost of single expansion

D = demand increase per year. $[L^3 t^{-1}]$

$t = t_{i+1} - t_i$ = number of years between each expansion. $[t]$

k, a = constants ($k > 0$; $0 < a < 1$)

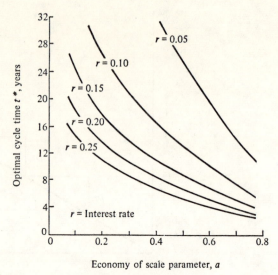

FIGURE 6.7.2
Optimal cycle time. (*After T. M. Rachford et al., Time Capacity Expansion of Waste Treatment Systems, J. Sanit. Eng. Div., ASCE, vol. 95, no. SA6, pp. 1063–1077, fig. 2, 1969.*)

The present-value costs of n expansions are given by the expression for compounding interest continually:

$$C(Dt) = k(Dt)^a + k(Dt)^a e^{-rt} + k(Dt)^a e^{-2rt} + \cdots + k(Dt)^a e^{-nrt}$$
$$= k(Dt)^a [1 + e^{-rt} + (e^{-rt})^2 + \cdots + (e^{-rt})^n] \qquad (6.7.2)$$

Since the values of the product rt must be greater than 0, the value of e^{-rt} is less than unity. Where such is the case, the terms in the geometric series enclosed in the brackets converge to $(1 - e^{-rt})^{-1}$. Therefore,

$$C(Dt) = \frac{k(Dt)^a}{1 - e^{-rt}} \qquad (6.7.3)$$

The minimum value of $C(Dt)$ can be computed by taking logarithms of both sides of Eq. (6.7.3), differentiating, and setting the result equal to 0:

$$\frac{d[\ln C(Dt)]}{dt} = \frac{a}{Dt^*} - \frac{re^{-rt^*}}{1 - e^{-rt^*}} = 0 \qquad (6.7.4)$$

where t^* = optimal time period between expansions

Equation (6.7.4) cannot be solved explicitly for t^*, and one must rely on numerical methods for its solution. Figure 6.7.2 gives a plot of the equation for various values of the interest rate r. The optimal time period between expansions t^* can be determined for any combination of the parameters a and r.

EXERCISES

6.1.1 Discuss the distinction between water-quality criteria and water standards.

6.1.2 List five beneficial uses made of water which involve quality criteria.

6.1.3 What types of limits are used in the U.S. Public Health Service Drinking Water Standards? Discuss each standard from the standpoint of its significance.

6.1.4 Discuss the difference between stream and effluent standards.

6.1.5 What classes of use are generally included in stream standards, and what key water-quality parameters are specified?

6.2.1 What are the chronic health effects of air pollution?

6.2.2 List the five classes of air pollutants. Discuss the major source of each as well as the way each manifests itself as an undesirable component of the air environment.

6.2.3 What is an adiabatic lapse rate, and how does it relate to an atmospheric temperature inversion?

6.2.4 What is an emission inventory, and for what purpose is it used?

6.2.5 What are the two general types of air-quality standards?

6.3.1 What is the nature of alpha, beta, and gamma radiation? What is the effect of each on living organisms? What shielding is necessary for protection from each?

6.3.2 Define each of the following terms:
(*a*) Dosimetry   (*b*) Roentgen unit   (*c*) Rad
(*d*) Rem   (*e*) Man-rems   (*f*) RBE
(*g*) $(MPC)_a$ (*h*) $(MPC)_w$

6.3.3 Discuss in detail the origin and relative importance of the radiation to which the earth's population is currently exposed.

6.3.4 Discuss the radiation hazards attendant to the operation of nuclear power reactors.

6.3.5 Discuss the radiation hazards attendant to the operation of nuclear-fuel reprocessing plants and the wastes they produce.

6.4.1 When and for what purpose should environmental impact statements be prepared?

6.4.2 List the four major sections that should constitute an environmental impact statement. What information should be contained in each section?

6.4.3 Discuss the form of the matrix that is part of the environmental assessment and the coding used for evaluating significant interactions.

6.4.4 To what factors should the text of the environmental impact assessment address itself?

6.4.5 Select a construction project currently in progress for which you have general information. Construct a reduced environmental-impact-assessment matrix for the project.

6.5.1 What shortcomings are inherent in predictive models based on population-growth curves?

6.5.2 The decennial census reports for three cities are given below. Estimate the 1980 population of each by means of graphic extrapolation.

	City A	City B	City C
1940	2,000	20,000	300,000
1950	2,200	60,000	600,000
1960	3,100	100,000	1,000,000
1970	6,000	140,000	1,200,000

6.5.3 Determine the growth constant characteristic of the population growth for each of the three cities listed in Exercise 6.5.2.

6.5.4 Using the growth constants determined in Exercise 6.5.3, estimate with the appropriate model the 1980 population of each of the three cities listed in Exercise 6.5.2.

6.5.5 An average correlation of water demand with size of population is given below. Using this correlation determine the constants of Eq. (6.5.8). Having determined the values of the constants, estimate the 1980 water demand of city C. Under what conditions should relationships such as Eq. (6.5.8) be used?

Population, thousands	Water demand, 10^6 gal/d
1	0.05
10	1.0
100	20.0
1,000	400.0

6.6.1 What is an input-output analysis, and what is the rationale for its application to estimates of water needs, waste-water quantities, and solid-waste generation?

6.6.2 Outline the format of the basic interindustry accounting matrix, and explain the significance of the entries. What is the relationship of the accounting matrix to a transaction table?

6.6.3 How are the entries of a table of technical coefficients computed?

6.6.4 How are the entries of a table of direct and indirect requirements computed? What is the significance of these entries? How can they be used to predict future water needs or waste generation?

6.6.5 What is the significance of water-use or waste coefficients?

6.6.6 Using Tables 6.6.2 to 6.6.8, check selected entries to establish a familiarity with the application of input-output analysis to the determination of a vector of future water needs.

6.7.1 In developing Eq. (6.7.3), what assumptions are made relative to the time-capacity relationship?

6.7.2 A correlation of the cost of a single expansion with size of expansion for a water-treatment facility is given below. Determine the values of the constants in Eq. (6.7.1).

Size of expansion, 10^3 gal	Cost, 10^3 dollars
1	15
10	50
100	160
1,000	500

6.7.3 If the interest rate on money borrowed for expansion is 0.5 percent compounded continually, what is the optimal time period between expansions if the cost relationship developed in Exercise 6.7.2 holds?

7
TIME SERIES

7.1 TREND, FREQUENCY, AND RANDOM COMPONENTS

A *time series* is a record of repeated observations made at a particular location in the environment. Each observation is a momentary summation of the effects of everything that is happening to the particular parameter being observed. Time series of interest to the environmental engineer include stream-flow records, sequential observations of water-quality measurements made in surface waters, waste-discharge flows, and time traces obtained from air monitoring devices. Analyses of such series provide insight as to the cause and effect of environmental phenomena as well as furnish means for forecasting future time series. Although the methodology of time-series analysis is extensive, only those methods of general use in environmental systems will be considered here. Data from which some time series are derived may be in a continuous record. However, data of this type can be converted to a discrete form prior to analysis. The average-temperature data found in Table 7.1.1 provide a typical example of a time series in tabular form.

A time series such as is shown in Fig. 7.1.1 can be considered as a sum of three components—a *trend component*, a *frequency* or *cyclic component*, and a

Table 7.1.1 AVERAGE TEMPERATURE, °F*

Year	Jan.	Feb.	Mar.	Apr.	May	June	July	Aug.	Sept.	Oct.	Nov.	Dec.	Annual
1931	44.2	46.2	47.7	59.7	66.7	77.8	81.9	77.8	77.6	61.8	57.5	52.4	62.6
1932	52.0	55.2	50.9	60.4	66.4	76.0	80.8	77.8	72.4	60.6	47.5	45.0	62.1
1933	47.6	44.3	52.4	59.8	73.2	77.2	76.7	77.2	76.4	63.4	52.1	51.8	62.7
1934	46.5	41.4	49.9	61.7	68.7	77.4	80.6	78.9	75.0	63.6	53.6	43.2	61.7
1935	45.0	46.4	58.8	61.5	69.2	75.8	78.6	78.3	73.8	63.3	55.4	38.2	62.0
1936	37.4	41.2	55.3	58.2	71.2	77.8	81.6	78.4	76.0	65.0	51.1	46.6	61.6
1937	51.8	44.3	49.6	57.5	68.8	79.2	79.5	79.0	70.8	58.9	48.0	41.3	60.7
1938	40.8	49.2	56.6	59.2	70.6	75.0	78.5	80.8	73.7	62.5	54.2	44.0	62.1
1939	45.6	49.7	55.4	60.6	68.2	79.6	80.1	78.3	74.0	63.8	48.8	45.0	62.4
1940	31.9	43.5	50.0	59.4	66.5	76.8	77.4	78.8	72.0	62.8	51.8	46.8	59.8
1941	43.7	40.6	47.3	63.1	71.2	76.7	78.8	79.6	75.7	67.8	51.2	45.6	61.8
1942	42.4	40.8	52.7	63.0	69.7	78.4	80.2	77.2	73.4	62.8	53.7	43.5	61.5
1943	46.6	46.9	50.6	63.0	73.2	80.7	79.0	80.6	72.0	59.6	49.0	45.4	62.2
1944	43.5	49.6	51.5	59.8	70.9	79.6	77.6	76.4	73.8	61.4	49.8	39.3	61.1
1945	43.2	46.8	60.6	64.0	66.3	76.8	78.1	77.0	75.6	61.2	52.9	38.2	61.7
1946	44.3	46.6	58.0	62.6	67.8	75.6	77.8	76.4	72.0	62.2	56.8	48.4	62.4
1947	46.8	38.8	44.6	64.7	69.4	75.5	76.2	79.1	74.5	65.6	49.4	43.4	60.7
1948	37.7	46.2	56.5	64.2	69.4	77.2	79.2	77.0	71.3	59.5	56.2	47.4	61.6
1949	52.6	51.2	53.1	59.6	69.6	76.2	79.8	76.7	69.3	67.0	49.3	45.5	62.5
1950	53.6	49.4	49.5	59.0	71.9	77.5	76.8	76.9	70.8	66.1	48.6	40.3	61.7
1951	43.4	46.3	53.2	59.2	67.9	76.3	79.3	80.1	74.1	65.2	47.2	45.6	61.5
1952	50.9	47.6	51.5	60.3	69.5	81.5	81.4	77.4	71.8	58.5	52.6	43.0	62.2
1953	46.5	48.0	54.1	60.2	73.3	78.6	79.2	78.6	72.6	64.7	52.0	44.0	62.7
1954	46.0	51.3	52.8	66.5	64.7	77.0	81.7	81.4	76.5	64.3	49.5	43.4	62.9
1955	43.5	46.6	54.2	64.0	70.0	71.4	78.2	79.3	73.1	60.2	50.4	42.0	61.1
1956	41.6	49.2	52.3	59.5	70.5	76.3	78.5	78.9	70.1	63.8	50.2	53.0	62.0
1957	46.0	52.2	51.2	64.0	70.3	76.8	79.2	77.2	73.9	57.9	52.7	44.9	62.2
1958	37.7	37.7	48.8	59.5	69.4	76.2	78.1	77.5	71.5	59.8	54.8	40.9	59.3
1959	42.4	46.8	49.3	61.2	70.6	75.3	77.6	78.9	71.6	63.2	50.9	45.3	61.1
1960	43.8	42.6	41.4	63.0	66.7	75.3	78.8	78.7	72.7	63.8	52.1	41.0	60.0
1961	39.5	48.5	54.7	55.5	64.9	72.9	75.9	75.7	73.5	60.6	56.6	43.9	60.2
1962	42.7	50.4	48.4	57.0	74.3	75.9	79.1	78.0	71.3	63.7	50.0	41.9	61.1
1963	39.6	40.3	55.3	63.8	68.6	74.1	76.0	77.2	70.3	61.6	50.5	36.1	59.4
1964	39.9	38.7	48.9	59.1	68.2	76.3	76.4	75.4	71.1	56.2	54.0	43.4	59.0
1965	42.6	42.2	47.5	61.4	71.3	72.1	77.7	77.7	72.4	60.0	52.9	45.4	60.3
1966	39.3	44.8	50.9	60.3	67.6	72.9	78.9	77.0	70.9	60.0	52.3	44.1	59.9

* From U.S. Weather Bureau, U.S. Dep. of Commerce, Climate of Clemson, S.C., *Agric. Weather Res. Ser.* 12, February 1967.

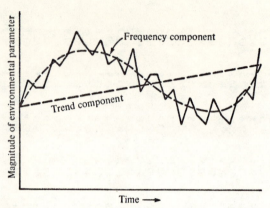

FIGURE 7.1.1
Time series demonstrating the components of parameter variability.

random component. Any one or a combination of these components may exist in a particular time series. A greater insight into time-series analysis is gained by considering the properties of each component separately.

The trend component, when it exists in a series, becomes discernable only when the record is long in duration. Care must be exercised in the identification of a long-term increase of the time-dependent parameter as a trend. A record of a relatively short length may exhibit an apparent trend that actually is a portion of a low frequency periodicity. A real trend might be identified in a record obtained from a monitoring station located in an area where air quality is in a process of being degraded by progressive build-up of industry. An apparent trend would be observed in the receding stream flows following a flood.

The trend portion of a particular time series may best be represented by a polynomial of second degree or higher. However, in most cases involving environmental phenomena the trend, if present, can be treated as a straight line. For a time series consisting of observations $x(t)$, made at equally spaced time intervals,

$$x_T(t) = a + b(t - \mu_t) \qquad (7.1.1)$$

where $x_T(t)$ = estimate of trend component

t = time expressed as an integer in a series of consecutive integers ranging from 1 through n (1, 2,..., n)

μ_t = mean of time variable

a, b = coefficients

The mean of the time variable can be computed with

$$\mu_t = \frac{\sum\limits_{i=1}^{n} t_i}{n} \qquad (7.1.2)$$

where n = number of observations

t_i = time of ith observation

For a given time series, the best estimate of the coefficients a and b can be computed using the expressions

$$a = \frac{\sum\limits_{i=1}^{n} x_i}{n} \qquad (7.1.3)$$

and

$$b = \frac{\sum\limits_{i=1}^{n} t_i x_i - n\mu_t \mu_x}{\sum\limits_{i=1}^{n} t_i^2 - n\mu_t^2} \qquad (7.1.4)$$

The coefficients a and b can be computed with the assistance of Algorithms 7.1.1 and 7.1.2.

It is to be noted that the coefficient a in Eqs. (7.1.1) and (7.1.3) represents the mean of the observations made of the dependent variable and not the intercept of the straight line of best fit at $t = 0$. The coefficient b is the slope or trend of such a line.

The frequency component of a time series may be composed of one or more harmonics. These harmonics can be described by the Fourier representation

$$x_F(t) = \sum_{k=1}^{m} (A_k \sin k\omega t + B_k \cos k\omega t) \qquad (7.1.5)$$

where $x_F(t)$ = estimate of frequency component

A_k, B_k = Fourier coefficients of kth harmonic

ω = angular frequency in radians per time unit

k = integer index identifying harmonic

m = total number of harmonics considered nonnegligible

The angular frequency is related to the frequency in cycles per time unit by

$$\omega = 2\pi f = \frac{2\pi}{P} \qquad (7.1.6)$$

where f = frequency in cycles per time unit

P = period measured in time units

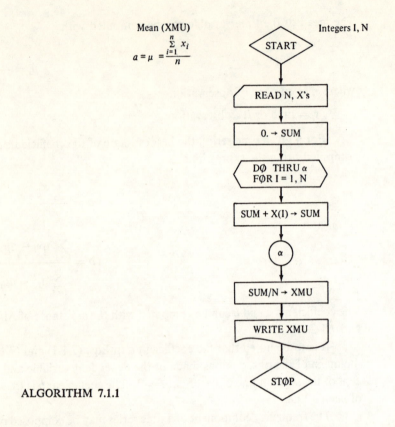

ALGORITHM 7.1.1

It is to be noted that the period P is referenced to the first harmonic. For other harmonics, the arguments of the trigonometric functions in Eq. (7.1.5) are $2\pi t/(P/k)$.

The frequencies observed in environmental phenomena may have periods anywhere from a fraction of a day to a year.

The best estimate of the Fourier coefficients is given by

$$A_k = \frac{2}{n} \sum_{i=1}^{n} x_i \sin k\omega t_i \qquad (7.1.7)$$

and

$$B_k = \frac{2}{n} \sum_{i=1}^{n} x_i \cos k\omega t_i \qquad (7.1.8)$$

Algorithm 7.1.3 can be used to compute the values of the Fourier coefficients.

Coefficient b in
Eq. (7.1.4).

$$b = \frac{\sum_{i=1}^{n} t_i x_i - n\mu_t \mu_x}{\sum_{i=1}^{n} t_i^2 - n\mu_t^2}$$

Initial computations:

μ_t Algorithm 7.1.1
μ_x Algorithm 7.1.1

where n = N
       x = X
      μ_t = TMU
      μ_x = XMU
      b = B

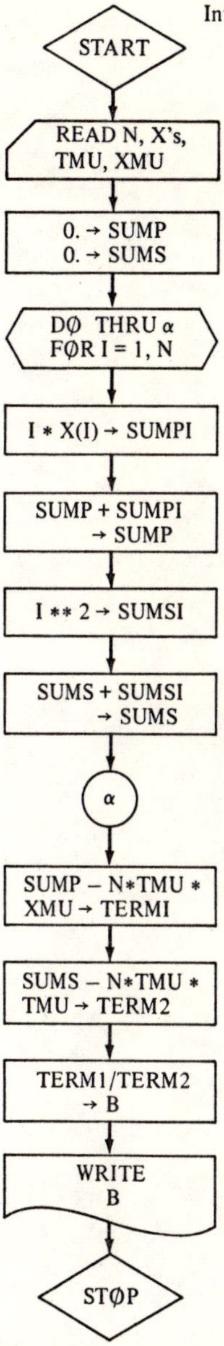

ALGORITHM 7.1.2

Integers I, N

START

READ N, X's, TMU, XMU

0. → SUMP
0. → SUMS

DØ THRU α
FØR I = 1, N

I * X(I) → SUMPI

SUMP + SUMPI → SUMP

I ** 2 → SUMSI

SUMS + SUMSI → SUMS

α

SUMP − N*TMU * XMU → TERM1

SUMS − N*TMU * TMU → TERM2

TERM1/TERM2 → B

WRITE B

STØP

Fourier coefficients in Eqs. (7.1.7) and (7.1.8) (AK and BK).

$$A_k = \frac{2}{n} \sum_{i=1}^{n} x_i \sin k\omega t_i$$

$$B_k = \frac{2}{n} \sum_{i=1}^{n} x_i \cos k\omega t_i$$

where n = N
k = K
ω = ØMEGA
x = X

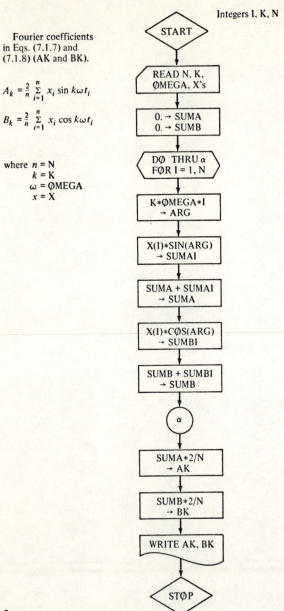

ALGORITHM 7.1.3

With use of a trigonometric identity, Eq. (7.1.5) can be modified to yield

$$x_F(t) = \sum_{k=1}^{m} C_k \cos(k\omega t - \theta_k) \quad (7.1.9)$$

All k harmonics need not be consecutive.

The amplitude and the phase shift of any particular harmonic can be computed with the following relationships:

$$C_k = (A_k^2 + B_k^2)^{1/2} \quad (7.1.10)$$

$$\theta_k = \arctan \frac{A_k}{B_k} \quad (7.1.11)$$

where C_k = amplitude of kth harmonic

θ_k = phase angle for kth harmonic in degrees relative to cosine term in Eq. (7.1.5)

A deviation about a mean value is inherent in data that fall into a frequency pattern. This deviation can be expressed by the *variance*. The variance accounted for in a given time series by a given harmonic can be computed with

$$\sigma_k^2 = \frac{C_k^2}{2} \quad \text{for } k < \frac{n}{2} \quad (7.1.12)$$

or

$$\sigma_k^2 = C_k^2 \quad \text{for } k = \frac{n}{2} \quad (7.1.13)$$

The random component of a time series is that component that has no deterministic pattern of behavior. Instead, it exhibits a behavior that is predictable only in its statistical properties. This component can be represented by

$$x_R(t) = z_n \sigma_R \quad (7.1.14)$$

where σ_R = standard deviation of random component of time series

z_n = standardized random, normal, and independently distributed variate with mean 0 and unit variance; values can be obtained from prepared tables or generated by computer

The random variate introduces a positive or negative variable that exceeds in magnitude the range of one standard error 32 percent of the time. Random variates with distributions other than normal may be substituted in Eq. (7.1.14). However, for most environmental phenomena of a random nature normal distributions provide a satisfactory representation.

A time series, therefore, may be represented by a relationship which combines all three components—trend, frequency, and random:

$$x(t) = x_T(t) + x_F(t) + x_R(t)$$

or

$$x(t) = a + b(t - \mu_t) + \sum_{k=1}^{m} C_k \cos(k\omega t - \theta_k) + z_n \sigma_R \quad (7.1.15)$$

The relative prominence of each component will vary with the particular time series being considered. In general, time series of environmental derivation fall into one of the following four categories:[1]

1 Time series that are composed of some periodicity, a certain degree of randomness, plus a mean with a time trend. Series of this type might be observed in cases where stream water quality is monitored over a relatively long period of time in an area experiencing industrial development.

2 Time series that are largely periodic and may include several distinct frequencies. Stream water temperature and tidal behavior generally result in time series of this type.

3 Time series that are composed of some periodicity and some degree of randomness. An example of this type can be found in the time records of dissolved oxygen in a river or estuary.

4 Time series that appear to be characterized almost entirely by random variation. Over a relatively short period of time, the average daily sewage flow to a waste treatment plant might yield this type of time series.

It must be emphasized that the categorization of a time series is dependent not only on the length of record but also on the particular statistic of the parameter of interest which is used. For example, although the average daily sewage flow may give a time series of type 4, the hourly flow may exhibit behavior that would assign it to type 2 or 3.

7.2 TIME-SERIES ANALYSIS

Time series provide useful information. Such information is extracted with techniques, the application of which is called *time-series analysis*. When applied to environmental systems, time-series analyses often yield information that provides an insight to cause-and-effect relationships influencing the parameter of concern. Furthermore, these techniques can be used to establish recursion

---

[1] R. V. Thomann, Time-series Analyses of Water-quality Data, *J. Sanit. Eng. Div., ASCE*, vol. 93, no. SA1, pp. 1–23, February 1967.

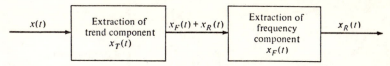

FIGURE 7.2.1
Filtering procedure for removing the trend and frequency components of a time series.

relationships that have application in the synthesis of predicted time series. The latter finds increasing use in simulation procedures and design.

The reduction of a time-series record into components is accomplished by the filtering procedure indicated in Fig. 7.2.1. The components are treated sequentially. Removal of the trend leaves the frequency and random components; then removal of the frequency leaves only the random. Details of the procedure are given by Algorithm 7.2.1. Figure 7.2.2 illustrates the result of extracting the frequency component from a time series of average daily dissolved-oxygen data. This particular series does not reveal a trend component.

The variance is of key interest in time-series analysis. This statistic is a measure of the variability of the data points in a series. If all the observations in a time record are treated statistically and a mean and variance computed, the latter is referred to as the *total variance* σ^2. All three components of a time series—trend, frequency, and random—contribute to the total variance:

$$\sigma^2 = \sigma_T^2 + \sigma_F^2 + \sigma_R^2 \qquad (7.2.1)$$

where $\sigma_T^2, \sigma_F^2, \sigma_R^2$ = variances contributed by trend, frequency, and random components, respectively

The variance contributed by each component can be isolated in a procedure that parallels the filtering procedure illustrated in Fig. 7.2.1. First, the estimate of the total variance can be computed by

$$\sigma^2 = \frac{\sum_{1=i}^{n}(x_i - \mu_x)^2}{n-1} \qquad (7.2.2)$$

Computation of the variance is facilitated with the use of Algorithm 7.2.2. The trend component of the observations is then removed, and the resulting data are treated with Eq. (7.2.2) to give the variance contributed by the frequency and random components $\sigma_F^2 + \sigma_R^2$. The last step is to remove the frequency component from the observations and treat the residual data with Eq. (7.2.2) to yield the random component σ_R^2. The random component will be truly random

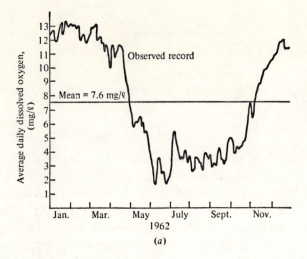

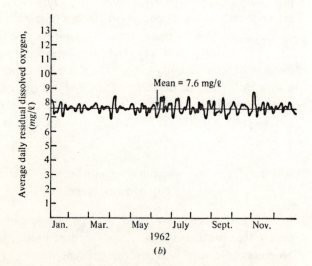

FIGURE 7.2.2
Average daily dissolved oxygen in the Delaware River Estuary at Torresdale, Pa. (*a*) Record containing both the frequency and random components; (*b*) record containing the random component after a frequency component of 20 harmonics has been removed. (*From R. V. Thomann et al., The Management of Time Variable Stream and Estuarine Systems, Chem. Eng. Prog. Symp. Ser., vol. 64, no. 90, 1968.*)

Random component of
time series, Eq. (7.1.15).

$$x_R(t) = x(t) - x_T(t) - x_F(t)$$
where $x_T(t) = a + b(t - \mu_t)$
$$x_F(t) = \sum_{k=1}^{m} C_k \cos(k\omega t - \theta_k)$$

$n = N$
$m = M$
$k = K$
$a = A$
$b = B$
$A_k = AK$
$B_k = BK$
$x = X$
$\omega = \text{ØMEGA}$
$t = I$
$x_R(t) = XRI$

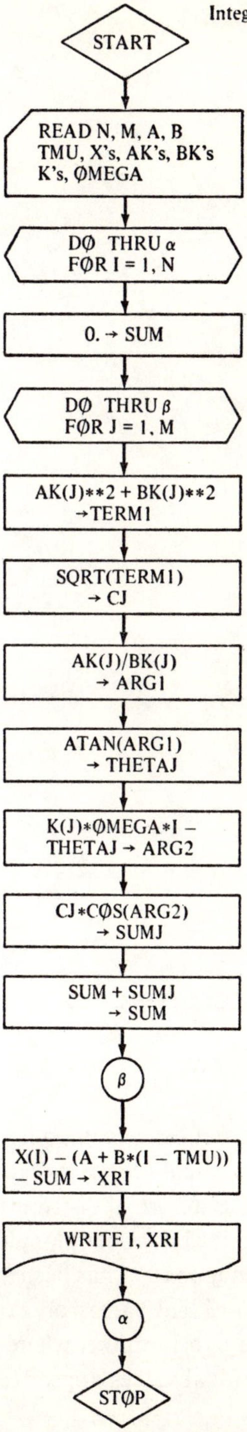

ALGORITHM 7.2.1

Integers I, J, K, M, N

⟨START⟩

READ N, M, A, B
TMU, X's, AK's, BK's
K's, ØMEGA

DØ THRU α
FØR I = 1, N

0. → SUM

DØ THRU β
FØR J = 1, M

AK(J)**2 + BK(J)**2
→ TERM1

SQRT(TERM1)
→ CJ

AK(J)/BK(J)
→ ARG1

ATAN(ARG1)
→ THETAJ

K(J)*ØMEGA*I −
THETAJ → ARG2

CJ*CØS(ARG2)
→ SUMJ

SUM + SUMJ
→ SUM

(β)

X(I) − (A + B*(I − TMU))
− SUM → XRI

WRITE I, XRI

(α)

⟨STØP⟩

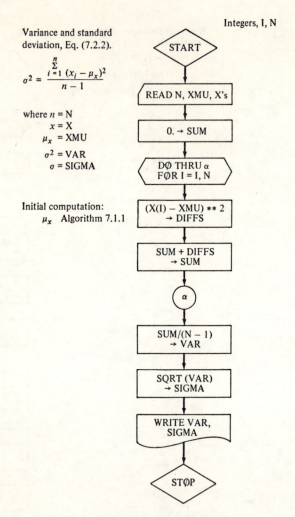

ALGORITHM 7.2.2

Variance and standard deviation, Eq. (7.2.2).

$$\sigma^2 = \frac{\sum\limits_{i=1}^{n}(x_i - \mu_x)^2}{n-1}$$

where n = N
x = X
μ_x = XMU
σ^2 = VAR
σ = SIGMA

Initial computation:
μ_x Algorithm 7.1.1

only if all the frequency component is removed. Such insurance is provided by plotting the residual record $x_R(t)$ on arithmetic probability paper. A straight-line plot reveals a normal distribution and, hence, a purely random residual. If the residual record does not plot along a straight line, a clue is obtained that not all the frequency content has been extracted from the residual data. An example of a plot of residual values on arithmetic probability paper is shown in Fig. 7.2.3. At the two locations where the observations plotted in the figure were made, the residual values appear to have a normal distribution.

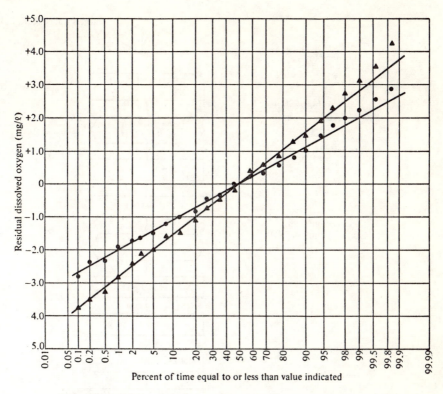

FIGURE 7.2.3
Distribution of hourly dissolved-oxygen series at two points along the Delaware River Estuary. (*From R. V. Thomann, Time-series Analysis of Water-quality Data, J. Sanit. Eng. Div., ASCE, vol. 93, no. SA*1, *February*, 1967.)

In most time series describing environmental behavior, the observations x_t are correlated with neighboring observations $x_{t-1}, x_{t-2}, \ldots$. For instance, in streams recovering from periods of high flow, daily values of flow exhibit marked correlation with the flows of the preceding days. A measure of such correlation is called the *autocovariance function*:

$$R(\tau) = \sum_{i=1}^{n-\tau} \frac{(x_i - \mu_x)(x_{i+\tau} - \mu_x)}{n - \tau} \quad (7.2.3)$$

where $x_i, x_{i+\tau}$ = dependent variable at time i and $i + \tau$, respectively

τ = time lag; $\tau = 0, 1, 2, \ldots, m$

Algorithm 7.2.3 can be used to facilitate the solution of Eq. (7.2.3).

Autocovariance as
a function of τ
where $\tau = 0, 1, 2, \ldots, m$
Eq. (7.2.3)

$$R(\tau) = \sum_{i=1}^{n-\tau} \frac{(x_i - \mu_x)(x_{i+\tau} - \mu_x)}{n - \tau}$$

where n = N
$m + 1$ = M
μ_x = XMU
x = X
R = SUM
τ = JJ

Initial computation:
μ_x Algorithm 7.1.1

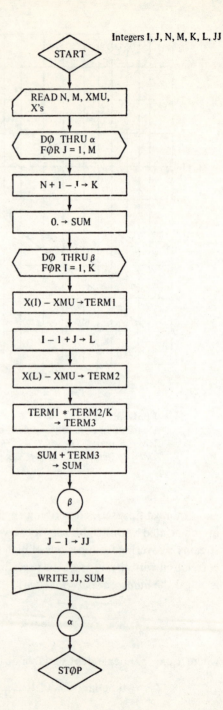

ALGORITHM 7.2.3

It is often convenient when comparing series with different scales of measurement to normalize the autocovariance function by dividing the autocovariance function $R(\tau)$ by the variance, which is simply the autocovariance with zero time lag $R(0)$. The result is called the *normalized autocorrelation function* $R(\tau)_0$. The autocorrelation function, therefore, is a dimensionless quantity that is equal to unity at zero lag. This function also provides an insight as to the *persistence* underlying the phenomenon of concern. If the function remains positive, it indicates that a persistence exists. For instance, if the autocorrelation function of an environmental variable, say water temperature, remains positive for lags up to approximately 24 h, then it can be concluded that the high (or low) temperatures will tend to persist for approximately a day. A negative function indicates that high (low) values of the variable will most likely be followed by low (high) values at some time τ later. No correlation, or persistence, is indicated when the function is 0. Two plots of the autocorrelation function are shown in Fig. 7.2.4. Plot (*a*) is typical of random time series in that the autocorrelation function falls away to 0 after a certain time lag and remains there. Plot (*b*) illustrates the behavior of a series which has some transient periodicity.

The autocovariance function finds use in the construction of a *variance spectrum*, or *power spectrum* as it is referred to in communications engineering, where its use first became popular. An analysis of a variance spectrum may reveal periodicities that are not easily discernible in a time series. Such analyses are becoming increasingly popular in environmental analysis, particularly in estuarine analysis.[1,2,3] The variance spectra for trend and periodic components of time series are illustrated in Fig. 7.2.5. Any random component present in a record may be concentrated in certain frequency bands of the record's spectrum. It is to be noted that a variance spectrum at frequency 0 includes all the record variance that does not reoccur in the period during which the record is analyzed. In other words, the spectrum at frequency 0 includes (1) any linear trends in the record, (2) any periodic components in the record of such low frequency that they appear as linear trends, and (3) any random component with a variance frequency of 0.

The first step in the construction of a variance spectrum is the Fourier transformation of the autocovariance function from the time domain to the fre-

---

[1] Ibid.
[2] T. A. Wastler, Application of Spectral Analysis to Stream and Estuary Field Surveys—I. Individual Power Spectra, *Publ.* 999-*WP*-7, U.S. Public Health Serv., Washington, 1963.
[3] C. G. Gunnerson, Optimizing Sampling Intervals in Tidal Estuaries, *J. Sanit. Eng. Div., ASCE*, vol. 92, no. SA2, pp. 103–125, 1966.

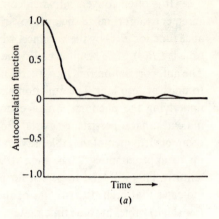

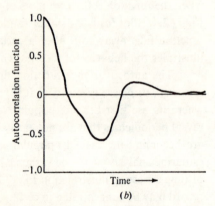

FIGURE 7.2.4
Autocorrelation functions.

quency domain. This step is performed using the expression

$$V_\tau = \frac{k}{m}\left(R_0 + R_m \cos\tau\pi + 2\sum_{l=1}^{m-1} R_l \cos\frac{l\tau\pi}{m}\right) \quad \text{for } \tau = 0, 1, 2, \ldots, m \quad (7.2.4)$$

where
V_τ = Fourier cosine transform of autocovariance at lag τ
R_0, R_m, R_l = autocovariances at lags $\tau = 0$, m, and l, respectively
k = constant
l = number having all integer values from 1 to $m - 1$

The value of the constant k is

$$k = \begin{cases} 1 & \text{for } \tau = 1, 2, \ldots, m-1 \\ \frac{1}{2} & \text{for } \tau = 0 \text{ and } m \end{cases} \quad (7.2.5)$$

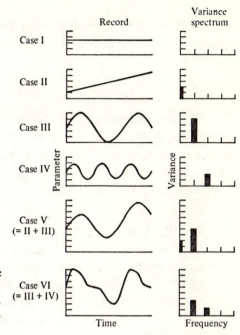

FIGURE 7.2.5
Variance spectra for trend and periodic components of time series. (*From C. G. Gunnerson, fig. 9, J. Sanit. Eng. Div., ASCE, vol. 92, no. SA2, pp. 103–125, 1966.*)

The second step is a weighting computation that smooths out some of the distortion of the spectrum resulting from the analysis of a single record. One such computation uses the relationships

$$U_0 = \tfrac{1}{2}(V_0 + V_1)$$
$$U_\tau = \tfrac{1}{4}V_{\tau-1} + \tfrac{1}{2}V_\tau + \tfrac{1}{4}V_{\tau+1} \quad \text{for } 1 \leq \tau \leq m-1 \quad (7.2.6)$$
$$U_m = \tfrac{1}{2}(V_{m-1} + V_m)$$

where $U =$ *smoothed* estimate of variance spectrum

The variance spectrum is formed by plotting the values of the spectral estimates as computed by Eq. (7.2.6) against corresponding values of the time lag. The dimensions of the spectral estimates are in variance per cycle per unit time, whereas the abscissa dimensions are in cycles per $2m\,\Delta t$, where Δt is the time interval between values in the series. For instance, if $\Delta t = 1$ d and m (the maximum number of lags) is 10, then the abscissa has the dimension of cycles per 20 d and has 11 discrete values ranging from 0 to 10, each corresponding to the spectral estimate for that particular time lag. Sometimes, it is more helpful to express the abscissa in terms of the periodicity of the variance, in which case the period corresponding to each lag is determined by

$$T_\tau = \frac{2m\,\Delta t}{\tau} \qquad (7.2.7)$$

The area under a variance spectrum is equal to the total variance of the record. Any dominant frequencies will appear as peaks in the spectrum, and the area under the curve within a frequency band represents the contribution of the frequency phenomena within the band to the total frequency component. The computation of the variance spectrum is facilitated by using Algorithm 7.2.4.

Solution procedure 7.2.1 As a rather obvious illustration of the utility of the variance spectrum, attention is directed to Fig. 7.2.6. Here is shown the time trace of the gauge height of the Potomac Estuary. The variance spectrum of the record is shown in Fig. 7.2.7. Most of the total variance is shown to occur at a period of 12 h, corresponding to the diurnal tidal variation. In this particular case such information can easily be deduced from the time trace itself. However, in many environmental records certain periodicities are not so obvious. Figure 7.2.8. shows time records of dissolved oxygen at several locations along the Potomac Estuary. The variance spectra for these records are shown in Fig. 7.2.9. Note is to be made of the fact that although the basic pattern of the spectra is similar the peak heights vary in magnitude from station to station along the estuary. The relative importance of the various environmental factors influencing the variation in dissolved oxygen varies with the location from which the record is taken.

Theoretically, the shortest period possible to resolve with a given sampling interval is the period that is two times the sampling interval. However, from a practical standpoint it has been found that a sampling interval equal to no more than one-third of the length of the shortest significant period is recommended. That is, to resolve a 12-h period, a sampling interval of no more than 4 h should be used.

The longest period resolved is determined by the number of lags used in the computation and by the sampling interval. Too many lags used in an analysis will reduce the precision of the spectral estimates, whereas too few will reduce the resolution of the spectral components. For environmental time series an optimum number of lags to use is approximately 10 percent of the total number of observations in the length of record analyzed. Furthermore, for such series it is recommended that the length

Integers J, L, M, JJ, MM

Smoothed estimate of the variance spectrum as a function of periodicity, computed with use of Eqs. (7.2.4)-(7.2.7)

$$V_\tau = \frac{k}{m}\left[R_0 + R_m \cos \tau\pi + 2 \sum_{\ell=1}^{m-1} R_\ell \cos\left(\frac{\ell\tau\pi}{m}\right)\right]$$

for $\tau = 0, 1, 2, \ldots, m$

$$k = \begin{cases} 1 & \text{for } \tau = 1, 2, \ldots, m-1 \\ \tfrac{1}{2} & \text{for } \tau = 0 \text{ and } m \end{cases}$$

$U_0 = \tfrac{1}{2}(V_0 + V_1)$
$U_\tau = \tfrac{1}{4} V_{\tau-1} + \tfrac{1}{2} V_\tau + \tfrac{1}{4} V_{\tau+1}$ for $1 \leq \tau \leq m-1$
$U_m = \tfrac{1}{2}(V_{m-1} + V_m)$

$$T_\tau = \frac{2m\Delta t}{\tau}$$

where $m + 1 = M$
$R = R$
$\Delta t = $ DELT
$U = U$
$T = $ PRD
$\tau = $ JJ
$k = $ K

Initial computation:
$R(\tau)$ Algorithm 7.2.3

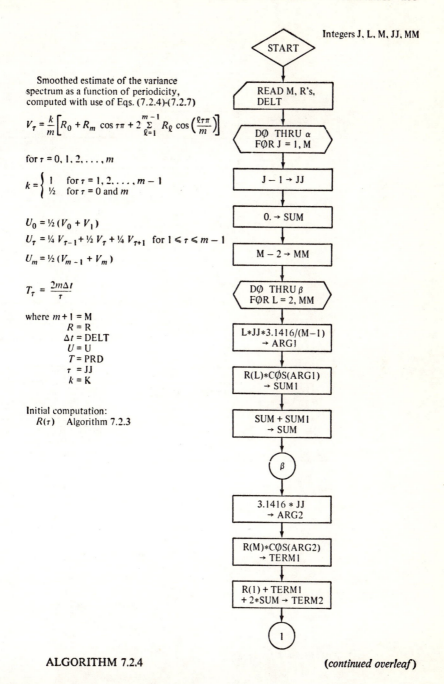

ALGORITHM 7.2.4

(continued overleaf)

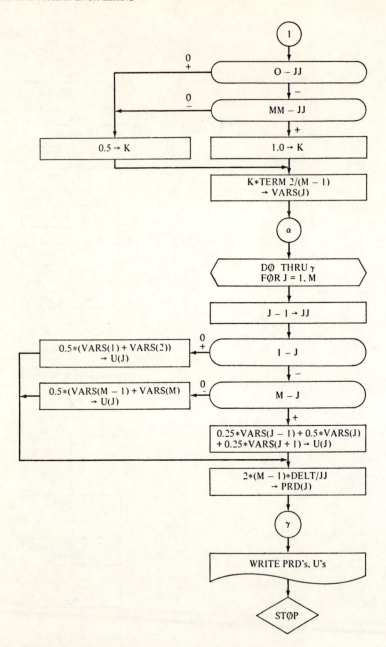

ALGORITHM 7.2.4 (*continued*)

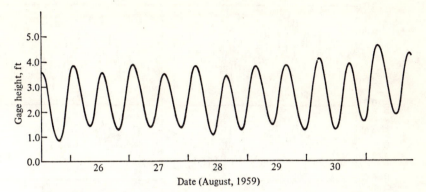

FIGURE 7.2.6
Portion of tide height record in the Potomac Estuary. (*From T. A. Wastler, Publ. 999-WP-7, U.S. Public Health Serv., Washington, 1963.*)

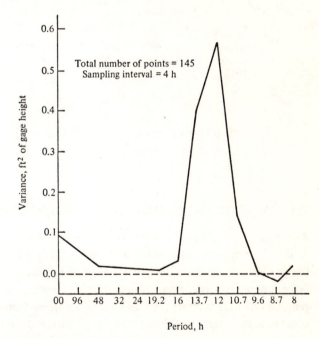

FIGURE 7.2.7
Spectrum of a tidal height record. (*From T. A. Wastler, Publ. 999-WP-7, U.S. Public Health Serv., Washington, 1963.*)

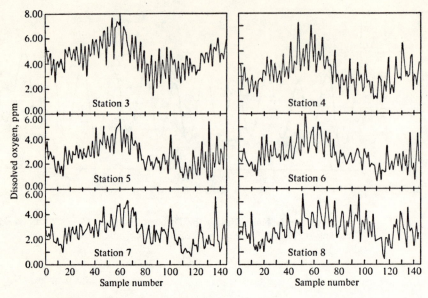

FIGURE 7.2.8
Dissolved-oxygen records obtained in the Potomac Estuary, August 1959. (*From T. A. Wastler, Publ. 999-WP-7, U.S. Public Health Serv., Washington, 1963.*)

of record to be analyzed be at least 10 times as long as the longest significant period to be resolved.[1] ////

7.3 SYNTHETIC STREAM–FLOW SEQUENCES

As mentioned in the previous section, time-series analysis of existing records can be used in synthesizing predicted time series. Such synthetic traces provide bases for procedures for estimating water storage requirements and low-stream-flow estimates, as well as for numerous other engineering uses. Furthermore, it was pointed out that in most time series describing environmental phenomena, the observations x_t are correlated with neighboring observations $x_{t-1}, x_{t-2}, \ldots$. In fact, the autocovariance function was described as a measure of this correlation. The present section will begin by exploring further the relationship between successive observations and how this relationship can be used in constructing a stream-flow model.

If the stream flow for a given time interval x_{i+1} is plotted as a function of

---

[1] T. A. Wastler, op. cit.

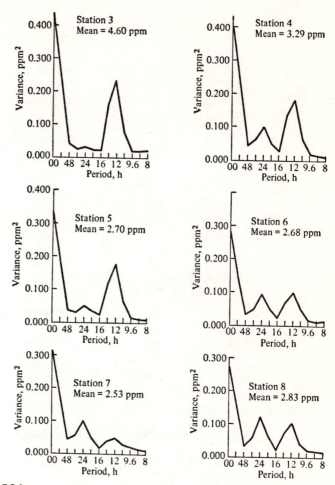

FIGURE 7.2.9
Dissolved-oxygen spectra in the Potomac Estuary. (*From T. A. Wastler, Publ. 999-WP-7, U.S. Public Health Serv., Washington, 1963.*)

the flow for the preceding interval x_i, the curve of best fit will most often suggest a linear relationship. However, as can be expected of all environmental phenomena, considerable scatter of points will occur about the straight line of best fit. If we assume that the curve of best fit for such a plot is linear, the model for the relationship between successive flows can be expressed by[1]

$$x_{i+1} = \mu + b(x_i - \mu) + z_{i+1}\sigma(1 - \rho^2)^{1/2} \qquad (7.3.1)$$

[1] M. B Fiering, A Markov Model for Low-flow Analysis, *Bull. Int. Assoc. Sci. Hydrol.*, vol. 9, p. 2, June 1964.

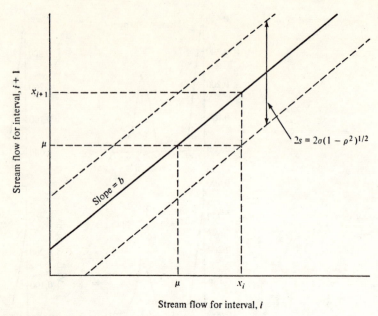

FIGURE 7.3.1
Definition sketch illustrating Eq. (7.3.1).

where μ = mean of flows

b = regression coefficient in $(i + 1)$st interval on values in ith interval

σ = standard deviation of flows

ρ = correlation coefficient between flows in successive time intervals

z_{i+1} = standardized random variate with mean 0 and unit variance

A definition sketch of Eq. (7.3.1) is shown in Fig. 7.3.1. The straight line, which has a slope equal to the value of the regression coefficient b, represents the deterministic component of the recursion relationship $\mu + b(x_i - \mu)$. The random component $z_{i+1}\sigma(1 - \rho^2)^{1/2}$ is represented by the vertical distance at which the values of x_{i+1} are located above or below the straight line. The term $\sigma(1 - \rho^2)^{1/2}$ is called the *standard error s*. The standard error defines a range in which approximately 68 percent of the correlation points are expected to fall. The correlation coefficient ρ is a measure of the degree of correlation between the successive flows and theoretically can assume all values between 0 for no correlation to ± 1 for perfect direct, or inverse, correlation. For the hypothetical extreme where the magnitude of the coefficient equals unity, the standard error becomes 0, and the recursion relationship expressed by Eq. (7.3.1) becomes completely deterministic. Note should be made of the similarity between the deterministic component of Eq. (7.3.1) and Eq. (7.1.1).

Stream flows may exhibit a variety of different distribution patterns about the mean. Among such patterns, the most common are the normal, logarithmic normal, and gamma distributions. For our purposes here, the normal and logarithmic normal distributions have the most significance. The random variate z_{i+1} in these cases must be normally distributed. If Eq. (7.3.1) is used to model other types of distributions, then the term z_{i+1} must conform to the particular distribution of concern.

The model expressed by Eq. (7.3.1) applies only when the population parameters μ, b, σ, and ρ do not change with time, such as could be assumed within a given season. However, when one is postulating a model that takes into account seasonal variations in the values of the population parameters, Eq. (7.3.1) must be modified:[1]

$$x_{i+1} = \mu_{j+1} + b_j(x_i - \mu_j) + z_i \sigma_{j+1}(1 - \rho_j^2)^{1/2} \qquad (7.3.2)$$

where x_i, x_{i+1} = discharges during ith and $(i + 1)$st seasons

μ_j, μ_{j+1} = mean seasonal discharges for jth and $(j + 1)$st seasons, respectively, within repetitive annual cycle

σ_{j+1} = standard deviation of discharges during $(j + 1)$st season

ρ_j = correlation coefficient between flows in jth and $(j + 1)$st seasons

b_j = regression coefficient for estimating flow in $(j + 1)$st season from that of jth season

The term "season" is a generic term which may be applied to time intervals varying from 1 to 6 months. The time interval used for a season will depend upon the particular use of the model. Algorithm 7.3.1 may be used in computations involving Eq. (7.3.2).

The application of the model expressed by Eq. (7.3.2) consists of three parts: First, the historical record is analyzed and the parameters μ_j, b_j, σ_j, and ρ_j for each season are computed. Second, either through a priori knowledge or specific testing the distribution of the data is determined; i.e., does the data, or

---

[1] H. A. Thomas, Jr., and Myron B Fiering, in Arthur Maass et al., "The Design of Water Resource Systems," chap. 12, Harvard, Cambridge, Mass., 1962.

Synthetic stream flow
sequence, Eq. (7.3.2).
(Seasonal flows.)

$$x_{i+1} = \mu_{j+1} + b_j(x_i - \mu_j) + z_i \sigma_{j+1}(1 - \rho_j^2)^{1/2}$$

where x = X (seasonal flow)
μ = XMU
b = B
z = Z
σ = VAR$^{1/2}$
ρ = RHϕ

and L = Years of synthetic record
N = Number of seasons per year

Initial computation:
μ Algorithm 7.1.1
b Algorithm 7.1.2
σ Algorithm 7.2.2
ρ Algorithm 7.3.2

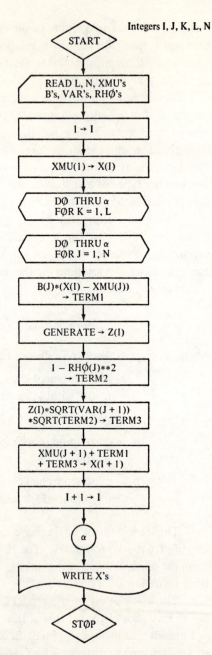

ALGORITHM 7.3.1

its logarithmic transform, have a normal distribution? Third, a flow sequence or a series of flow sequences are generated using the model with the computed parameters. The goal in generating each sequence is that by all statistical tests of significance the sequence cannot be distinguished from the original historical record.

The computation of the means μ_j and μ_{j+1} in Eq. (7.3.2) is facilitated by the use of Algorithm 7.1.1. The computation of b_j can be performed by substituting in Algorithm 7.1.2 x_i for t_i, x_{i+1} for x_i, μ_i for μ_t, and μ_{j+1} for μ_x. The standard deviation σ_{j+1} can be computed using Algorithm 7.2.2. The correlation coefficient ρ_j can be computed with the relationship

$$\rho_j = \frac{\sum_{i=1}^{n-1} x_i x_{i+1}}{(n-1)\sigma_j \sigma_{j+1}} \qquad (7.3.3)$$

where x_i, x_{i+1} = flows in seasons j and $j+1$

n = number of years of record

j = serial index identifying season

A solution of Eq. (7.3.3) is facilitated by using Algorithm 7.3.2 (see p. 243).

In the synthesis of flow sequences, negative values will often occur. When such values occur, they can in most cases be replaced by 0s without the sacrifice of too much statistical integrity.

It should be emphasized that an observed time series is but one realization (or event) of a stochastic process. The analysis of time series has as its purpose the determination of the structure of the time series, i.e., the components that make up the time series that has been realized. The generation, or synthesis, of a time series is accomplished by combining the components derived from the analysis of the time series in such a way that the generated series are other possible realizations of the observed time series.

Solution procedure 7.3.1 Table 7.3.1 (p. 242) lists parameters of monthly flows at a given point along a stream calculated from 32 years of observed record. The season interval in this case was 1 month. Using these parameters a synthetic record of flow was computed which extended 510 years. In the computation, the index j ran cyclic from 1 to 12, and the index i sequentially from 0 to 6,119 (6,120 values, or 510 years of 12 monthly flows).[1] The magnitudes of the mean monthly flows of the synthesized

[1] Ibid.

Table 7.3.1 SERIAL CORRELATION PARAMETERS OF MONTHLY FLOWS AT A POINT ALONG A STREAM CALCULATED FROM 32 YEARS OF OBSERVED RECORD*

Month	Regression coefficient b_j, $(10^2 \text{ acre-ft})^{-1}$	Correlation coefficient ρ_j	Standard deviation σ_j, 10^2 acre-ft	Mean flow μ_j, 10^2 acre-ft
January	0.5470	0.7411	1,219	1,698
February	1.0231	0.6367	899	1,778
March	1.4901	0.6388	1,445	3,066
April	0.3700	0.2347	3,372	8,939
May	0.1825	0.1515	5,317	18,100
June	0.2281	0.7804	6,404	12,618
July	0.1637	0.8755	1,873	3,469
August	0.3876	0.4668	350	1,079
September	2.0764	0.8511	291	845
October	2.1884	0.8328	710	1,264
November	0.6815	0.7842	1,868	1,813
December	0.6464	0.8612	1,623	1,989
Average				4,722

* After H. A. Thomas, Jr., and Myron B Fiering, in Arthur Maass et al., "The Design of Water Resource Systems," table 12.1, Harvard, Cambridge, Mass., 1962. (Copyright 1962 by the President and Fellows of Harvard College.)

Table 7.3.2 COMPARISON OF THE MEAN MONTHLY FLOWS OF 510 YEARS OF SYNTHESIZED RECORD WITH THOSE OF 32 YEARS OF OBSERVED RECORD*

	Magnitude of mean monthly flows, 10^2 acre-ft	
Month	Observed, 32 years	Synthesized, 510 years
January	1,698	1,720
February	1,778	1,780
March	3,066	3,127
April	8,939	9,044
May	18,100	17,794
June	12,618	11,992
July	3,469	3,348
August	1,079	1,059
September	845	843
October	1,264	1,250
November	1,813	1,783
December	1,989	2,007
Total	56,658	55,747
Average	4,722	4,645

* After H. A. Thomas, Jr., and Myron B Fiering, in Arthur Maass et al., "The Design of Water Resource Systems," table 12.2, Harvard, Cambridge, Mass., 1962. (Copyright 1962 by the President and Fellows of Harvard College.)

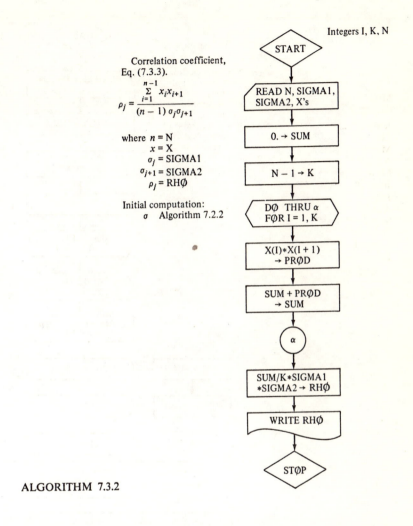

ALGORITHM 7.3.2

record are compared with those of the observed record in Table 7.3.2. The weak correlations between the monthly spring runoffs in Table 7.3.1 are attributed to the change in time and duration of the spring thaw from year to year. ////

7.4 STORAGE–YIELD RELATIONSHIPS

Synthetic stream-flow sequences find application in the establishment of storage-yield relationships for water supply. Given sequences derived from stream-flow records obtained at the site of a proposed dam along with desired drafts, the

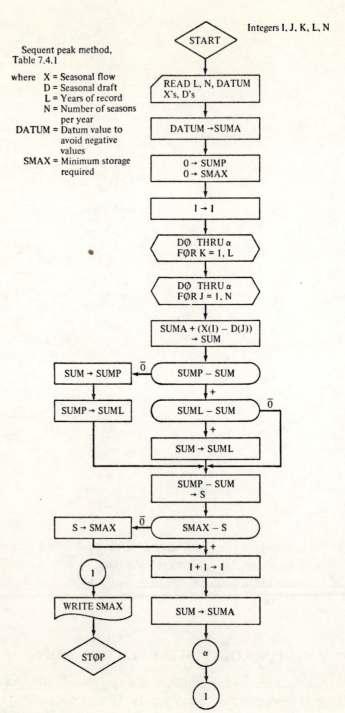

ALGORITHM 7.4.1

Table 7.4.1 STORAGE-YIELD DETERMINATION USING THE SEQUENT PEAK METHOD*†

Year no. (1)	Season no. (2)	Inflow x_i (3)	Percent annual draft (4)	Draft D_i (5)	$x_i - D_i$ (6)	$\sum (x_i - D_i)$ (7)	Storage S_i (8)
						100.000	
1	1	10	4	5.976	4.024		
						104.024	4.024
	2	12	4	5.976	6.024		
						110.048	10.048
	3	21	5	7.470	13.530		
						123.578P_1	14.784
	4	8	6	8.964	−0.964		
						122.614	13.820
	5	6	9	13.446	−7.446		
						115.168T_1	6.374
	6	19	10	14.940	4.060		
						119.228	10.434
	7	22	15	22.410	−0.410		
						118.818	10.024
	8	29	11	16.434	12.566		
						131.384P_2	14.784
	9	11	9	13.446	−2.446		
						128.938	12.338
	10	3	8	11.952	−8.952		
						119.986	3.386
	11	14	7	10.458	3.542		
						123.528	6.928
	12	11	12	17.928	−6.928		
						116.600T_2	0.000
2	1	10	4	5.976	4.024		
						120.624	4.024
	2	12	4	5.976	6.024		
						126.648	10.048
	3	21	5	7.470	13.530		
						140.178P_3	14.784
	4	8	6	8.964	−0.964		
						139.214	13.820
	5	6	9	13.446	−7.446		
						131.768T_3	6.374
	6	19	10	14.940	4.060		
						135.828	10.434
	7	22	15	22.410	−0.410		
						135.418	10.024
	8	29	11	16.434	12.566		
						147.984P_4	14.784
	9	11	9	13.446	−2.446		
						145.538	12.338
	10	3	8	11.952	−8.952		
						136.586	3.386
	11	14	7	10.458	3.542		
						140.128	6.928
	12	11	12	17.928	−6.928		
						133.200	0.000

* After H. A. Thomas, Jr., and Myron B Fiering, The Nature of the Storage-yield Function, table I, in "Operations Research in Water Quality Management," chap. 2, Harvard University Water Program, 1963.
† All terms in 10^3 acre-ft.

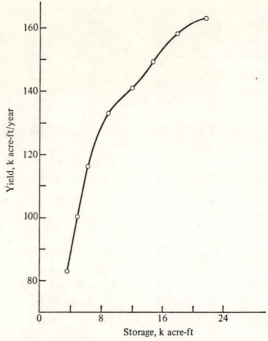

FIGURE 7.4.1
Storage-yield relationship. (*From H. A. Thomas and Myron B Fiering, The Nature of the Storage-Yield Function, "Operations Research in Water Quality Management," chap. 2, Harvard University Water Program, 1963.*)

minimum capacity of the reservoir can be determined so that no shortage of draft should occur. The basic procedure, called the *sequent peak method*, is as follows:[1] Consider a flow sequence that is sufficiently long to include at least two cycles of repetitive hydrologic behavior. The cycles are generally annual. In Table 7.4.1 (p. 245) monthly flows x_i and monthly drafts D_i are tabulated over a 2-year period. The differences between the two are computed for each month (column 6) and then summed. An adequately large datum avoids negative values in the cumulative net inflows tabulated in column 7. Reviewing column 7, the local maximum values (sequent peaks) P_k are identified such that $P_{k+1} > P_k$. The lowest value T_k falling between each pair of maximum values P_k and P_{k+1} is then identified. The minimum storage to meet the stated draft will be

$$S_m = \max (P_k - T_k) \qquad (7.4.1)$$

The procedure is outlined in Algorithm 7.4.1 (see p. 244).

[1] H. A. Thomas, Jr., and M. B Fiering, The Nature of the Storage-Yield Function, in "Operations Research in Water Quality Management," chap. 2, Harvard University Water Program, 1963.

Obviously, draft cannot exceed inflow, and the maximum draft that can be imposed in a given situation is

$$\sum_{i=1} D_i = \sum_{i=1} x_i \qquad (7.4.2)$$

Within the limits described by Eq. (7.4.2), the sequent peak method can be used with varying draft rates to provide a relationship such as is presented in Fig. 7.4.1. This figure identifies the yield that is associated with a particular storage value.

The sequent peak method has been explained in the context of a record of minimum length, i.e., two annual cycles. In application, the method should be applied to records of much greater length. This is done by analyzing first the historical stream-flow record to determine the parameters in the model expressed by Eq. (7.3.2). Then, using the computed parameters a family of synthetic records can be generated, the length of each record reflecting the economic life of the proposed reservoir. The sequent peak method can be applied to each record to determine the minimum storage indicated by that particular record. The distribution of the minimum-storage values derived from the family of synthetic records can then be examined on probability paper to determine the risk involved in using any particular value of storage.

7.5 PREDICTING MINIMUM STREAM FLOWS

The waste loading that may be imposed upon a stream without deleterious effect is closely related to the low-flow regime of the stream. Obviously, for a given waste discharge, the lower the stream flow, the higher will be the concentration of the waste in the stream. For this reason, low-flow determinations are important in the design of waste treatment facilities. Unlike floods, in which severity is described by a flow magnitude without specific reference to duration, the severity of a drought is intimately related to its duration.

When dealing with low-flow sequences, use is made of daily flow data. Such sequences can be synthesized with a model that is a modification of Eq. (7.3.1):[1]

$$x_{i+1} = \mu_j + b_j(x_i + \mu_{j-1}) + z_{i+1}\sigma_j(1 - \rho_j^2)^{1/2} \qquad (7.5.1)$$

where x_{i+1} = estimated value of flow in $(i + 1)$st day

x_i = flow in ith day

j = serial index identifying season

Inasmuch as stream flows are bounded at the lower end by zero flow and, theoretically, are unbounded at the upper end, daily stream-flow data exhibits

---

[1] M. B Fiering, op. cit.

considerable skewness. Consequently, a suitable transformation must be made of parameters μ and σ before Eq. (7.5.1) can be applied to the generation of synthetic daily flows. A logarithmic transformation has been used for this purpose. The mean and standard deviation of the logarithms are

$$\mu_{\log} = \log\left[\mu\left(1 + \frac{\sigma^2}{\mu^2}\right)^{-1/2}\right] \quad (7.5.2)$$

and

$$\sigma_{\log} = \left[\frac{\log(1 + \sigma^2/\mu^2)}{2.31}\right]^{1/2} \quad (7.5.3)$$

where μ, $\mu_{\log}$ = means of untransformed and transformed data, respectively

σ, $\sigma_{\log}$ = standard deviations of untransformed and transformed data, respectively

Seasonal values of $\mu_{\log}$, $\sigma_{\log}$, and ρ are required to describe a yearly cyclic pattern. As successive years of daily flows are generated using Eq. (7.5.1), the January values are used for the first 31 d, the February values for the next 28 d, and simiiarly until the last 31 d are computed using the December values.

Solution procedure 7.5.1 One of many techniques that can be used for predicting low flows is as follows: The model described in Eq. (7.5.1) is used to generate m replicate sets of N years of daily flows. For each set, the annual n-d low flow $q_{i,j}$ is extracted from each year of the set. See Table 7.5.1. Then the minimum value of $q_{i,j}$ in each set is identified. The probability distribution of these minimum values $q_{k,j}$ is identified and analyzed by plotting the values on the appropriate probability paper.

The n-d period selected will depend upon the objective of the analysis. For stream-pollution work, however, 7-d and 15-d low flows are of usual interest. The record length selected N again depends upon the particular goal of the analysis. For stream-pollution purposes this may vary from 5 to 25 years.

Table 7.5.1 TABULATED ANNUAL n-DAY LOW FLOWS

Year	Set				
	1	2	3	4	...m
1	$q_{1,1}$	$q_{1,2}$	$q_{1,3}$	$q_{1,4}$	$q_{1,m}$
2	$q_{2,1}$	$q_{2,2}$	$q_{2,3}$	$q_{2,4}$	$q_{2,m}$
3	$q_{3,1}$	$q_{3,2}$	$q_{3,3}$	$q_{3,4}$	$q_{3,m}$
⋮	⋮	⋮	⋮	⋮	⋮
N	$q_{N,1}$	$q_{N,2}$	$q_{N,3}$	$q_{N,4}$	$q_{N,m}$
Minimum value	$q_{k,1}$	$q_{k,2}$	$q_{k,3}$	$q_{k,4}$	$q_{k,m}$

Algorithms 7.5.1 and 7.5.2 can be used to generate synthetic streamflow traces, which in turn can be analyzed for the lowest n-d flow for each year and the lowest for the period of record.   ////

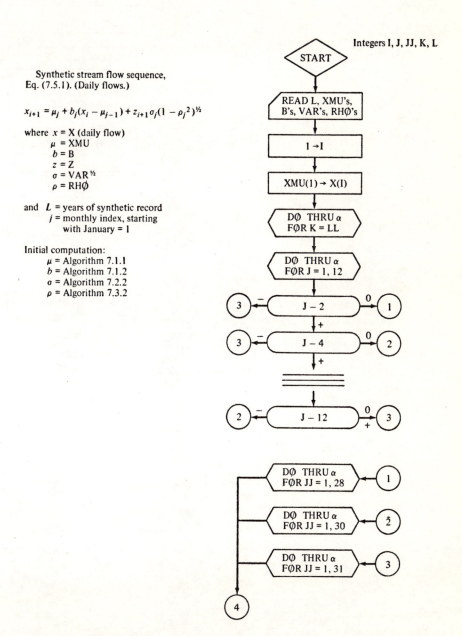

Synthetic stream flow sequence, Eq. (7.5.1). (Daily flows.)

$$x_{i+1} = \mu_j + b_j(x_i - \mu_{j-1}) + z_{i+1}\sigma_j(1 - \rho_j^2)^{1/2}$$

where $x = $ X (daily flow)
 $\mu = $ XMU
 $b = $ B
 $z = $ Z
 $\sigma = $ VAR$^{1/2}$
 $\rho = $ RHØ

and $L = $ years of synthetic record
 $j = $ monthly index, starting with January = 1

Initial computation:
 $\mu = $ Algorithm 7.1.1
 $b = $ Algorithm 7.1.2
 $\sigma = $ Algorithm 7.2.2
 $\rho = $ Algorithm 7.3.2

ALGORITHM 7.5.1 (*continued overleaf*)

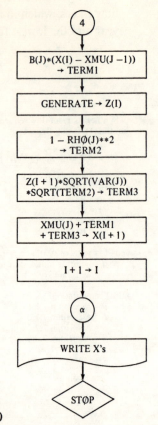

ALGORITHM 7.5.1 (*continued*)

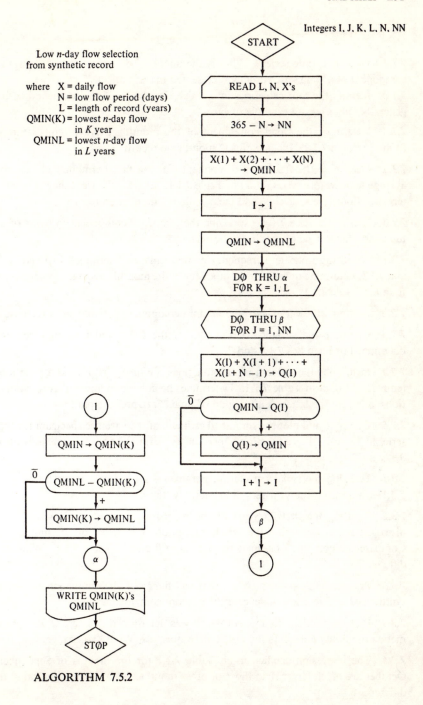

ALGORITHM 7.5.2

EXERCISES

7.1.1 What is a time series? What is the anatomy of a time series? How is each component of a time series usually represented mathematically?

7.1.2 Identify the four categories into which time series of environmental derivation generally fall. Give an example of each.

7.1.3 Determine the trend component of the monthly average temperature series found in Table 7.1.1. Include all 432 months on record.

7.1.4 Determine the frequency component (for the first harmonic) of the monthly average temperature series found in Table 7.1.1. Express the frequency component in terms of Fourier coefficients, and include all 432 months on record.

7.1.5 Repeat Exercise 7.1.4, expressing the frequency component in terms of amplitude and phase angle of the first harmonic.

7.2.1 Using the trend and frequency components as determined in Exercises 7.1.3 and 7.1.5, determine the random component of the monthly average temperature series found in Table 7.1.1.

7.2.2 Compute the variance of the random component determined in Exercise 7.2.1.

7.2.3 Compute the variance of the sum of the trend and frequency components determined in Exercises 7.1.3 and 7.1.5.

7.2.4 Determine the autocovariance functions for time lags 0 to 30 for the monthly temperature data in Table 7.1.1. Determine the corresponding autocorrelation functions, and plot the latter as a function of time. Interpret the plot.

7.2.5 Compute and plot the smoothed estimate of the variance spectrum for the time series in Table 7.1.1 as a function of periodicity. Construct the estimate over the range $0 \leq \tau \leq 30$.

7.3.1 Utilizing the serial correlation parameters of monthly flows found in Table 7.3.1, synthesize a 100-year record of monthly flows.

7.3.2 Treating the first 25 years of the record synthesized in Exercise 7.3.1 as a historical record, compute the serial correlation parameters of the monthly flows. Compare the computed parameters with those found in Table 7.3.1. Why are they different?

7.3.3 Prepare a computer program that will filter out all components of the record synthesized in Exercise 7.3.2 except the random component.

7.3.4 Plot the random component which was determined in Exercise 7.3.3 on arithmetic probability paper. Is the distribution normal? Would you expect it to be?

7.3.5 The regression coefficients in Table 7.3.1 for the months of September and October are much larger than those of other months. What is the significance of this difference?

7.4.1 Using the first 4 years of the 100-year record of monthly flows synthesized in

Exercise 7.3.1 and the monthly percentages of the annual draft tabulated in column 4 of Table 7.4.1, determine without the assistance of a computer the storage requirements of a reservoir designed to meet an annual draft of 90 percent of the average inflow.

7.4.2 Using the complete 100-year record of monthly flows synthesized in Exercise 7.3.1 and the monthly percentages of the annual draft tabulated in column 4 of Table 7.4.1, determine the storage requirements of a reservoir designed to meet an annual draft of 90 percent of the average inflow. Compare the answer with that of Exercise 7.4.1.

7.4.3 Repeat Exercise 7.4.2 to determine the storage requirements of a reservoir designed to meet annual drafts of 60, 70, 80, and 100 percent. With this information construct a plot similar to that in Fig. 7.4.1.

7.4.4 Repeat Exercise 7.4.2 for 20 complete, 100-year records, each synthesized using the serial correlation parameters found in Table 7.3.1. Plot results on arithmetic probability paper. Do the annual draft values have a normal distribution?

7.4.5 From the information gained from Exercise 7.4.4 determine the appropriate mean and standard deviation of the annual draft values.

7.5.1 Assume that the regression coefficients and correlation coefficients in Table 7.3.1 are serial correlation parameters of daily flows (in cubic feet per second) at a point along a stream. Assume further that the values of the mean flow and standard deviation in that table are 100 times larger than the mean daily flow and the corresponding standard deviation (also in cubic feet per second). Synthesize ten 5-year daily stream-flow records with Eq. (7.5.1) using the logarithmic transforms of the daily flow means and standard deviations.

7.5.2 Extract from each of the daily stream-flow records synthesized in Exercise 7.5.1 the annual 7-d low flow.

7.5.3 Plot the annual 7-d low-flow values determined in Exercise 7.5.2 on the appropriate probability paper, and determine the mean and standard deviation.

8
MANAGEMENT SYSTEMS

8.1 WATER–QUALITY MANAGEMENT

Water-quality management, in the sense that it will be discussed in the present chapter, refers to the control of water quality in a surface water system to achieve a desired water-use goal in an optimal fashion. Such control can be exercised either through a set of operational procedures such as stream-flow regulation and the regulation of waste discharge or by the construction of permanent waste treatment facilities. Water-use goals are arrived at through a decision-making process which takes into account the needs for water by municipalities, industry, agriculture, and the general public as they relate to the use of water for fishing and other recreation. Once water-use goals are established, water-quality criteria are selected which reflect these goals.

The difference between water-quality goals and stream standards should be noted. Goals are used as targets in a program for *improving* water quality. Standards, on the other hand, are minimum limits imposed for the purpose of *preventing further degradation* of water quality.

Water-quality criteria selected to attain a particular set of water-use goals may be stated in terms of maximum, allowable concentration levels of several water-quality parameters. However, the minimum, allowable concentration of dissolved oxygen is one of the most important criteria. This is because fairly high levels of dissolved oxygen (i.e., greater than 4 mg/l) are required to maintain a normal, balanced community of aquatic organisms whereas low concentrations of dissolved oxygen (i.e., less than 1.0 mg/l) permit the establishment of conditions which limit fish life and promote the evolution of odors and the discoloration of the water. Furthermore, there is a direct cause-and-effect relationship between the organic wastes discharged into the water system and the dissolved oxygen therein.

The control of water quality in an optimal fashion implies control achieved at least cost. However, the burden of cost incurred in attaining water-use goals varies with the method used in allocating costs.

Three methods of cost allocation have been identified;[1] The *uniform-treatment method* requires that an identical percentage of the *raw BOD load* be removed from each source before discharge into the surface water system. The term "raw BOD load" refers to the BOD of the waste before any removal by treatment. Uniform treatment is the method commonly used in present water-quality management programs. The method has the advantage of being convenient to administer. However, this advantage is outweighed by the disadvantages of being economically inefficient and inequitable. Many sources of waste discharge are required to increase the degree of treatment at noncritical locations in the system simply because it is necessary to do so at critical locations. Although each source is required to remove the same percent of raw BOD load, no allowance is made for differences in treatment costs.

The *cost-minimization method* selects the level of BOD removal to be attained at each source to achieve the desired dissolved-oxygen level at a minimum cost to the region. No unnecessary treatment is required, and only those removals are called for which produce an increment of improved quality at the lowest cost. The method is equitable in the sense that a source which does not lower the dissolved oxygen incurs no cost. Yet in another sense the method is inequitable. Suppose that two industries are located next to each other along a river and are discharging the same type of waste. The marginal effects of the wastes on the river might be expected to be the same. However, if the treatment cost at one industry is lower than that at the other, the industry with the lower treatment cost will be required to attain a higher degree of BOD removal than will the one with the high treatment cost. In fact, the latter may not be required to provide any additional removal. Obviously the cost-minimization method would be difficult to implement administratively because of the unequal cost burden.

[1] E. T. Smith and A. R. Morris, Systems Analysis for Optimal Water Quality Management, *J. WPCF*, vol. 41, no. 9, pp. 1635–1646, 1969.

The *zoned-optimization method* combines elements of the uniform-treatment and cost-minimization methods. The surface water system is divided into geographic zones. The treatment level in each zone is chosen so as to attain the dissolved-oxygen goal at minimum overall cost. If only one zone is established, the method becomes one of uniform treatment. If zones are established for each source, the method becomes one of cost minimization. The zoned-optimization method results in a measure of equity in that waste sources located near each other are treated similarly. Zones do not have to be established on a geographic basis. They can be established on a categorical basis as well. For instance, all paper mills can be placed in one category, municipalities in another, and so forth. The zoned-optimization method has the advantage of being almost as easy to administer as the uniform-treatment method, yet the requirement of similar treatment for sources located near each other (or located in the same category) tends to minimize the objections of the individual dischargers regarding their treatment requirements as compared with those of their neighbors.

Certain constraints are imposed in all three methods. The dissolved-oxygen concentration in any part of the system is not permitted to be lowered from the level at which it exists prior to implementation of the control program, even if the level is greater than that imposed by the water-quality criteria. Furthermore, no source may discharge more BOD than it had been discharging prior to program implementation. When a particular treatment level is established, no source treating its waste to a higher level may lower the degree of treatment.

Basic to the development of any method of water-quality control and cost allocation is the requirement for a quantitative formulation of the cause-and-effect relationships existing in the surface water system. The finite-volume transport model discussed in Secs. 5.1 and 5.4 provides such a formulation. The finite-volume model relates through the steady-state transfer function ϕ_{ij} the dissolved-oxygen response in any volume to the BOD input in any other volume. The transfer function describes mathematically the effects of the system parameters on transport through the system. The collection of steady-state transfer functions describing the system constitute the physical model with which a cost-allocation method can be developed.

Once a physical model for the system has been developed, the cost-minimization method can be formulated as a problem in linear programming. Such a formulation is as follows:[1] Minimize

$$\sum_{k=1}^{m} B_k W_k \qquad (8.1.1)$$

---

[1] R. V. Thomann and D. H. Marks, Results from a Systems Analysis Approach to the Optimum Control of Estuarine Water Quality, *Proc. Third Int. Conf. Water Pollut. Res.*, sec. III, paper 2, Munich, September 1966.

subject to the constraints

$$\sum_{j=1}^{n} \phi_{ij} W_j \geq \Delta c_i \qquad i = 1, 2, 3, \ldots, n \qquad (8.1.2)$$

in which
$$W_j = \sum_{k \in j} W_k \qquad (8.1.3)$$

and
$$0 \leq W_k \leq U_k \qquad k = 1, 2, \ldots, m \qquad (8.1.4)$$

where B_k = unit cost of effluent modification at source k, dollars/(kg/d)

W_k = effluent modification at source k, kg/d

ϕ_{ij} = steady-state transfer in volume i per unit input of waste-load discharge in volume j, (mg/l)/(kg/d)

Δc_i = increase in concentration of dissolved oxygen sought in volume i, mg/l

U_k = upper limit to amount of waste to be removed at source k, kg/d

Equation (8.1.1) is the sum, which is to be minimized, of the costs incurred in waste removal at each of the k sources. Equation (8.1.2) is the physical model introducing the cause-and-effect relationships that characterize the system. Equation (8.1.4) specifies that no waste source will treat less than at its present level of treatment and imposes an upper bound to the degree to which treatment will be required. The upper bound may reflect the limitations of available technology.

Equation (8.1.1) is referred to as the *objective function*. A basic requirement in the formulation of a problem to be solved by linear programming is that the objective function be linear. Unfortunately, the relationship between the degree of waste treatment and treatment cost is not linear but a convex curve. The relationship, however, can be assumed to be linear within certain treatment ranges, and the problem of nonlinearity can be dealt with by approximating the curve relating cost with treatment level with a series of linear segments. Each segment can then be considered as a separate source of waste discharge. In most instances, the cost curves need be approximated by no more than two or three segments.

The dissolved-oxygen concentration, being the key water-quality parameter of concern here, establishes the treatment requirements at the different sources. Consequently, the concentration levels existing in the component finite volumes of the system are determined for the critical period of the year (i.e., that period when the flows are the lowest and temperature is the highest). The differences between these levels and that concentration established as the goal to be attained determine the improvement that must be sought. These differences provide the values for Δc_i in Eq. (8.1.2).

The uniform-treatment method employs Eqs. (8.1.2) to (8.1.4) along with the relationship

$$W_k = \begin{cases} (R - P_k)T_k & \text{for } P_k < R \\ 0 & \text{for } P_k \geq R \end{cases} \quad (8.1.5)$$

where R = uniform percent removal

P_k = current percent removal at source k

T_k = BOD of waste at source k prior to any treatment, kg/d

A search technique is used in which a low treatment level R is selected and its resulting DO levels determined. If the DO levels thus calculated do not satisfy the goal, the treatment level is increased by small increments until the goal is reached in all parts of the system. The objective function, Eq. (8.1.1) is not involved in the search technique.

The zoned-optimization method does not lend itself to a linear-programming formulation. Nonlinearities occur in the constraint functions as a result of discontinuities appearing at points where new sources enter the problem. A method of solution is found elsewhere.[1]

Solution procedure 8.1.1 The cost-minimization method requires the following basic information:

1 The steady-state transfer-function matrix
2 The differences between the existing DO values in each finite volume during the critical period of the year and the DO concentration goal to be attained
3 The BOD loads currently discharged at each source
4 The slope and upper breakpoint of each segment of the cost curve for the BOD removal at each source

The steady-state transfer-function matrix $\mathbf{X}$ is computed by the matrix manipulations discussed in Solution Procedure 5.4.2. The transfer-function matrix is also called the *unit-loading matrix*. The values of the elements of this matrix ϕ_{ij}, which are dependent solely on the system's parameters, are determined from the relationships given in Sec. 5.4.

---

[1] Delaware Estuary Comprehensive Study, "Preliminary Report and Findings," U.S. Dept. of Interior, Federal Water Pollution Control Administration, Philadelphia, 1966.

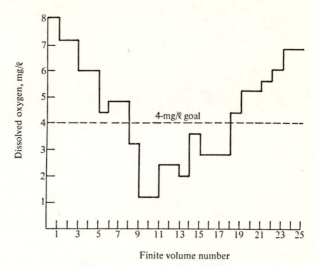

FIGURE 8.1.1
Average dissolved-oxygen concentration in an estuary during the critical period of the year.

The existing DO values in each finite volume during the critical period of the year are determined from a field survey. A plot of such information is shown in Fig. 8.1.1. The difference between the existing values and that value selected to be attained (4 mg/l in Fig. 8.1.1) gives the vector of concentration values to be used in the constraint equations. Since the dissolved-oxygen concentration in any part of the system is not permitted by the method to be lowered from the existing level, all negative values of the concentration difference are converted to 0.

The BOD loads currently being discharged at each source, as well as the cost curves for removing additional BOD, usually must be obtained by an inventory and survey of the sources. For illustrative purposes, a cost curve for one hypothetical source discharging into an estuarine system is shown in Fig. 8.1.2. The ultimate BOD of the waste prior to any treatment is 10,000 kg/d. Current treatment operations reduce the BOD by 30 percent, leaving 70 percent to be discharged into the estuary. It is assumed that it would be reasonable to require up to 90 percent removal for this particular waste. Consequently, the cost curve extending from the current level of treatment (30 percent) to the reasonable maximum treatment (90 percent) is nonlinear.

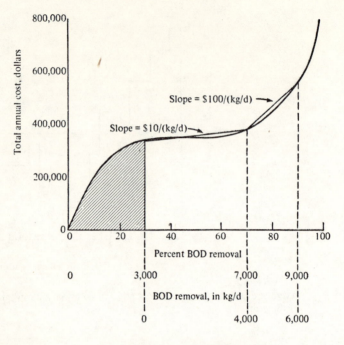

FIGURE 8.1.2
Cost curve for a hypothetical source discharging into an estuarine system.

To linearize the problem, the cost curve is approximated by two linear segments between 30 and 90 percent removal. The first segment has a slope of $10/(kg/d)$ and has an end point at 70 percent removal. The second segment extends from 70 percent to 90 percent and has a slope of $100/(kg/d)$. In setting up the problem, the source is treated as two separate sources, each with a linear cost curve given by a segment of the total cost curve. For example, let the waste source be numbered 12 and 13. The unit costs of effluent modification for this source are $B_{12} = \$10/(kg/d)$ and $B_{13} = \$100/(kg/d)$. The upper limits to the amount of waste to be removed are $U_{12} = 7{,}000 - 3{,}000 = 4{,}000$ kg/d and $U_{13} = 9{,}000 - 7{,}000 = 2{,}000$ kg/d. There is no danger of any segment being taken out of order since each segment of the curve has a higher unit cost than the one preceding. The linear-programming method in looking for the least cost will choose the lowest unit cost available to it.

Each source is treated in a similar manner. Equations (8.1.1) to

(8.1.4) are then provided with the proper values of ϕ_{ij}, Δc_i, B_k, and U_k and solved for the decision variables W_k. ////

Solution procedure 8.1.2 The uniform-treatment method does not require a linear-programming solution. Instead a search technique is used in which a low treatment level R is selected and its resulting DO levels determined. If the DO levels thus calculated do not satisfy the goal set for the system, the treatment level is increased by small increments (say 5 percent) until the goal is reached. Algorithm 8.1.1 illustrates the procedure. Note should be made of the fact that two indices are used (j and k) to reference the values of W_{jk}, P_{jk}, T_{jk}, and U_{jk} in Eqs. (8.1.2) to (8.1.5). These indices identify the finite volume and the cost-curve segment for which the variables apply. Equation (8.1.1), although not contributing to the determination of the decision variables W_{jk}, does give the total cost for the treatment required by the uniform-treatment method. ////

Solution procedure 8.1.3 Equations (8.1.1) and (8.1.2) constitute a problem in linear programming. In its general application, linear programming deals with the problem of allocating limited resources among competing activities in an optimal manner. In the context of Eqs. (8.1.1) and (8.1.2), dissolved oxygen in the finite volumes is the limited resource, and the treatment plants are the competing activities. The mathematical statement of the linear-programming problem consists of two types of functions: The function which is being minimized (or maximized) is called the *objective function*. The functions which establish restrictions to the solution of the problem are commonly called *constraints*. The latter may be expressed in terms of equalities or inequalities. All the mathematical functions—objective and constraint—must be linear.

A typical statement of a linear-programming model takes the form of the following: Maximize

$$C = c_1 x_1 + c_2 x_2 + \cdots + c_n x_n \quad (8.1.6)$$

subject to the constraints

$$\begin{aligned}
a_{11} x_1 + a_{12} x_2 + \cdots + a_{1n} x_n &\le b_1 \\
a_{21} x_1 + a_{22} x_2 + \cdots + a_{2n} x_n &\le b_2 \\
&\cdots \cdots \cdots \cdots \cdots \cdots \cdots \cdots \cdots \\
a_{m1} x_1 + a_{m2} x_2 + \cdots + a_{mn} x_n &\le b_m
\end{aligned} \quad (8.1.7)$$

$$x_1, x_2, \ldots, x_n \ge 0$$

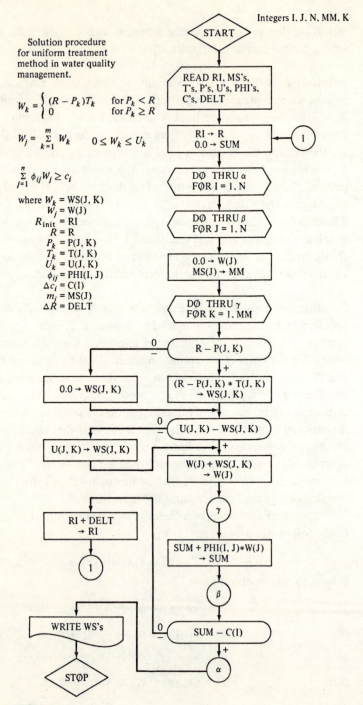

ALGORITHM 8.1.1

where C = variable chosen for overall measure of effectiveness, i.e., variable to be maximized (or minimized)

c_j = increase in effectiveness resulting from each unit increase in x_j

x_j = level of activity j

b_i = limit to availability of resource i

a_{ij} = amount of resource i consumed by each unit of activity j

m, n = number of available resources and competing activities, respectively

The terms a_{ij}, b_i, and c_j are known constants, and the term x_j is the variable, called the *decision variable*. The nonnegativity constraints $(x_j \geq 0)$ rule out negative activity levels, a requirement dictated by the real-world situation modeled by the linear-programming functions. As indicated before, the objective function can be minimized, if that is the objective, and the inequalities can be greater than or equal to b_i or equal to b_i instead of less than or equal to b_i.

A graphic solution of a simple, two-dimensional linear-programming problem serves to illustrate the basic nature of the solution to the linear-programming problem. Consider the following problem: Maximize

$$C = 6x_1 - 5x_2 \qquad (8.1.8)$$

subject to the constraints

$$\begin{aligned} x_1 &\leq 3 \\ x_2 &\leq 4 \\ 2x_1 - x_2 &\leq 8 \\ x_1, x_2 &\geq 0 \end{aligned} \qquad (8.1.9)$$

The problem is solved graphically in Fig. 8.1.3. The area located within the plotted constraints is indicated by the arrows. The straight lines labeled *a*, *b*, and *c* are *possible solutions*. Straight lines *a* and *b* are *feasible solutions* because they pass through or are tangent to the area within the plotted constraints. Straight line *b* is the *optimum feasible solution* because it is the feasible solution with the maximum value of *C*.

The graphic method cannot be used for more than two or three variables. For this reason linear-programming problems are commonly solved

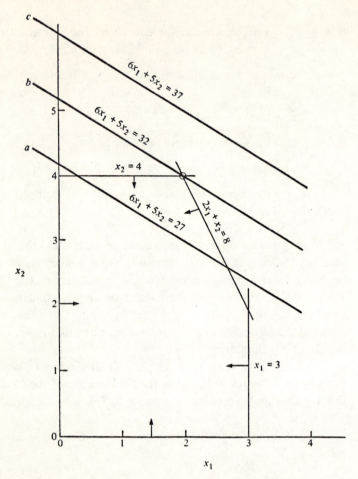

FIGURE 8.1.3
Graphic solution of a two-dimensional linear-programming problem.

by the *simplex method*. This method is an algebraic procedure which progressively approaches the optimum feasible solution through an iterative process until optimality is finally reached.

Many user-oriented computer programs are available for the solution of linear-programming problems by the simplex method. The programs reduce the computer-programming procedure to a matter of following a simple set of instructions. MPS/360 is such a program.[1] ////

---

[1] IBM Mathematical Programming System/360 Version 2, Linear and Separable Programming, *User's Manual* H20-0476-2.

8.2 SOLID-WASTE MANAGEMENT

The term *solid-waste management* is used to describe the collection and disposal of solid waste materials. Implied in the term is that such activities are optimized with respect to economics, the public health, and the environment. Historically, however, solid-waste management has been relegated to the lowest levels of responsibility. As a result, most practices are the result of cut-and-try techniques handed down from one generation to another.

Both the collection activity and disposal are influenced by the quantity of solid waste generated. For municipal wastes, factors such as population density, social-economic status of the population, climate, and frequency of collection service are important factors to be considered. Consequently, average values of solid-waste generation have little significance, and for any particular city the quantity currently being generated is best determined through surveys. One such survey is reported in detail elsewhere.[1]

A model can be developed on a regional basis for forecasting solid-waste quantities. Such a model incorporates the input-output economic description of the region as well as the predictive description of the household sector. The generation of wastes by the industrial sector is given by Eq. (6.6.9):

$$\mathbf{X}_W = \mathbf{W}\mathbf{Y}_F \qquad (6.6.9)$$

where $\mathbf{X}_W = n \times 1$ matrix of future solid-waste generation by each sector

$\mathbf{Y}_F = n \times 1$ matrix of future final demands

$\mathbf{W} = n \times n$ waste-generation matrix

and where

$$\mathbf{W} = {}_T(\mathbf{I} - \mathbf{A})^{-1}\mathbf{L} \qquad (6.6.8)$$

where $\mathbf{L} = n \times n$ diagonal matrix with solid-waste coefficients l_i appearing in main diagonal and with 0 in all off-diagonal locations

$\mathbf{A} = n \times n$ matrix of technical coefficients defined by Eq. (6.6.3)

The increase in the final-demand vector for future years can be calculated from

$$\mathbf{Y}_{F(t+n)} = \frac{\mathbf{Y}_{F(t)}}{(1 + p)^n} \qquad (8.2.1)$$

where $\mathbf{Y}_{F(t)}, \mathbf{Y}_{F(t+n)}$ = final-demand vectors for years t and $t + n$, respectively

p = percent increase in gross regional product per year

---

[1] J. E. Quon, Refuse Quantities and Frequency of Service, *J. Sanit. Eng. Div., ASCE*. vol. 94, no. SA2, pp. 403–420, April 1968.

The quantity of solid waste generated by the households in the region can be expressed in terms of the type of housing:

$$x_h = h_s n_s + h_m n_m \quad (8.2.2)$$

where x_h = total waste generated by household sector of region

n_s, n_m = number of single-family units and multiple-family units, respectively, in region

h_s, h_m = household-waste coefficients for single-family and multiple-family housing units, respectively

The number of single- and multiple-housing units must be determined from surveys and from housing forecasts for the region.

The total waste estimated to be generated for any one year will be

$$x_T = x_h + \sum_{i=1}^{n} x_{Wi} \quad (8.2.3)$$

where x_{Wi} = elements of vector $\mathbf{X}_W$

Solid wastes may be classified in a number of different ways. One system of classification includes garbage, commercial swill (semiliquid food wastes from restaurants), domestic refuse, commercial refuse, rubbish, and ashes. However, it is generally most convenient to classify solid waste as to the activity by which it is generated (domestic, agricultural, industrial, etc.) or in terms that have significance as far as the prevailing (or planned) method of disposal is concerned. For instance, in an area serviced by a refuse incinerator it might be desirable to classify the domestic and commercial wastes as garbage, combustible refuse, and noncombustible refuse. In the broadest sense, the term "solid waste" includes all solid materials generated by man and his activities for which man has no further need. Sludge from waste treatment plants, ashes from incinerators, junked car bodies, cans, bottles, etc., all can be considered as solid wastes.

Disposal Systems

Disposal systems are of two types: disposal without provision for salvage or energy recovery and disposal with partial recovery of salvageable material and/or energy.[1]

The first type, those disposal systems without provision for salvage or energy recovery, includes open dumps, sanitary landfills, and incineration, both

---

[1] H. F. Ludwig and R. J. Black, Report on the Solid Waste Problem, *J. Sanit. Eng. Div., ASCE*, vol. 94, no. SA2, pp. 355-370, April 1968.

central and on-site. Open dumps, although in general use throughout the country, give rise to a variety of health and environmental problems. Such problems include rodent infestations, mosquitoes, odors, and fires. At relatively little additional expense, open dumps can be replaced by sanitary landfills.

Sanitary landfills are engineered operations in which solid wastes are covered with compacted earth. This type of disposal is relatively inexpensive and, if managed properly, does not result in the problems generated by open dumps. When completed, a landfill site can be used for recreation, parking, or light construction. Details for the design, construction, and operations of a landfill operation are found elsewhere.[1]

Many cities in the United States use incinerators to reduce the volume of solid wastes. Noncombustible wastes are either collected separately or passed through the incinerator along with other refuse. Magnetic devices may be used to separate ferrous metals from the ashes for salvage. Incineration reduces the volume of wastes to a fraction of the original volume. Provision must be made for disposal of incinerator ash. To avoid air-pollution problems, incinerators must be equipped with electrostatic precipitators. Good equipment and intelligent operation are essential to satisfactory incineration.

Efficient incinerators used in the home, apartment buildings, and for commercial establishments would reduce the solid-waste handling problem. However, present designs of on-site incinerators are generally inadequate and contribute significantly to air-pollution problems. Open burning of refuse is not allowed in most areas. The principles and practices of incineration are discussed in detail elsewhere.[2]

Those systems in which there is partial recovery of salvageable materials and/or energy include garbage grinders discharging into sewers, swine feeding operations, and composting. When garbage is ground and discharged into the sewer, most of it is treated anaerobically in digesters at the waste-water treatment plant. As a result, additional methane is produced by the digesters which can be used as a supplemental or primary source of fuel. Residual solids in the form of digested sludge can be used as a soil conditioner.

At one time several cities disposed of their garbage by feeding it to swine. However, to prevent the spread of vesicular exanthema in hogs and trichinosis in humans, laws were enacted which required that garbage be sterilized by cooking prior to its being used for feeding swine. Because widespread use of household garbage grinders has reduced the amount of garbage and the cost of separate

[1] Berkeley: Sanit. Eng. Res. Lab., University of Calif., An Analysis of Refuse Collection and Sanitary Landfill Disposal, *Tech. Bull.* 8, December 1952.
[2] R. C. Corey, "Principles and Practices of Incineration," Interscience-Wiley, New York, 1969.

garbage collection has become excessive, swine feeding as a method of disposal has declined.

Composting of solid wastes in large operations has been widely practiced in Europe, primarily because agricultural practices in that part of the world have long used compost as fertilizer. Because of the rather limited fertilizer values of compost, the market in the United States has never been substantial. The separation of metals, rags, and glass from refuse is an integral part of the composting operation. The salvage value of these materials is barely sufficient to pay for the cost of separation. A detailed study of the composting process is reported elsewhere.[1]

Collection Systems

Approximately 80 percent of the expenditures for solid-waste management in the United States support the collection activity. Consequently, any significant reductions in the cost of solid-waste management must be gained primarily by improving the efficiency of collection.

The collection activity consists of two operations: collection and haul. The collection operation consists of removing the refuse from the point of generation. It extends in time from the moment the collection vehicle arrives at the refuse generation area until the vehicle finishes a collection route. The haul operation begins when the collection vehicle departs for the disposal site from the point where the last container of refuse was loaded and includes the time required to reach the disposal site, to unload, and to return to the first container on the next collection route. A rational method for the analysis and design of the collection activity is found elsewhere.[2] Such a method, however, does not include an optimization of the system from the standpoint of assigning collection routes to the disposal sites.

The nature of the haul operation will depend upon the size of the area to be served. The service area may be an entire metropolitan area or a region. On the other hand, it may consist only of a small municipality. Whatever the size, the area can be divided into collection districts. Each collection district can consist of a single collection route or a group of contiguous routes serviced by a single transfer facility. In either case, the collection district can be thought of, and expressed as, a collection point. Furthermore, there may be several locations available for the final disposal of the collected wastes. These locations, or

---

[1] Berkeley: Sanit. Eng. Res. Lab., University of Calif., Reclamation of Municipal Refuse by Composting, *Tech. Bull.* 9, June 1953.

[2] Berkeley: Sanit. Eng. Res. Lab., An Analysis of Refuse Collection and Sanitary Landfill Disposal, op. cit.

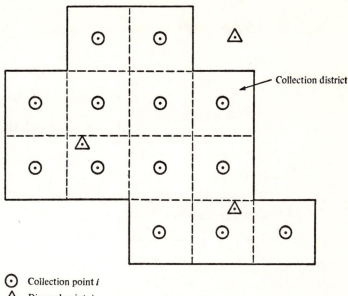

- ⊙ Collection point i
- △ Disposal point j

FIGURE 8.2.1
Representation of a solid-waste-management service area.

disposal points, are sites where an incinerator, sanitary landfill, or some other disposal operation is located. A representation of a solid-waste-management service area is shown in Fig. 8.2.1. In this representation of a highly uncomplicated service area, thirteen collection districts are serviced by three disposal points. The centroids of the collection districts are identified as the collection points.

The cost of the haul operation per unit weight of solid waste can be stated as a function of the distance between the collection point and disposal point:

$$c_{ij} = f(l_{ij}) \quad (8.2.4)$$

where c_{ij} = unit transportation cost of hauling wastes from collection point i to disposal point j

l_{ij} = distance from collection point i to disposal point j

With minimum cost as the objective, the optimal haul arrangements can be determined from the following relationships: Minimize

$$\sum_{i=1}^{m} \sum_{j=1}^{n} c_{ij} x_{ij} \quad (8.2.5)$$

subject to the constraints

$$\sum_{j=1}^{n} x_{ij} = a_i \qquad i = 1, 2, \ldots, m \qquad (8.2.6)$$

$$\sum_{i=1}^{m} x_{ij} = b_j \qquad j = 1, 2, \ldots, n \qquad (8.2.7)$$

$$x_{ij} \geq 0 \qquad \text{for all } i \text{ and } j \qquad (8.2.8)$$

where x_{ij} = quantity of wastes transported from collection point i to disposal point j
a_i = total quantity of waste transported from collection point i
b_j = capacity of disposal point j
m, n = number of collection points and disposal points, respectively

Implicit in the development of the constraints Eqs. (8.2.6) and (8.2.7) is that

$$\sum_{i=1}^{m} a_i = \sum_{j=1}^{n} b_j \qquad (8.2.9)$$

In fact, Eq. (8.2.5) does not have a feasible solution unless Eq. (8.2.9) applies.

Equations (8.2.5) to (8.2.8) constitute a special type of problem in linear programming, called the *transport problem*. The solution can be found by the simplex method. However, the fact that the constraint coefficients equal 0 or 1 permits the use of more efficient methods of solution.

Solution procedure 8.2.1 For the purpose of demonstrating how the terms in Eqs. (8.2.5) to (8.2.7) can be evaluated, let it be assumed that the area depicted in Fig. 8.2.1 is the solid-waste-management service area of concern. The distance matrix, which is given in Table 8.2.1, was determined from maps of the service area. Also given in Table 8.2.1 is the waste-generation vector, the values of which were obtained from waste surveys of representative collection districts. Let it be assumed further that the capacities of disposal points 1 to 3 are limited to 296, 320, and 344 tons/d, respectively.

A transportation-time matrix can be computed by multiplying the elements of the distance matrix by a factor equal to the reciprocal of the average velocity at which the waste is transported to the disposal site. Generally this factor will be on the order of 0.05 h/mi. An empirically derived relationship can then be used to convert the time matrix into a unit-cost matrix. One such relationship is[1]

$$c_{ij} = \frac{\$(2.55 + 0.325S + 5.40M + 0.95)t_{ij}}{60S} \qquad (8.2.10)$$

where c_{ij} = unit transportation costs, dollars/ton

S = size of truck, tons of haul

M = size of crew

t_{ij} = transportation time, h

The cost based on a standard 6-ton truck with a two-man crew calculates out to be

$$c_{ij} = \$0.04514 t_{ij} \qquad (8.2.11)$$

////

[1] F. R. Dair, Time, Crew Size, and Costs, *Solid Wastes Manage. Refuse Removal J.*, vol. 10, no. 8, pp. 6–10, August 1967.

Table 8.2.1 DISTANCE MATRIX AND QUANTITY VECTOR FOR SOLID-WASTE-MANAGEMENT SERVICE AREA SHOWN IN FIG. 8.2.1

Collection point i	Distance to disposal point j, mi			Waste quantities generated at point i, tons/d
	1	2	3	
1	10	8	20	72
2	5	12	18	84
3	17	6	22	84
4	12	4	17	72
5	8	8	14	78
6	5	12	8	60
7	20	5	15	90
8	16	2	11	60
9	12	7	7	108
10	10	12	3	60
11	17	10	6	60
12	15	12	2	78
13	17	18	5	54
				960

8.3 WASTE-WATER REUSE SYSTEMS

Traditionally, the first stage of water resource development has involved a situation in which a primary water supply is available in large excess of the water demand. In such a situation, planning focuses on providing storage and distribution systems in order to regulate the water, both in time and space, to satisfy the overall demand. When primary sources are not sufficient to furnish the quantity required to meet the total demand, then secondary and/or supplemental sources of water must be considered. Possible secondary sources include treated effluents from municipal and industrial waste treatment plants and agricultural return flows. Supplemental sources might be either water imported from other areas or water derived through the desalinization of seawater.

There are two types of reuse: *Recycle* is the reuse of effluent water within a demand sector. For instance, the effluent from the treatment plant of a particular industry might be recycled for use as process water, or irrigation return flows may be desalinated for further agricultural usage. *Sequential reuse* is the reuse of effluent water from one sector by another sector. In the past, water-reuse efforts have been directed toward the improvement of treatment technology so as to provide water for a specific demand sector. However, a more effective plan for reuse incorporates an optimal allocation of water from various origins in which the quality and quantity requirements of each use, or demand, sector can be satisfied at minimum cost. The methodology for achieving such a plan is discussed below.

Table 8.3.1 is a tableau in which are listed a variety of possible sources of supply and demand. The sources of demand which are labeled "destinations" in the tableau are, with the exception of "system outflow," diversion requirements. When the total diversion requirements exceed the primary supply available, secondary and/or supplemental supply sources must be sought. System outflow is simply the difference between the available primary supply and the consumptive use.

The problem of allocating water from various sources of supply (origins) to different categories of demand (destinations) can be solved by a method similar to that used in solving the classic transportation problem discussed in the preceding section. The problem is formulated as follows:[1] Minimize

$$\sum_{i=1}^{m} \sum_{j=1}^{n} c_{ij} x_{ij} \quad (8.3.1)$$

---

[1] A. B. Bishop and D. W. Hendricks, Water Reuse Systems Analysis, *J. Sanit. Eng. Div., ASCE*, vol. 97, no. SA1, pp. 41–57, 1971.

Table 8.3.1 MATRICES OF ELEMENTS USED IN WATER-REUSE SYSTEMS ANALYSIS

		Destinations j						
Origins i	Municipal (1)	Industrial (2)	Agricultural (3)	Recreation (4)	Wildlife (5)	Hydropower (6)	System outflow (7)	Availabilities (8)
Primary sources								
Surface water (1)	x_{11} Δq_{11} c_{11}	x_{12} Δq_{12} c_{12}	· · ·	· · ·	· · ·	· · ·	x_{17} Δq_{17} c_{17}	a_1
Groundwater (2)	x_{21} Δq_{21} c_{21}	x_{22} Δq_{22} c_{22}	· · ·	· · ·	· · ·	· · ·	x_{27} Δq_{27} c_{27}	a_2
Secondary sources								
Municipal effluent (3)	· · ·	· · ·	· · ·	· · ·	· · ·	· · ·	· · ·	·
Industrial effluent (4)	· · ·	· · ·	· · ·	· · ·	· · ·	· · ·	· · ·	·
Agricultural return flow (5)	· · ·	· · ·	· · ·	· · ·	· · ·	· · ·	· · ·	·
Supplemental sources								
Imported water (6)	· · ·	· · ·	· · ·	· · ·	· · ·	· · ·	· · ·	·
Desalinization of seawater (7)	x_{71} Δq_{71} c_{71}	x_{72} Δq_{72} c_{72}	· · ·	· · ·	· · ·	· · ·	x_{77} Δq_{77} c_{77}	a_7
Demand-sector requirements (8)	b_1	b_2	·	·	·	·	b_7	

subject to the constraints

$$\sum_{j=1}^{n} x_{ij} = a_i \qquad i = 1, 2, \ldots, m \qquad (8.3.2)$$

$$\sum_{i=1}^{m} x_{ij} = b_j \qquad j = 1, 2, \ldots, n \qquad (8.3.3)$$

$$x_{ij} \geq 0 \qquad \text{for all } i \text{ and } j \qquad (8.3.4)$$

$$\sum_{i=1}^{m} a_i - \sum_{j=1}^{n} b_j > 0 \qquad (8.3.5)$$

where x_{ij} = quantity of water from origin i allocated to destination j

c_{ij} = cost of treating (if required) and delivering unit quantity of water from origin i to destination j

a_i = quantity of water available from origin i

b_j = quantity of water required by destination j

Since system outflow is not strictly a diversion requirement but rather depends on the consumptive use and losses in the system, Eq. (8.3.5) takes the form of an inequality rather than an equality. System outflow, therefore, is actually a dummy destination where water is not reused, and has an allocation cost of 0. In the case of reuse, effluent quality constraints can be imposed by assigning penalty costs to elements in the system-outflow vector. For instance municipal and industrial waste water must be treated at cost before it can be discharged to the system outflow. Water from other origins, already being of sufficient quality, can be released to the system outflow at no cost.

Table 8.3.2 WATER SUPPLY AND DEMAND CHARACTERISTICS OF AGROURBAN SYSTEM*†

System	Primary supply	Diversion requirement	Effluent or return flow	Consumptive use
Surface water	83			
Groundwater	48			
Jordan River water	270			
Municipal		88	74	14
Industrial		124	87	37
Agricultural		270	103	167
Wildlife refuge		141	65	76
Total	401	623	329	294

System outflow = primary supply − consumptive use and losses
= 401 − 294 = 107

* A. B. Bishop and D. W. Hendricks, Water Reuse Systems Analysis, *J. Sanit. Eng. Div.*, ASCE, vol. 97, no. SA1, pp. 41–57, 1971.
† *Note:* All data in acre-feet per year.

The cost incurred in providing water from any source for any use depends on the water quality of the source, the quality required by the use, and the facilities required to treat, transport, and deliver the water. Therefore, the unit costs c_{ij} will include two components—the cost of treatment and the cost of transportation.

Treatment costs are developed from matrices of *water-quality differentials* Δq_{ij}. The water-quality differential is the difference between the requirement quality with respect to a given parameter and the quality of the available water. A different matrix must be used for each quality parameter of significance. These matrices are used as a basis for determining the degree of treatment necessary to match each source with its possible uses.

Unfortunately, unit treatment costs as well as unit transportation costs will vary with the quantity of water involved. Because the quantities of water involved are unknowns which are to be determined, the relationships employed in allocation should be used in an iterative procedure.

Solution procedure 8.3.1 A case study of an integrated urban and agricultural system located in an arid climate will be considered. The supply and demand characteristics of the system are given in Table 8.3.2. Matrices for the water-quality differentials Δq_{ij} for BOD and total dissolved solids (TDS) are given in Tables 8.3.3 and 8.3.4. The zero and less-than-

Table 8.3.3 MATRIX OF WATER-QUALITY DIFFERENTIALS FOR BOD*

Origin	Destinations					
	Municipal requirement	Industrial requirement	Agricultural requirement	Bird refuge	System outflow	BOD effluent quality, mg/l
Surface water supply	20†	<0	<0	0	0	20
Groundwater supply	0	<0	<0	20	20	0
Jordan River supply	20	<0	<0	0	0	20
Municipal effluent	300	<0	0	280	280	300
Industrial effluent	300	<0	0	280	280	300
Agricultural return flow	5	<0	<0	<0	<0	5
Bird refuge	20	<0	<0	0	0	20
BOD influent requirement, mg/l	0	600	300	20	20	

* A. B. Bishop and D. W. Hendricks, Water Reuse Systems Analysis, *J. Sanit. Eng. Div., ASCE*, vol. 97, no. SA1, pp. 41–57, 1971.
† BOD differential between requirement quality and availability quality, in milligrams per liter.

zero values present in these matrices indicate that the effluent quality from a particular origin equals or is better than the quality requirement of a particular destination.

From unit-cost treatment-plant-capacity data and from information concerning the cost of pumping, transporting, and delivering the water, the matrix of unit costs found in Table 8.3.5 was computed. Since unit costs will vary with the scale of the development, an iterative procedure must be used to ensure that the allocated capacities match the size of the facilities from which the cost data were derived. It is to be noted that the right-hand column of Table 8.3.5 contains the water supply or availabilities at each origin and the bottom row contains the diversion requirements for each use sector. These data were taken from Table 8.3.2.

The minimum cost allocation is shown in Table 8.3.6. Such allocation can be determined either by solving the set of Eqs. (8.3.1) to (8.3.5) by the simplex method or by the transportation-problem algorithm, a special case of the simplex method. For either method, a variety of user-oriented computer programs are available. Figure 8.3.1 is a flow diagram constructed from the data given in Table 8.3.6. ////

Table 8.3.4 MATRIX OF WATER-QUALITY DIFFERENTIALS FOR TDS*

Origin	Destination					
	Municipal requirement	Industrial requirement	Agricultural requirement	Bird refuge	System outflow	TDS effluent quality, mg/l
Surface water supply	<0†	<0	<0	<0	<0	300
Groundwater supply	<0	<0	<0	<0	<0	300
Jordan River supply	300	0	<0	<0	<0	800
Municipal effluent	400	100	0	<0	<0	900
Industrial effluent	400	100	0	<0	<0	900
Agricultural return flow	600	200	100	<0	<0	1,000
Bird refuge	1,000	700	600	<0	0	1,500
TDS influent requirement, mg/l	500	800	900	3,000	1,500	

* A. B. Bishop and D. W. Hendricks, Water Reuse Systems Analysis, *J. Sanit. Eng. Div., ASCE*, vol. 97, no. SA1, pp. 41–57, 1971.
† TDS differentials between requirement quality and availability quality in milligrams per liter.

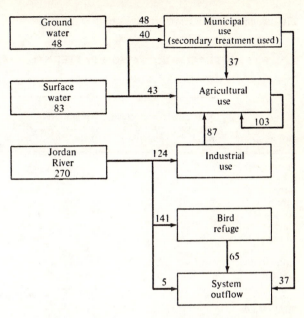

FIGURE 8.3.1
Flow diagram of optimal allocation of water in an integrated urban and agricultural system located in an arid climate. All figures are in thousands of acre-feet.

Table 8.3.5 MATRIX FOR UNIT COSTS*

Origin	Municipal requirement	Industrial requirement	Agricultural requirement	Bird-refuge requirement	System outflow	Availabilities, thousands of acre-ft
Surface water supply	38†	38	5	5	0	83‡
Groundwater supply	23	23	10	10	0	48
Jordan River supply	108	10	5	0	0	270
Municipal effluent	135	56	10	51	46	74
Industrial effluent	115	39	10	29	29	87
Agricultural return flow	108	93	5	5	0	103
Bird-refuge outflow	108	93	93	193	0	65
Requirements, thousands of acre-ft	88‡	124	270	141	107	730

* A. B. Bishop and D. W. Hendricks, Water Reuse Systems Analysis, *J. Sanit. Eng. Div., ASCE*, vol. 97, no. SA1, pp. 41–57, 1971.
† Cost in dollars per acre-foot.
‡ Water quantities in right column and bottom row are from Table 8.3.2.

Table 8.3.6 OPTIMAL ALLOCATION PATTERN*†

| | Destination | | | | | |
Origin	Municipal require-ment	Industrial require-ment	Agricultural require-ment	Bird refuge	System outflow	Supply availabilities, thousands of acre-ft
Surface water supply	40‡		43			83
Groundwater supply	48					48
Jordan River supply		124		141	5	270
Municipal effluent			37		37	74
Industrial effluent			87			87
Agricultural return flow			103			103
Bird refuge					65	65
Demand requirements, thousands of acre-ft	88	124	270	141	107	

* A. B. Bishop and D. W. Hendricks, Water Reuse Systems Analysis, *J. Sanit. Eng. Div., ASCE,* vol. 97, no. SA1, pp. 41–57, 1971.
† *Note:* Total cost = $9,053,000 or $14.50/acre-ft; total diversion = 623,000 acre-ft.
‡ Allocation amount in thousands of acre-feet.

EXERCISES

8.1.1 What is the difference between water-quality goals and stream standards?

8.1.2 What three methods of cost allocation have been proposed for water-quality control? What are the principal characteristics of each?

8.1.3 What is the basic requirement in the formulation of any problem for solution by linear programming?

8.1.4 Exercise 5.4.4 required the determination of the DO-BOD unit-loading matrix and the dissolved-oxygen concentrations for the estuarine system, the characteristics of which are tabulated in Table 5.4.1. The waste discharges are tabulated in the same table. Assume the cost of waste-discharge reduction at all sources to be $10,000/(1,000 lb of BOD reduction) and that the upper limit to the amount of waste to be removed at each source is 90 percent of the present discharge. Using the cost-minimization method of cost allocation, what will be the cost of maintaining a minimum of 4 mg/l of dissolved oxygen in all parts of the system?

8.1.5 Repeat Exercise 8.1.4 using the uniform-treatment method. Assume that no treatment is currently being given to the waste discharges. Compare the cost with that determined by the cost-minimization method.

8.2.1 What activities are defined by the term "solid-waste management," and what conditions are implied by the term?

8.2.2 What factors influence the quantity of solid wastes generated in a municipality?

8.2.3 Describe how a regional-growth model such as is discussed in Sec. 6.6 can be used to forecast solid-waste quantities.

8.2.4 Prepare a detailed discussion of solid-waste disposal systems, emphasizing the advantages and shortcomings of each method of disposal.

8.2.5 What is the relative significance of the collection activity in a solid-waste management system? What operations are included in this activity?

8.2.6 Figure 8.2.1 and Table 8.2.1 define the conditions for a solid-waste collection activity. With minimum cost as the objective, assign the collection districts to the disposal sites. Base the cost on a standard 6-ton truck with a two-man crew.

8.3.1 What do the elements x_{31}, x_{42}, and x_{53} in Table 8.3.1 have in common? What do the elements x_{32}, x_{33}, x_{41}, x_{43}, x_{51}, and x_{52} have in common?

8.3.2 Under what conditions can a solution to a linear-programming problem be sought using the transportation-problem algorithm in place of the simplex method?

8.3.3 What problem is confronted in computing the unit-cost matrix such as the one in Table 8.3.5? How is the difficulty resolved?

8.3.4 Using Eqs. (8.3.1) to (8.3.5) and the information contained in Table 8.3.5, check the values contained in the optimal allocation matrix found in Table 8.3.6.

9
ENGINEERED TRANSPORT SYSTEMS

9.1 PIPE NETWORK ANALYSIS

A definition sketch for the isothermal flow of an incompressible fluid through a pipe, with friction occurring but with no addition of work or heat, is shown in Fig. 9.1.1. Using the notations in the figure, the mechanical-energy equation is written as

$$z_1 + \frac{p_1}{\gamma} + \frac{\bar{V}_1^2}{2g} = z_2 + \frac{p_2}{\gamma} + \frac{\bar{V}_2^2}{2g} + h_f \qquad (9.1.1)$$

where z = elevation head above arbitrary datum. $[L]$
p = pressure. $[FL^{-2}]$
γ = specific weight. $[FL^{-3}]$
$\bar{V}$ = mean velocity of flow. $[Lt^{-1}]$
g = acceleration of gravity. $[Lt^{-2}]$
h_f = permanent head loss due to friction. $[L]$

The head loss h_f results from the conversion of mechanical energy to heat energy by friction. In an isothermal system this term represents a permanent loss to the energy of the system. Pipe-flow computations commonly involve the determination of the term h_f.

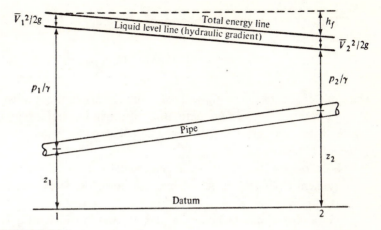

FIGURE 9.1.1
Definition sketch for the flow of an incompressible fluid through a pipe, with friction losses occurring but with no addition of work or heat.

When dealing with large hydraulic systems, it is common practice to determine the frictional head loss from one of several exponential formulas empirically derived from turbulent-flow experiments. Most of the equations relate frictional head loss in feet of water to the hydraulic radius of the conduit, length of conduit, and roughness of the interior surface. Application of the exponential equation has been confined primarily to water systems.

The *Hazen-Williams equation* was derived originally for turbulent flow in both pipes and open channels, but now it is more commonly used for pipe flow. The equation is generally written as

$$\bar{V} = 1.318 C_H R^{0.63} G^{0.54} \qquad (9.1.2)$$

where $\bar{V}$ = velocity of flow, ft/s

R = hydraulic radius, ft

G = hydraulic gradient, ft/ft

C_H = Hazen-Williams coefficient

The hydraulic radius is the ratio of the cross-sectional area of a conduit to its wetted perimeter. For pipes, $R = D/4$, where D is the pipe diameter. The hydraulic gradient G is the ratio of the frictional head loss h_f to the length of pipe L. The slopes of both the total-energy and liquid-level lines in Fig. 9.1.1 can be used as the hydraulic gradient. The liquid-level line is frequently referred to as the *hydraulic grade line.*

Through substitution and rearrangement, Eq. (9.1.2) can be solved for the frictional head loss:

$$h_f = 3.03 \frac{L}{D^{1.17}} \left(\frac{V}{C_H}\right)^{1.85} \quad (9.1.3)$$

The value of the coefficient C_H will vary with the material out of which the pipe is made, with the degree of deterioration of the inside surface, and, to a lesser degree, with pipe size. Values for different conduit materials are listed in Table 9.1.1.

Solution of the Hazen-Williams equation is facilitated by the use of the alignment chart in Sec. A.2. Here the head loss is expressed in feet per 1,000 ft of pipe.

For a pipe of specified diameter, length, and roughness, Eq. (9.1.3) can be reduced to

$$h_f = kQ^{1.85} \quad (9.1.4)$$

where Q = volumetric rate of flow

k = constant

Water distribution systems should be designed to furnish at all times sufficient flow at satisfactory pressures. To ensure greatest reliability in sustaining such an objective, it is desirable to supply a point of demand from at least two directions. Water distribution systems designed as grid systems accomplish this purpose.

There is no direct method of designing water distribution grids. Generally, a grid network is located to coincide with the street network of the community. Pipe sizes are selected initially to provide flow velocities of 3 to 5 ft/s. For the purpose of analysis, the grid is generally reduced to trunk mains; the small service mains are neglected. The reduced grid is then analyzed hydraulically to obtain the correct flow distribution pattern. If the corrected flow pattern

Table 9.1.1 VALUES OF THE HAZEN-WILLIAMS COEFFICIENT C_H FOR VARIOUS KINDS OF PIPE

Kind of pipe	C_H
Cast iron, new	130
Cast iron, old	100
Cast iron, lined with cement or bituminous enamel	140
Steel, welded, lined with cement or bituminous enamel	140
Steel, riveted, coated with coal tar	110
Asbestos cement	140
Concrete	120
Wood stave	120

produces undesirable pressure contours, then appropriate adjustments are made in pipe sizes, and the process of hydraulic analysis is repeated. Flow withdrawals at the junction points of the reduced grid are determined by the demands to be satisfied at these points. Miscellaneous losses for fittings and valves are usually disregarded in water-distribution-system analyses. These losses are insignificant at flow velocities normally found in the system.

The hydraulic analysis of distribution grids is generally performed using the *Hardy Cross method of network analysis*.[1] The method consists of assuming a particular flow distribution pattern throughout the grid and then determining the head loss that would occur between the source of supply and critical points in the system. If the head loss via one path to a particular point is equal to that via another path, the flows are correctly distributed. If the head losses differ, then the assumed distribution is incorrect and must be corrected. Basic to the method are the principles that (1) the total flow reaching any junction of two or more pipes must equal the total flow leaving the junction and (2) the change in pressure between any two points in a closed network is the same by any and all paths connecting the points.

The relationship for determining the flow correction is developed in the following way: Consider a single circuit of several pipes. If the assumed distribution of flow were correct, the head loss around the circuit according to Eq. (9.1.4) would be

$$\sum h_f = \sum k Q^{1.85} = 0 \qquad (9.1.5)$$

If the assumed distribution were not correct, the correction for each pipe would be

$$Q = Q_0 + \Delta Q \qquad (9.1.6)$$

where Q = correct flow

Q_0 = assumed flow

ΔQ = flow correction

Substituting Eq. (9.1.6) into the expression for the head loss in each pipe yields

$$k Q^{1.85} = k(Q_0 - \Delta Q)^{1.85} = k(Q_0^{1.85} + 1.85 Q_0^{0.85} \Delta Q + \cdots) \qquad (9.1.7)$$

If ΔQ is small compared with Q_0, the remaining terms in the expansion can be neglected. Then for all pipes in the circuit

$$\sum k Q^{1.85} = \sum k Q_0^{1.85} + 1.85 \Delta Q \sum k Q_0^{0.85} \qquad (9.1.8)$$

---

[1] Hardy Cross, Analysis of Flow in Network of Conduits or Conductors, *Bull.* 286, University of Ill., November 1936.

By substituting Eq. (9.1.5) into (9.1.8) and rearranging the latter to solve for the flow correction, one has

$$\Delta Q = - \frac{\sum k Q_0^{1.85}}{1.85 \sum k Q_0^{0.85}} = - \frac{\sum h_{f_0}}{1.85 \sum (h_{f_0}/Q_0)} \qquad (9.1.9)$$

where h_{f_0} = frictional head loss computed on basis of Q_0

Care must be exercised regarding the sign of the head loss in both the numerator and denominator. By convention, flow in the clockwise direction around a circuit is considered of one sign and flow in the counterclockwise direction is considered of the other sign. Consequently, the head losses in the numerator are computed with due regard given to the signs. On the other hand, the summation of head losses in the denominator is computed using the absolute values of the individual head losses. The flow correction ΔQ has a single direction for all pipes in the circuit.

A simple pipe network is shown in Fig. 9.1.2. Shown also are the flows assumed to exist in each pipe. These flows are used as the initial values in the Hardy Cross analysis. The computations involved in performing the first trial are presented in Table 9.1.2. The table serves to illustrate the methodology used when making the computations with a slide rule or hand calculator.

Solution procedure 9.1.1 When performed with a slide rule or hand calculator, the Hardy Cross analysis, even of networks as simple as the one shown in Fig. 9.1.2, is laborious and time-consuming. However, the method lends itself readily to solution by digital computers. Such a solution can be obtained using a program similar to the one outlined in Algorithm 9.1.1.

A solution to the network shown in Fig. 9.1.2 was obtained using the Fortran IV language. The results, in terms of flow distributions and head losses, are listed in the last two columns in Fig. 9.1.3. Twenty-eight trials were required to reduce ΔQ to 5 gal/min. When using flow units of gallons per minute and head-loss units of feet per 1,000 ft, the Hazen-Williams expression for the latter becomes

$$G = \left[\frac{Q}{4.63 C_H D^{2.63}} \right]^{1.85} \qquad (9.1.10)$$

The final head losses were used in Fig. 9.1.3 to compute the elevation of the total-energy line at each pipe junction. These elevations were

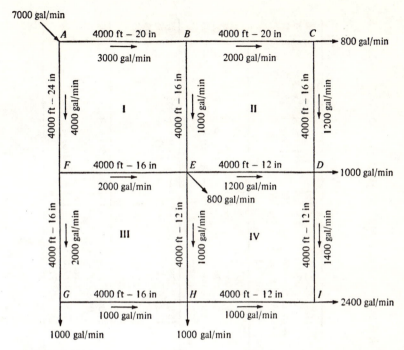

FIGURE 9.1.2
Definition sketch for pipe network analyzed by the Hardy Cross method.

referenced to an elevation of the total-energy line of 100.00 ft at point A, the inflow point to the network. The elevation of the total-energy line at each junction was computed by adding the head losses in all the pipes in a single path from point A to the junction. Within the round-off error and the error introduced by the finite value of ΔQ still remaining, the head losses from point A to a particular junction will be the same regardless of the path followed. The difference between the elevation of the total-energy line at point A and the sum of the head losses in a path to the junction yields the elevation of the total-energy line at that junction.

The elevations of the total-energy line at the junctions were used to construct 10-ft contours over the sketch of the network in Fig. 9.1.3. The spacing between the contours indicates proportionately high head losses in the lower right-hand side of the network. The designer should investigate the substitution of larger pipes for those initially selected between H and I, and between D and I.   ////

Table 9.1.2 COMPUTATIONS INVOLVED IN THE HARDY CROSS ANALYSIS OF THE PIPE NETWORK SHOWN IN FIG. 9.1.2

Pipe	D, in	L, ft	Q_1, gal/min	G, ft/1,000 ft	$\sum h_f$	$\sum h_f/Q_1$	ΔQ, gal/min		Q_2, gal/min
AB	20	4,000	3,000	2.8	11.2	0.00373	+240		3,240
BE	16	4,000	1,000	1.0	4.0	0.00400	+240 − (−153) = +393		1,393
EF	16	4,000	−2,000	−3.7	−14.8	0.00740	+240 − (−213) = +453		−1,547
FA	24	4,000	−4,000	−2.0	−8.0	0.00200	+240		−3,760
					$\sum = -7.6$	0.01713			
BC	20	4,000	2,000	1.3	5.2	0.00260	−153		1,847
CD	16	4,000	1,200	10.0	40.0	0.03333	−153		1,047
DE	12	4,000	−1,200	−6.0	−24.0	0.02000	−153 − (−151) = −2		−1,202
EB	16	4,000	−1,000	−1.0	−4.0	0.00400	−153 − (+240) = −393		−1,393
					$\sum = +17.2$	0.05993			
FE	16	4,000	2,000	4.0	16.0	0.00800	−213 − (+240) = −453		1,547
EH	12	4,000	1,000	4.3	17.2	0.01720	−213 − (−151) = −62		938
HG	16	4,000	−1,000	−1.0	−4.0	0.00400	−213		−1,213
GF	16	4,000	−2,000	−3.7	−14.8	0.00740	−213		−2,213
					$\sum = +14.4$	0.03660			
ED	12	4,000	1,200	6.0	24.0	0.02000	−151 − (−153) = +2		1,202
DI	12	4,000	1,400	8.0	32.0	0.02280	−151		1,249
IH	12	4,000	−1,000	−4.3	−17.2	0.01720	−151		−1,151
HE	12	4,000	−1,300	−4.3	−17.2	0.01720	−151 − (−213) = +62		−938
					$\sum = +21.6$	0.07720			

ENGINEERED TRANSPORT SYSTEMS 287

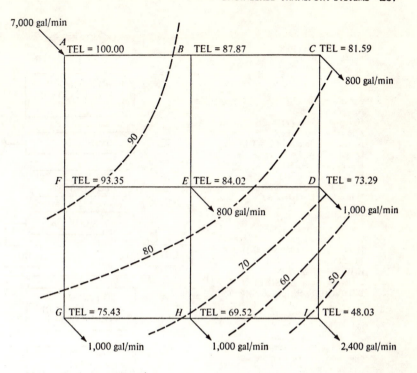

Pipe	Diameter D, in	Length L, ft	Flow Q, gal/min	Head Loss h_f, ft
AB	20	4,000	3,227	12.13
BE	16		963	3.85
EF	16		1,557	9.28
FA	24		3,773	6.65
BC	20		2,264	6.28
CD	16		1,464	8.30
DE	12		789	10.78
EH	12		931	14.50
HG	16		1,217	5.91
GF	16		2,217	17.92
DI	12		1,253	25.26
IH	12		1,147	21.40

FIGURE 9.1.3
Results of the Hardy Cross analysis of the pipe network in Fig. 9.1.2.

288 ENVIRONMENTAL SYSTEMS ENGINEERING

Procedure for pipe
network analysis
using the Hardy Cross
method.

No. of circuits = NC
No. of pipes = NP
Hazen-Williams Coef. = CH
Allowable error = EPS
No. of pipes in circuit L = J(L)
Length of pipe I = PL(I)
Dia. of pipe I = D(I)
Flow in pipe I = Q(I)
*Index identifying pipe k
 in circuit L = ID(L, K)

Σ absolute values
 of corrections = SCØRR
Σ Q(I) = SQ(I)
Head loss in ft / 1000 ft = G(I)
Head loss in pipe I = H(I)
H(I)/Q(I) = HØVQ(I)
Σ H(I) in circuit L = SH(L)
Σ HØVQ(I) in circuit L = SHØVQ(L)

*ID(L, K) serves as an
index which includes the
pipe number and a sign.
The sign is positive for all
pipes not common to more
than one circuit. For pipes
found in two circuits, the
sign is positive for the
first circuit in which the
pipe appears and negative
for the second circuit.

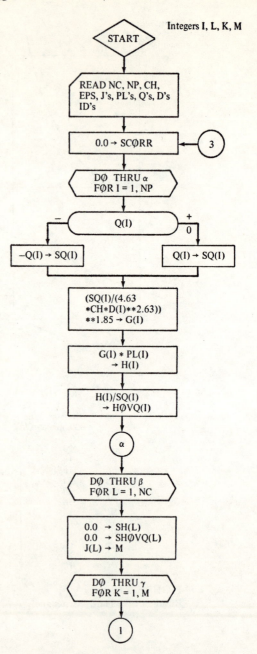

ALGORITHM 9.1.1

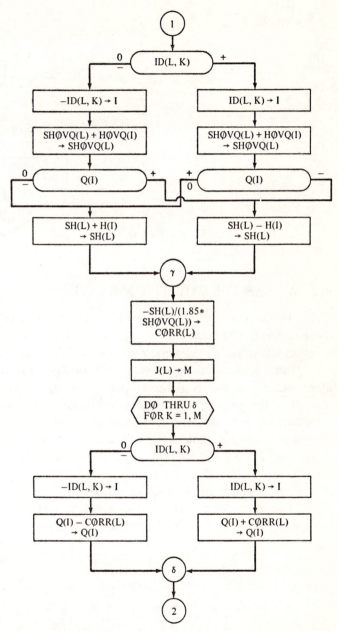

ALGORITHM 9.1.1 (*continued*)   (*continued overleaf*)

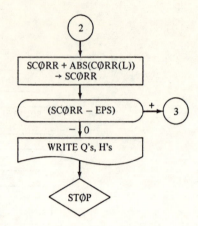

ALGORITHM 9.1.1 (*continued*)

9.2 WATER DISTRIBUTION SYSTEMS

The water distribution system includes, in addition to the pipe network, storage and pumping facilities. The design of all three components requires information, or assumptions, on the rates and pressures at which water must be supplied.

Water consumption rates vary considerably. The average daily rates, along with the normal variations from the average daily rates, are given in Table 9.2.1 for typical communities in the United States. Average consumption rates will vary with climate, cost of water, extent of metering, standard of living, and other factors. Variations from the average daily consumption will depend on the size of the community. The smaller the community, the larger will be the variations.

Pipe networks are normally designed on the basis of an estimated fire demand plus coincident draft. Fire demand can be estimated from a formula recommended by the American Insurance Association.[1] The formula is

$$Q = 1{,}020\sqrt{P}\,(1 - 0.01\sqrt{P}) \qquad (9.2.1)$$

where Q = rate of fire flow, gal/min

P = population, thousands

This expression is considered applicable to communities with populations up to 200,000 persons. For larger populations, additional quantities are required

---

[1] National Board of Fire Underwriters (now the Am. Insur. Assoc.), Standard Schedule for Grading Cities and Towns of the United States, New York, 1956.

because of the possibility of a second fire. The coincident draft is generally equated to the average consumption rate for the maximum day.

Distribution systems should be designed to maintain fairly uniform pressures. Pressures between 40 and 75 lb/in^2 in the main trunk system are normally desirable. However, during periods when a fire demand is being met, pressures down to 20 lb/in^2 may be acceptable. Pressures in excess of 75 lb/in^2 may result in excessive leakage and malfunction of hot-water heaters.

Storage is needed in the distribution system to accommodate rates of demand in excess of the capacity of the water treatment plant and to equalize operating pressures. Emergency reserve storage may also be included in the system. Storage may take the form of surface reservoirs, standpipes, and elevated tanks. The storage units should be located near the center of greatest demand so that optimal pressure control can be derived.

That storage needed to accommodate fluctuating rates of demand is the sum of the equalizing storage and fire reserve requirements. Since water treatment plants are generally designed on the basis of the demand occurring during the maximum day, the fluctuations in demand which need to be accommodated by equalizing storage are those occurring during the maximum day.

Attention is directed to Fig. 9.2.1. Shown there are three diagrams, each constructed by accumulating the flow demand for the maximum day. Shown with each diagram is a straight line representing the accumulated flow resulting

Table 9.2.1 WATER CONSUMPTION RATES FOR UNITED STATES COMMUNITIES

Category	Average daily consumption	
	Normal range, gal/capita/d	Average, gal/capita/d
Domestic	15–70	50
Commercial and industrial	10–100	65
Public	5–20	10
Water unaccounted for	10–40	25
Total	40–230	150

Normal variations from average daily consumption		
Ratio	Normal range	Average
Maximum day/average day	(1.2–2.0):1	1.5:1
Maximum hour/average hour	(2.0–3.0):1	2.5:1

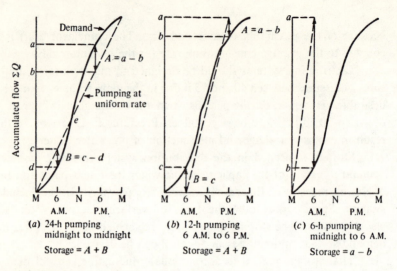

FIGURE 9.2.1
Determination of equalizing storage from accumulative flow diagram for maximum day.

from pumping to storage at a uniform rate. In Fig. 9.2.1a, water is pumped to storage over a 24-h period. The storage requirement for such a pumping schedule is equal to the sum of the vertical distances A and B. A and B are the maximum vertical distances between the demand and pumping curves. At some hour of the morning the accumulated pumping exceeds the accumulated demand by B gal. If this excess is stored, it is used up by the time that the demand curve crosses the pumping curve at point e. At some hour in the afternoon, the accumulated demand exceeds the accumulated pumping by A gal, a quantity that must be drawn from storage and replenished by midnight.

In Fig. 9.2.1b, pumping extends from 6 A.M. to 6 P.M. For such a schedule, the storage requirement is equal to the sum of the vertical distances A and B. Here the distances are measured from the terminal points on the pumping curve to the demand curve. The accumulated demand exceeds the accumulated pumping by B gal by 6 A.M. Such a quantity must be provided by storage. At 6 P.M., the accumulated pumping exceeds the demand by A gal, but this excess is required to furnish water from storage between 6 P.M. and midnight.

When pumping extends from midnight to 6 A.M. as is shown in Fig. 9.2.1c, the storage required will be equal to the vertical distance between the pumping curve at the end of the pumping day and the demand curve. From 6 A.M. until midnight, the entire demand must be met from storage.

The pumping schedule used will depend upon the economics of the particular situation. Twenty-four-h pumpage to storage will allow the use of smaller pumping capacity. However, it is sometimes desirable to plan to pump for shorter time periods. For example, for small communities it may be advantageous to pump only during the working day, or it may be more economic to pump to storage only during those hours when electric-power rates are the lowest.

Fire reserve storage requirements are based primarily upon the size of the community served. The American Insurance Association recommends that fire reserve storage be large enough to supply water for fighting fire for 4, 6, and 8 h in communities of 1,000, 2,000, and 4,000 people, respectively. For communities of 6,000 and larger, 10-h fire reserve storage is recommended.

Emergency storage is used to take care of all water needs when, for some reason, normal supply service is interrupted. The American Insurance Association recommends an emergency storage of 5 d at the maximum daily flow.

Pumping-capacity requirements are based on the *system-head curves*. The latter are derived from a knowledge of the relationship between the flow capacity of the system and the total dynamic head:

$$h = h_s + h_v + h_f \qquad (9.2.2)$$

where h = total dynamic head

h_s = static head

h_v = velocity head

h_f = frictional head loss

The *static head* is equal to the difference in elevation between the water level in the pump suction pool and the elevation to which the water is to be raised. Therefore, the static head will vary as the water levels in the storage tank and pump suction pool fluctuate. For this reason, two system-head curves are considered when selecting pumping equipment. A maximum curve is used to identify the system head when storage is full and the water level in the pump suction pool is low. A minimum curve is used to identify the reverse situation— an empty storage tank and a high suction-pool level. The vertical distance between the maximum and minimum curves represents the range within which the system head varies. Representative system-head curves are shown in Fig. 9.2.2a. It should be pointed out that the velocity head of water discharged to the surface of a storage tank is considered negligible when computing the system-head curves.

The performance of pumps is predicted on the basis of *pump-characteristic curves*. These curves relate flow capacity to discharge head, efficiency, and brake horsepower. Typical pump-characteristic curves are shown in Fig. 9.2.2b.

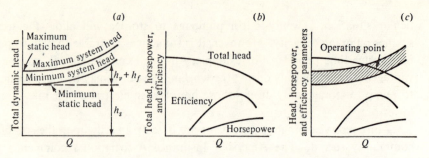

FIGURE 9.2.2
(a) System-head, (b) pump-characteristic, and (c) operation curves.

Pump-characteristic curves are used in conjunction with system-head curves in selecting pumping equipment. See Fig. 9.2.2c. An operating point is located at some point along that portion of the pump-head curve which intersects the range between the maximum and minimum system-head curves. The operating point identifies the discharge flow and head at which the pump is to operate. A pump is selected that will function at the operating point at an efficiency that is close to maximum. When variable flow capacity is required, it is most practical to provide two or more pumps in parallel so that the discharges of each pump are maintained at close to peak efficiency.

Solution procedure 9.2.1 The procedure for determining the elevation required for a storage tank servicing a water distribution system is applied to the pipe network shown in Figs. 9.1.2 and 9.1.3. The same network is shown in Fig. 9.2.3. Included are the ground surface elevations (above sea level) at the pipe junctions where inflow and discharge take place. Junction I has been established as the optimal location for a storage tank. Such a tank must be high enough to furnish at least 40 lb/in² pressure at the highest plumbing fixture in the area served by the system. Therefore, the required height will be

$$h_r = h_p + \Delta h_e + h' \qquad (9.2.3)$$

where h_r = height of minimum water level in storage tank above ground level

h_p = desired pressure

Δh_e = difference in ground elevation at tank site and that at highest point in system

h' = height of plumbing fixtures at highest point

ENGINEERED TRANSPORT SYSTEMS 295

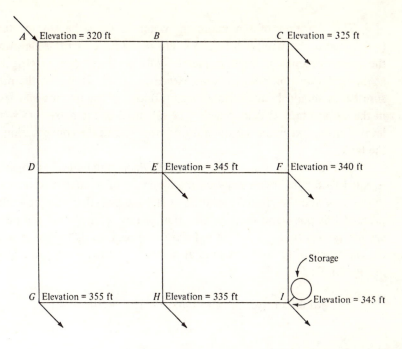

Junction	Ground elevation, ft	Max. elevation of plumbing fixture	Minimum head, ft	Minimum pressure,* lb/in²
A	320			
C	325	337	122	53
E	345	357	102	44
F	340	352	107	46
G	355	367	92	40
H	335	347	112	49
I	345	357	102	44

FIGURE 9.2.3
Definition sketch for pressure head calculations. (Computed on the basis that the minimum elevation of the water level in the storage tank will be 345 + 114 = 459 ft above sea level.)

The highest point in the system is found to be at G. If the plumbing fixtures in the area served by this junction are assumed to be located no higher than 12 ft above the ground, then

$$h_r = 2.31(40) + (355 - 345) + 12$$
$$= 114 \text{ ft}$$

The elevation of the minimum water level in the storage tank (referenced to sea level) will be $345 + 114 = 459$ ft. Using this elevation, the pressure heads at highest plumbing-fixture levels are investigated at the other junction points. None exceed 75 lb/in^2. However, the pressure heads should be investigated again, once the maximum water level in the storage tank is determined. The elevation of the maximum water level will depend upon the quantity of storage and the configuration of the tank.

The minimum system head against which pumping must operate is equal to the difference between the elevation of minimum water level in the tank and the ground elevation at A (assuming that pumping takes place at this point and that the elevation of the water level in the pump suction pool is equal to the ground elevation) plus the difference between the elevations of total-energy lines at points A and I (the latter are given in Fig. 9.1.3):

$$\text{Minimum system head} = (459 - 320) + (100 - 48)$$
$$= 191 \text{ ft} \qquad ////$$

9.3 OPEN-CHANNEL FLOW

Open-channel flow differs from flow in pipes in that a free surface is present in the former. Open channels of direct concern to the environmental engineer include flumes, short weir channels, and partly filled pipes. Flow in open channels can be either *steady* or *unsteady* depending on whether or not the depth at a point remains constant. In addition, flow can either be *uniform* or *nonuniform*. Flow is uniform when the depth of the liquid is the same at all points along the channels and nonuniform when the depth varies.

A definition sketch is shown in Fig. 9.3.1 for the flow of a liquid in an open channel. The energy balance can be written as

$$z_1 + y_1 + \frac{\overline{V}_1^2}{2g} = z_2 + y_2 + \frac{\overline{V}_2^2}{2g} + h_f \qquad (9.3.1)$$

where y = depth. $[L]$

As a rule, interest centers around the determination of the velocity produced by a given slope or, conversely, around the determination of a slope to give a desired velocity. Flow may be either laminar or turbulent.

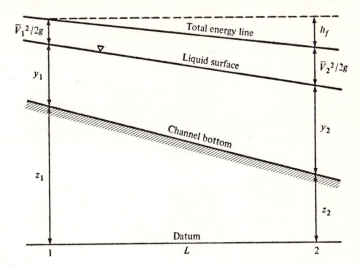

FIGURE 9.3.1
Definition sketch for flow in open channels.

The energy per unit weight of a liquid with respect to the bottom of the channel in which it is contained is called the *specific energy*:

$$E = y + \frac{\bar{V}^2}{2g} \quad (9.3.2)$$

A plot of the specific energy E as a function of the depth y is shown in Fig. 9.3.2. From the curve it can be seen that for a given discharge there exists a depth y_c at which the specific energy is a minimum. The velocity of flow at the critical depth is called the *critical velocity* $\bar{V}_c$. When for the given discharge the depth of flow is greater than the critical depth, flow conditions are said to be *subcritical*; when depths occur which are less than the critical depth, *supercritical* flow conditions are said to prevail. For any value of specific energy except the minimum, there are two depths, called *conjugate* depths, at which flow is possible. It is to be noted that the upper leg of the curve merges with a straight line inclined 45° with the horizontal. All points along the latter represent the potential-energy component of the specific energy. Subcritical and supercritical flow conditions are possible both in laminar- and turbulent-flow regimes.

If an obstruction which increases the flow depth is placed suddenly across a channel in which a liquid is flowing, a surge wave will form which will tend to move upstream against the motion of the oncoming liquid. When flow is subcritical, the flow velocity is so low that a small disturbance such as a surge wave can be transmitted upstream and, thereby, change upstream conditions.

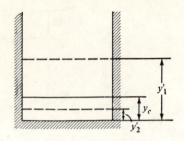

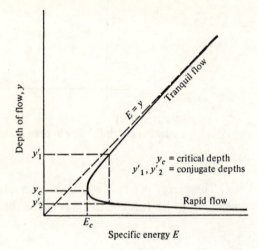

FIGURE 9.3.2
Specific energy as a function of depth, discharge remaining constant.

In such instances flow is controlled by conditions downstream. When flow is supercritical, the liquid velocity is greater than that of a surge wave, and any such disturbance formed is carried downstream. At critical flow, the velocity is equal to the velocity of a surge wave.

There is a criterion applicable to all channels which gives the relationship existing between the discharge and cross section of flow when flow conditions are critical. Attention is directed to Fig. 9.3.3. The specific energy of the liquid flowing through the trapezoidal section is

$$E = y + \frac{\overline{V}^2}{2g} = y + \frac{Q^2}{2gS^2} \qquad (9.3.3)$$

where Q = volumetric flow rate. $[L^3 t^{-1}]$

S = channel cross section. $[L^2]$

For a given discharge, the cross section at the critical depth is found when the specific energy is a minimum. By differentiating Eq. (9.3.3) with

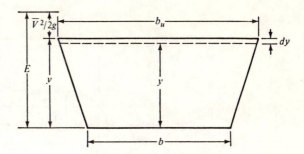

FIGURE 9.3.3
Specific energy of liquid flowing through a trapezoidal channel.

respect to the depth y and equating to 0, one has

$$\frac{dE}{dy} = 1 - \frac{Q^2}{gS^3}\frac{dS}{dy} = 0 \qquad (9.3.4)$$

Since at critical flow

$$Q = S\bar{V}_c \qquad (9.3.5)$$

and the change in S is given by

$$dS = b_u\, dy \qquad (9.3.6)$$

Eq. (9.3.4) can be written as

$$\frac{\bar{V}_c^2}{g} = \frac{S}{b_u} \qquad (9.3.7)$$

where $\bar{V}_c$ = flow velocity at critical depth. $[Lt^{-1}]$
b_u = top width of channel. $[L]$

By multiplying both sides by S^2, Eq. (9.3.7) becomes

$$\frac{Q^2}{g} = \frac{S^3}{b_u} \qquad (9.3.8)$$

When the relationship given in Eq. (9.3.8) is obtained, flow is critical. The equation holds for any cross section as long as the term b_u expresses the width at the top of the channel.

A specific relationship can be developed for rectangular channels. For such channels Eq. (9.3.8) can be expressed as

$$y_c = \left(\frac{Q^2}{gb^2}\right)^{1/3} \qquad (9.3.9)$$

where y_c = critical depth. $[L]$
b = channel width. $[L]$

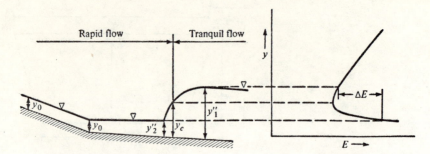

FIGURE 9.3.4
Specific-energy relationships occurring in a hydraulic jump.

When supercritical flow conditions are induced in water flowing in a channel with characteristics which promote subcritical flow, flow is unstable, and a hydraulic jump will result. Attention is directed to Fig. 9.3.4. Here, water flowing initially in a channel with a slope such that flow is supercritical flows into a second reach of channel with a slope less than critical. The flow depth increases gradually from y_0 to y_2''. At the latter depth, the depth changes abruptly to y_1'', which either is normal for the new slope or is controlled by down-channel conditions. The abrupt change is called a *hydraulic jump*. Flow conditions down channel from the jump are subcritical.

The problem usually encountered in connection with the hydraulic jump is that of determining from a knowledge of the down-channel depth y_1'' the depth from which the jump takes off y_2''. The turbulence produced in a jump results in a sizable loss of energy. Consequently, a specific-energy balance across the jump cannot be used. Instead, resort is made to the momentum relationships involved.

The change in the linear momentum of water with respect to time is equal to the resultant force acting on the water. Neglecting frictional forces, the forces active over the jump are the hydrostatic forces at both ends of the water in the jump (Fig. 9.3.5):

$$F_1 - F_2 = \frac{\gamma}{g} Q(\overline{V}_2 - \overline{V}_1) \qquad (9.3.10)$$

$$\frac{\gamma b}{2}(y_1''^2 - y_2''^2) = \frac{\gamma Q^2}{gb}\left(\frac{1}{y_2''} - \frac{1}{y_1''}\right) \qquad (9.3.11)$$

Solving for y_2'' gives

$$y_2'' = \frac{y_1''}{2}\left[\left(1 + \frac{8Q^2}{gb^2 y_1''^3}\right)^{1/2} - 1\right] \qquad (9.3.12)$$

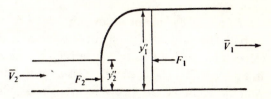

FIGURE 9.3.5
Hydrostatic forces acting at a hydraulic jump.

The energy lost as the result of turbulence produced in the jump can be computed from

$$\Delta E = \left(y_2'' + \frac{\bar{V}_2^2}{2g}\right) - \left(y_1'' + \frac{\bar{V}_1^2}{2g}\right) \quad (9.3.13)$$

Surface profiles often are of interest in the solution of problems involving steady, nonuniform flow in open channels. By multiplying both sides of Eq. (9.3.1) by the distance between points 1 and 2, L, and rearranging terms one has

$$L = \frac{(y_2 + \bar{V}_2^2/2g) - (y_1 + \bar{V}_1^2/2g)}{(z_1 - z_2)/L - h_f/L} \quad (9.3.14)$$

where $(z_1 - z_2)/L$ = slope of channel invert

h_f/L = slope of energy-grade line

The slope of the energy-grade line can be computed by means of one of the equations commonly used for determining the velocity of turbulent, uniform flow in open channels. [See Eq. (9.3.16).] Starting at a known depth the distance L is computed to a slightly different depth. If the increments of depth are small, a satisfactory profile of the water surface can be obtained.

In most practical situations, when water flows in an open channel, the flow regime is turbulent. For turbulent, uniform flow in open channels, the velocity of flow can be estimated with one of several empirical formulas. These formulas, referred to as *uniform-flow formulas*, have the general form

$$\bar{V} = CR^a G^b \quad (9.3.15)$$

where $\bar{V}$ = mean velocity, ft/s

R = hydraulic radius, ft

G = hydraulic gradient

C = coefficient

a, b = exponents

For steady, uniform flow, the slope of the total-energy line is parallel to both the liquid surface and the bottom of the channel. Consequently, the slope of any one of these characteristics can be used as the hydraulic gradient G in Eq. (9.3.15). The coefficient C is a term expressing flow resistance; it varies with the hydraulic radius, channel roughness, viscosity, and many other factors.

The *Manning equation* is widely used in engineering practice. This relationship is commonly expressed as

$$\bar{V} = \frac{1.486}{n} R^{2/3} G^{1/2} \qquad (9.3.16)$$

where n = coefficient of roughness

The similarity of Eq. (9.3.16) to the Hazen-Williams expression in Eq. (9.1.2) should be noted as well as the similarity of both to Eq. (9.3.15). The values of the coefficient of roughness for different types of conduit materials are listed in Table 9.3.1.

Solution procedure 9.3.1 In open-channel flow computations, it is often necessary to determine the depth of flow. The determination is not straightforward, and resort is made to trial-and-error solution.

For a rectangular channel, Eq. (9.3.16) can be expressed as

$$\frac{(by)^{5/3}}{(b + 2y)^{2/3}} = \frac{nQ}{1.486 G^{1/2}} \qquad (9.3.17)$$

Table 9.3.1 VALUES OF THE ROUGHNESS COEFFICIENT n FOR DIFFERENT KINDS OF CONDUITS*

	Condition of interior surface			
Kind of conduit	Best	Good	Fair	Bad
Vitrified-clay sewer pipe	0.010	0.013	0.015	0.017
Unglazed-clay drainage tile	0.011	0.012	0.014	0.017
Concrete pipe	0.012	0.013	0.015	0.016
Cement-mortar surfaces	0.011	0.012	0.013	0.015
Concrete-lined channels	0.012	0.014	0.016	0.018
Semicircular metal flumes, smooth	0.011	0.012	0.013	0.015
Semicircular metal flumes, corrugated	0.023	0.025	0.028	0.030
Wood-stave pipe	0.010	0.011	0.012	0.013

* Selected from values found in H. W. King, "Handbook of Hydraulics," table 90, p. 268, McGraw-Hill, New York, 1939.

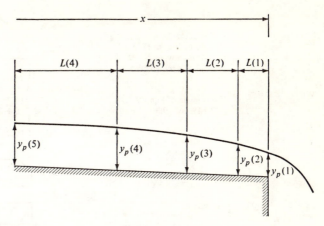

FIGURE 9.3.6
Definition sketch for drop-down curve at free-discharge end of an open channel.

Given the value of the channel width b and the quantities on the right-hand side of the equation, the value of the depth y can be determined through iteration. The value of depth thus found is called the *normal depth*; i.e., it is that depth which normally occurs at a slope G.

When a free discharge occurs at the end of an open channel, however, the depth of flow decreases as the end is approached. The depth at the end is approximately $0.7y_c$, and the critical depth y_c has been found to occur at a distance of $3y_c$ to $4y_c$ from the end. The water-surface profile in such cases is referred to as a *drop-down curve*, and it is sometimes desirable to compute the coordinates of the curve with respect to the channel invert.

Reference is made to Fig. 9.3.6. Let it be assumed that the open channel sketched therein is rectangular. The critical depth $y_c = y_p(2)$, which can be computed using Eq. (9.3.9), is assumed to exist at a distance $L(1) = 3.5y_c$ from the end. The depth at the end $y_p(1)$ can be assumed to be equal to $0.7y_c$. From the point of critical depth $y_p(2)$, going upstream the depths at different points can be computed using Eq. (9.3.14) in a stepwise procedure. In such computations, the slope of the energy-grade line can be determined with Eq. (9.3.16) using average values in the reach for the quantities $\overline{V}$ and R.

A computer procedure is given in Algorithm 9.3.1. The procedure yields the depths of flow at points upstream from the discharge end until a point is reached where the depth of flow is normal. ////

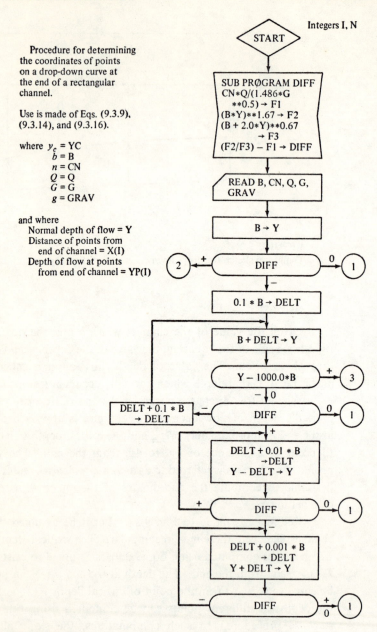

Procedure for determining the coordinates of points on a drop-down curve at the end of a rectangular channel.

Use is made of Eqs. (9.3.9), (9.3.14), and (9.3.16).

where y_c = YC
b = B
n = CN
Q = Q
G = G
g = GRAV

and where
Normal depth of flow = Y
Distance of points from end of channel = X(I)
Depth of flow at points from end of channel = YP(I)

ALGORITHM 9.3.1

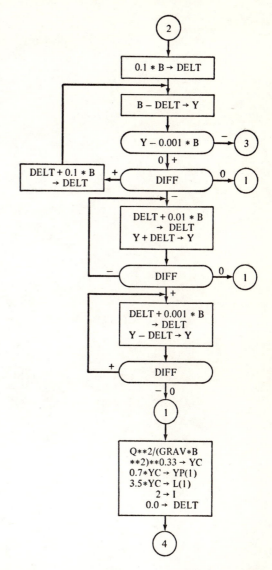

ALGORITHM 9.3.1 (*continued*) (*continued overleaf*)

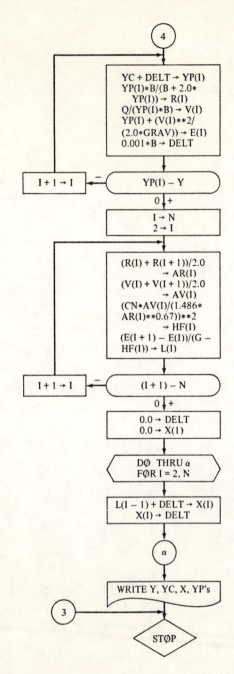

ALGORITHM 9.3.1 (*continued*)

9.4 DOMESTIC WASTE-WATER COLLECTION SYSTEMS

Domestic waste-water collection systems are generally composed of pipes buried underground at elevations that permit the collection of wastes by gravity. It is common practice to design on the basis that the pipes will be completely full when accommodating peak flows. For this reason, open-channel flow conditions prevail most of the time the pipes are in service. Although several empirical equations have been used in the selection of pipes for collection systems, the Manning equation (9.3.16) has come into more general use for such purposes. A nomogram of the Manning equation is given in Sec. A.3. A chart in Sec. A.4 gives the hydraulic elements of partially filled circular sections. It is to be noted that the velocity of flow at one-half depth is equal to the velocity when the pipe is flowing full.

Waste-water collection systems are classified as follows: *Separate systems* are designed to transport domestic and other waste waters only. Storm drainage, surface water, and groundwater are not intentionally admitted to such systems. *Storm drainage systems* are designed to transport storm water, surface water, street wash or other drainage, domestic and industrial waste waters being excluded. *Combined systems* are systems which receive both domestic waste waters and storm and surface water drainage. Combined systems, although still in existence, are now obsolete. The use of separate systems eliminates the necessity of treating storm waters prior to their discharge into the receiving streams. Such use greatly reduces the cost of treatment facilities and their operation.

Separate systems are composed of several elements: *Building sewers* are used to transport waste waters from individual buildings and residences. These pipes generally have a diameter of 6 in or more and are placed at grades no less than 1 or 2 percent. Building sewers discharge into *collector sewers*. The collector sewers are located along the streets and alleys of the area being served and discharge into *interceptor sewers*. The latter transport the waste water from the area served by the system to the treatment plant. Vitrified clay, asbestos cement, and concrete pipe are all used in the construction of sewers. Cast-iron pipe is often used when the waste must be transported under pressure.

Waste-water collection systems are normally designed to accommodate the maximum hourly flow rates. These are determined in the following manner: The quantity of domestic waste water will vary in different sections of a community. Table 9.4.1 gives the average population densities for different types of districts. From such a table and community planning studies, future population densities for the areas to be served can be estimated. This information along with that of anticipated average daily water demands and the expected percent of the latter being discharged into the collection system provides a working value of the future, average, daily waste-water flows. For collector sewers, the maximum hourly flow rate often is predicted by multiplying the future, average, daily waste flow by a factor of 4. For interceptors, the average flow is multiplied by a factor of 2.5. As a safety factor, the design flow usually includes an allowance for unavoidable infiltration due to faulty pipe joints. If significant quantities of industrial wastes are to be discharged into the system, an additional increment of flow increase must be considered. Normally, conditions at the 30- and 50-year horizon are assumed for the future.

Pipe size is determined from the quantity of waste water expected to be handled and either of two other factors—slope and velocity. If the ground surface is very flat, the minimum velocity will be used in order that the pipe does not lie too far underground. If the ground slope is greater than that necessary to produce the minimum velocity, then slope is the controlling factor, and the pipe is laid approximately parallel to the ground surface. Sometimes the slope of collector pipes coming into a manhole from streets other than the one under consideration will govern the elevation of the bottom of a manhole. Unless the manhole is specifically designed as a drop manhole, the inverts of all pipes coming into a manhole should be located at the elevation of the bottom of the manhole.

A minimum velocity of 2.5 ft/s is generally adopted to prevent the deposition of organic solids, unless the ground is very flat. In the latter case, a minimum velocity of 2.0 ft/s may be acceptable. Velocities greater than 10 ft/s should be avoided to minimize abrasion. Where excessive velocities would otherwise occur, resort is often made to drop manholes to reduce slopes.

System design ordinarily begins with the collector system. A minimum depth of cover is adopted sufficient to permit the pipe to drain an ordinary basement. Design proceeds from the pipe farthest away from the discharge

Table 9.4.1 AVERAGE POPULATION DENSITIES

Character of district	Development	Density of population/acre	Impervious surface percentage
Dense residential	Two-family houses and six-family apartment buildings	55	34
Medium residential	Mostly single-family houses	35	27
Light residential	Single-family houses only, some on double lots	15	20
Mercantile		14	100
Light commercial		30	80
Industrial		10	60

end of the system. Interceptor design is governed by the collector-system requirements. Table 9.4.2 is an outline of a procedure followed for the design of a domestic waste-water collection system.

Solution procedure 9.4.1 Figure 9.4.1 is a plan of a collector portion of a separate waste-water collection system. Basic data for system design are as follows:

1. Average water consumption, gallons per capita = 150
2. Percent of water used discharged to collection system = 85
3. Maximum hourly discharge, percent of average discharge = 400
4. Population density, capita per acre = 55
5. Infiltration, gal/(acre)(d) = 2,500
6. Minimum velocity, ft/s = 2
7. Minimum cover, ft = 7
8. Increment of area increase at each manhole, acres = 2.5
9. As a special requirement, industrial waste discharge at manhole 6, 10^6 gal/d = 2

Table 9.4.2 PROCEDURE OUTLINE FOR THE DESIGN OF A DOMESTIC WASTE-WATER COLLECTION SYSTEM

---

1. Select basic data.
   a. Estimate future density of population in each area to be served.
   b. Estimate maximum hourly rate of flow on a per capita basis.
   c. Estimate rate of groundwater infiltration.
   d. Adopt minimum size of pipe. (Eight-inch minimum is preferable and required by many city codes.)
   e. Adopt minimum velocity of flow of 2.5 ft/s, or 2.0 ft/s if ground is very flat. Use no velocities greater than 10 ft/s.
   f. Adopt depth of cover sufficient to allow the system to drain ordinary basements. Seven ft is a common value.
2. Prepare a contour map of the area to be served, and draw a line to represent the pipeline in each street. Place arrows on lines to indicate direction of flow.
3. Locate and number all manholes. Place manholes at:
   a. All changes of direction.
   b. All changes of slope.
   c. All pipe junctions except house connections.
   d. Upper end of all collector lines for flushing.
   e. Every 300 or 400 ft for cleaning.
4. From a contour map, or from profile levels made in the field, prepare a surface profile for each street along the centerline of the proposed pipeline.
5. Sketch on the map the limits of area tributary to each manhole, i.e., the area served by each collector line between manholes. Planimeter and record the increments of area.
6. Starting at the upper end of the system, design the system from manhole to manhole. Unless drop manholes are required by excessive differences in invert elevations, the inverts of all pipes at manholes should be at the same elevation. A pipe of one diameter should never discharge into another with a smaller diameter.

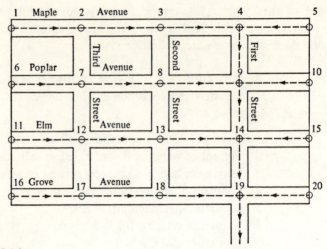

FIGURE 9.4.1
Plan of collector portion of a separate waste-water collection system.

A computation sheet for system design, completed for all pipes up system from manhole 9, is found in Table 9.4.3. For the portion of the system which has been designed, ground slope controls. For this reason, invert elevations although computed to 0.01 ft are expressed in terms computed to 0.1 ft. In computing the invert elevations from ground elevations and required cover, allowance has been made for approximate pipe thickness. ////

Table 9.4.3 COMPUTATION SHEET OF DESIGN OF WASTE-WATER COLLECTION

From manhole no.	To manhole no.	Line	Street	Length, ft	Area, acres Increment	Total	Waste discharge, 10^6 gal/d	Infiltration, 10^6 gal/d
1	2	1	Maple Ave.	400	2.5	2.5	0.069	0.006
2	3	2	Maple Ave.	400	2.5	5.0	0.138	0.012
3	4	3	Maple Ave.	400	2.5	7.5	0.207	0.018
5	4	4	Maple Ave.	400	2.5	2.5	0.069	0.006
4	9	5	First St.	300		10.0	0.276	0.024
6	7	6	Poplar Ave.	400	2.5	2.5	2.069	0.006
7	8	7	Poplar Ave.	400	2.5	5.0	2.138	0.012
8	9	8	Poplar Ave.	400	2.5	7.5	2.207	0.018
10	9	9	Poplar Ave.	400	2.5	2.5	0.069	0.006

9.5 STORM-WATER COLLECTION SYSTEMS

The *rational method* for estimating runoff from land areas is widely used in the design of storm-water collection systems, storm drains, and culverts which must transport the runoff resulting from single storms. The method is developed from the consideration that the quantity of storm water will vary directly with the size of area on which the rain falls, the intensity of the rainfall, and the perviousness of the surface on which the rain falls. Two basic assumptions are made when the method is applied:

1 The rate of runoff to any point in the system is a function of the average rainfall rate during the time required for the storm water to flow from the remotest part of the drainage area to the point under consideration. The time of flow is called the *time of concentration*.
2 The highest intensity of rainfall in any storm occurs within the time of concentration.

The rational method uses the relationship

$$Q = CiA \qquad (9.5.1)$$

where Q = maximum rate of runoff, ft³/s

A = drainage area tributary to point under consideration, acres

i = average rainfall intensity, in/h

C = average runoff coefficient

SYSTEM

Total flow		Diameter of pipe, in	Slope, ft/ 1,000 ft	Velocity, ft/s	Capacity, ft³/s	Surface elevation		Invert elevation	
10⁶ gal/d	ft³/s					Upper end	Lower end	Upper end	Lower end
0.075	0.116	8	10.0	2.3	1.3	228.2	224.2	220.40	216.40
0.150	0.232	8	5.0	2.1	0.9	224.2	222.2	216.40	214.40
0.225	0.348	8	8.0	2.9	1.1	222.2	219.0	214.40	211.20
0.075	0.116	8	16.5	2.5	1.4	225.6	219.0	217.80	211.20
0.300	0.464	8	12.0	3.4	1.4	219.0	215.4	211.20	207.00
2.08	3.22	12	9.0	5.0	3.5	225.4	221.8	217.20	213.60
2.15	3.34	15	3.0	3.4	3.6	221.8	220.6	213.60	212.40
2.23	3.46	15	18.5	7.0	8.5	220.6	215.4	212.40	207.00
0.075	0.116	8	19.0	2.7	1.7	222.4	215.4	214.60	207.00

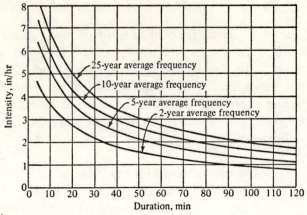

FIGURE 9.5.1
Typical intensity-duration curves.

The units for the terms of Eq. (9.5.1) are assumed to be consistent because 1.008 ft^3/s is equal to 1 in/h of rainfall falling on an area of 1 acre.

The term i is the average rainfall intensity for the period of maximum rainfall of a given frequency of occurrence having a duration equal to the time required for the runoff originating during said period of maximum rainfall to flow from the remotest part of the drainage area to the point under consideration. Typical curves relating the rainfall intensity to duration of intensity are shown in Fig. 9.5.1. Such curves should be derived empirically from rainfall records accumulated for the specific area of concern. At locations where records are unavailable, resort has to be made to regional curves.[1,2,3]

The value of rainfall intensity to be used in Eq. (9.5.1) is found by equating the corresponding value of duration to the time of concentration t_c. The time of concentration is the sum of the *inlet time* and the *time of flow* in the pipe above the point under consideration. Inlet time depends upon the slope, the nature of the surface cover, and the distance of travel. It is defined as the time it takes for a particle of water to reach the system inlet from the most remote point on the drainage area. For densely developed areas, where impervious

---

[1] U.S. Weather Bureau, U.S. Dep. of Commerce, Rainfall Intensities for Local Drainage Design in the United States, *Tech. Pap.* 24, Washington, 1955.
[2] U.S. Weather Bureau, U.S. Dep. of Commerce, Rainfall Intensity-Duration-Frequency Curves for Selected Stations in the United States, Alaska, Hawaiian Islands and Puerto Rico, Washington, 1955.
[3] U.S. Weather Bureau, U.S. Dep. of Commerce, Rainfall Intensities for Local Drainage Design in Western United States, *Tech. Pap.* 28, Washington, 1955.

surfaces shed their water to storm drains through closely spaced inlets, an inlet time of 5 min is often used. For well-developed areas with relatively flat slopes, the use of an inlet time of 10 to 15 min is customary. For flat residential areas with widely spaced inlets, the use of inlet times of 20 to 30 min is common. The time of flow is determined from velocity and distance in the section of the collection system extending from the farthest inlet to the point under consideration.

Values of the appropriate runoff coefficient to be used in Eq. (9.5.1) can be estimated from published data such as are presented in Table 9.5.1, or they can be derived from a rational evaluation of the actual nature of the surface area and its composite runoff characteristics. In the latter case, the percent composition of the various types of surfaces included in the drainage area is determined, and then a weighted average of the average runoff coefficients for the different types of surfaces is computed.

In general, the procedure for the layout of a storm-water collection system is similar to that of a domestic waste-water system. Manholes are

Table 9.5.1 AVERAGE RUNOFF COEFFICIENTS*

Description of area	Runoff coefficients
Business:	
Downtown areas	0.70–0.95
Neighborhood areas	0.50–0.70
Residential:	
Single-family areas	0.30–0.50
Multiunits, detached	0.40–0.60
Multiunits, attached	0.60–0.75
Residential (suburban)	0.25–0.40
Apartment dwelling areas	0.50–0.70
Industrial:	
Light areas	0.50–0.80
Heavy areas	0.60–0.90
Parks, cemeteries	0.10–0.25
Playgrounds	0.20–0.35
Railroad: yard areas	0.20–0.40
Unimproved areas	0.10–0.30

* From *Am. Soc. Civ. Eng. Man. Eng. Pract.*, no. 37, p. 48, 1960.

located at street intersections unless such intersections are more than approximately 400 ft apart. In the latter case, manholes are located at intermediate points as well. It is assumed that all the storm water flows downgrade to the manhole at the lowest end of the area under consideration.

Storm-water collector pipes are generally designed for a minimum velocity of 3 ft/s, flowing full, in order to prevent the deposition of solids. Pipe diameters of 15 in are generally taken as minimum values in storm drainage system design.

Solution procedure 9.5.1 Table 9.5.2 is an illustrative computation sheet for the design of a storm-water collection system for the residential area depicted in Fig. 9.4.1. Basic data for the design are as follows:

1 Minimum velocity, ft/s = 3
2 Minimum cover, ft = 5
3 Average runoff coefficient = 0.6
4 Storm frequency, average number of years = 5

Rainfall intensities were obtained from the 5-year curve in Fig. 9.5.1. It is to be noted that fewer manholes and pipes are required in the storm drainage system than are required in the domestic waste-water system. This is because it is assumed that all storm water flows to the lower end of the drainage area before it enters a manhole. Also, it is to be noted that invert elevations of pipes converging at a manhole do not have to be at the same elevation. It is important, however, that the invert of the pipe transporting drainage from a manhole be at no higher elevation than that of any pipe discharging into the manhole. ////

Table 9.5.2 COMPUTATION SHEET FOR DESIGN OF STORM-WATER COLLECTION

From manhole no.	To manhole no.	Line	Street	Length, ft	Area, acres Increment	Total	Time of flow, min To upper end	In pipe section above	Time of concentration	Rainfall intensity, in/h
2	3	1	Maple Ave.	400	2.5	2.5	10.0		10.0	5.1
3	4	2	Maple Ave.	400	2.5	5.0	10.0	1.5	11.5	4.9
4	9	3	First St.	300	5.0	10.0	11.5	1.1	12.6	4.8
7	8	4	Poplar Ave.	400	2.5	2.5	10.0		10.0	5.1
8	9	5	Poplar Ave.	400	2.5	5.0	10.0	1.5	11.5	4.9
9	14	6	First St.	300	5.0	20.0	12.6	0.6	13.2	4.6

EXERCISES

9.1.1 A, B, C, and D identify the four corners of a rectangular system of pipes, the diameter and length of which are indicated below:

Pipe	Diameter, in	Length, ft
AB	6	1,200
BC	6	2,000
CD	4	1,200
DA	8	2,000

Water flowing at the rate of 500 gal/min flows into the system at pipe junction A and out of the system at junction B. The elevations at junctions A and B are 2,000 and 2,050 ft, respectively. If the pressure in the pipe at A is 40 lb/in^2, what will be the pressure at B, and how will the flow divide itself through the system?

9.1.2 With Algorithm 9.1.1 as a guide, prepare in some computer language a procedure for pipe-network analysis using the Hardy Cross method.

9.1.3 After replacing the pipes DI and IH in the network shown in Fig. 9.1.3 with pipes having a diameter of 16 in, reanalyze the flow distribution in the network.

9.1.4 After analyzing the network in Fig. 9.1.3 according to instructions given in Exercise 9.1.3, construct 10-ft pressure contours for the system.

9.1.5 Reanalyze the flow distribution in the network in Fig. 9.1.3 after the discharge at C is increased to 1,800 gal/min and the discharge at D is decreased to 0.

9.2.1 A water distribution system is being designed for a community of 100,000 people. For what rate of flow and pressure should the pipes in the main trunk system be selected?

9.2.2 For what rate of flow should the water-treatment plant for the community in Exercise 9.2.1 be designed?

SYSTEM

Runoff acre, ft^3/(s)(acre)	Total runoff, ft^3/s	Pipe diameter, in	Slope, ft/ 1,000 ft	Velocity, ft/s	Capacity, ft^3/s	Surface elevation Upper end	Surface elevation Lower end	Invert elevation Upper end	Invert elevation Lower end
3.06	7.68	18	5.0	4.3	7.7	224.2	222.2	217.60	215.60
2.94	14.70	21	8.0	6.0	14.8	222.2	219.0	215.35	212.15
2.88	28.80	24	12.0	8.6	29.0	219.0	215.4	211.90	208.30
3.06	7.68	21	3.0	4.3	9.0	221.8	220.6	214.95	213.75
2.94	14.70	21	13.0	7.9	18.0	220.6	215.4	213.75	208.55
2.76	55.20	42	4.0	7.3	62.0	215.4	214.2	206.80	205.60

9.2.3 What types of storage should be included in a water distribution system? How are the storage capacities determined?

9.2.4 The following water meter readings for a small community were recorded for the maximum day:

Time	Gallons	Time	Gallons
Midnight	0	1 P.M.	425,000
1 A.M.	22,500	2	452,000
2	46,800	3	484,400
3	72,000	4	526,400
4	99,000	5	564,200
5	128,600	6	594,200
6	159,800	7	623,000
7	203,000	8	642,800
8	248,000	9	659,000
9	295,400	10	674,000
10	324,200	11	689,000
11	360,200	Midnight	710,000
Noon	403,400		

Determine the equalization storage required if water is pumped to storage over a 24-h span. Determine the storage required if pumps are operated only from 6 A.M. to 6 P.M.

9.2.5 Determine how high a storage tank must be to serve the system described in Solution procedure 9.2.1 if the tank is to be located at point F instead of point I.

9.2.6 Discuss the procedure followed in selecting pumping equipment to be used in a water distribution system.

9.3.1 Design the cross section of an unlined canal to carry 1,000 ft^3/s of water at a slope of 0.4 ft/mi in a material the nature of which requires a side slope of $1\frac{1}{2}$:1 (horizontal/vertical) and imposes a maximum flow velocity of 2 ft/s. Use an n value of 0.0225.

9.3.2 At what depth will 80 ft^3/s of water flow in a rectangular channel 3 ft wide and lined with concrete if the bottom of the channel is constructed on a slope of 7 ft/1,000 ft? Assume flow to be steady and uniform.

9.3.3 Determine the critical depth for water flowing at a rate of 4 ft^3/s in a trapezoidal channel 2 ft wide at the bottom and with side slopes of 1:2 (horizontal/vertical).

9.3.4 A jump is to be induced in a rectangular channel 2 ft wide transporting water at a rate of 6 ft^3/s. If the downstream depth is maintained at 2 ft, from what depth will the jump take off? How much energy will be lost in the jump?

9.3.5 The rectangular channel described in Exercise 9.3.2 terminates in a free discharge. Determine the coordinates of points on the drop-down curve.

9.4.1 List the components of a separate waste-water collection system. What flow rate is each designed to accommodate?

9.4.2 Discuss the factors that determine whether slope or velocity is considered in selecting pipe sizes for a waste-water collection system.

9.4.3 Two circular sewers join at a manhole. One has an 84-in diameter, and at maximum flow it carries 55 ft³/s on a slope of 0.0002. The smaller sewer has a 30-in diameter and is on a slope of 0.0015. The latter carries a maximum rate of flow of 9.5 ft³/s. At what height must the invert of the smaller pipe be above the invert of the larger sewer so that during maximum flow there will be no backing up of the sewage in the smaller sewer. Use $n = 0.012$.

9.4.4 Two 36-in-diameter circular sewers on a slope of 0.004 converge at an invert elevation of 50.00 ft and at a horizontal angle of 45° from each other. They discharge into a 48-in sewer laid on a slope of 0.005. Compute the invert elevation of the 48-in sewer so that when the two 36-in sewers are flowing full there will be no surcharging at any sewer and the water surface in all sewers will coincide at the point of junction. Assume $n = 0.015$ and a turbulence loss of 0.01 in the manhole.

9.4.5 Complete the design of the separate waste-water collection system for the residential area depicted in Fig. 9.4.1.

9.5.1 What basic assumptions are made in applying the rational method to the estimation of runoff?

9.5.2 On what factors does the time of concentration depend?

9.5.3 Three areas A, B, and C, each 1 acre in area, discharge runoff to three manholes: A to manhole 1, B to manhole 2, and C to manhole 3. The inlet time from each area is 7 min, 8 min, and 15 min, respectively. All areas have an average runoff coefficient of 0.8. The direction of flow is from manhole 1 to manhole 2 to manhole 3 and out. The time of flow between manholes is 5 min. The relation of rainfall intensity to time of concentration is given by $i = 105/(t + 15)$ in/h.

Using the rational method, (*a*) determine the maximum flow out of manhole 3, and (*b*) determine what size of pipe out of manhole 3 is required for the maximum discharge if $n = 0.015$ and $s = 0.0092$.

9.5.4 Complete the design of the storm-water collection system described in Solution procedure 9.5.1.

10
WATER TREATMENT AND RENOVATION SYSTEMS

10.1 TREATMENT TRAINS

The treatment of water and waste water involves the removal of undesirable materials by one process or a combination of several processes which can be physical, chemical, or biologic in nature. The materials which are removed range in size from simple ions to large floating materials. Such materials can be either inorganic, organic, or both. Furthermore, these materials may find their way into the water naturally with little or no help from man, or they may result from one or several uses to which man subjects water. Traditionally, the term "water treatment" has been applied to activities in which water is treated prior to use, whereas "waste-water treatment" has been used to describe the processing of water after use. As water renovation practices for direct reuse become more widely employed, the separate terminology describing treatment systems, i.e., water treatment systems and waste-water treatment systems, becomes less significant in the descriptive sense.

A treatment or renovation system consists of one or more *treatment trains*. A treatment train is defined as a sequence of one or more processes used to accomplish a particular treatment objective. If treatment is to accomplish several objectives, then the system will consist of several trains, each designed for a specific treatment objective. For present purposes, treatment objectives will be categorized in terms of the size of materials removed and the subsequent treatment of the removed materials.

Figure 10.1.1 shows a size spectrum of particles removed by treatment systems. The unit pL is analogous to pH and is defined as

$$pL = -\log L \quad (10.1.1)$$

where L = characteristic size dimension, m

The pL unit provides a convenient scale for classifying a wide range of sizes. The pL scale ranging from pL 0 to 10 is partitioned into three sectors, the boundaries between which are not precise and even overlap. These sectors are identified here by the terms "gross particulates," "suspended particulates," and "dissolved materials." The size sectors are established in such a way that they correspond roughly to the range of particle sizes removed by three types of treatment trains used in ordered sequence. A *gross-particulate removal* (GPR) *train* will remove most of the particles of pL 4 or larger. A treatment system consisting of a gross-particulate removal train followed by a *suspended-particulate removal* (SPR) *train* will remove most of the particles larger than pL 6. By adding a *dissolved-materials removal* (DMR) *train* to the end of the system, almost complete removal of all materials in the water can be accomplished. The sequence in which the treatment trains are employed is important; a suspended-particulate treatment train very seldom should be located before a gross-particulate removal train, and a dissolved-materials removal train nearly always should be located last. Certain exceptions may occur occasionally but not often enough to destroy the generalization. Finally, it is important to emphasize that a system may include only one or two of the removal trains, depending upon (1) the size range of particles in the water to be treated and (2) the size range of particle acceptable in the effluent.

The materials removed from the water by the gross- and suspended-particulate removal trains may require treatment prior to disposal. Such treatment is provided by a *sludge treatment* (ST) *train*. The sludge may include, in addition to the removed particulates, solids generated by one or more processes within the removal trains.

Component processes of removal and sludge trains are depicted in Tables 10.1.2 to 10.1.5. In these tables the treatment processes are correlated with the treatment result. As with the sequence of trains within a system, order is important within the trains themselves. In general, the sequence in which the processes are employed within a treatment train is from left to right in the tables. Alternative processes to accomplish a given result are found in the same column. A train may or may not consist of processes to accomplish all the results listed in the tables. Modern treatment technology includes a much broader selection

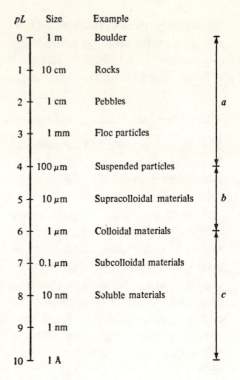

FIGURE 10.1.1
Categorization of particle size spectrum. (*a*) Gross particulates; (*b*) suspended particulates; (*c*) dissolved materials. (*Adapted from T. Helfgott et al., Analytical and Process Classification of Effluents, table 1, J. Sanit. Eng. Div., ASCE, vol. 96, no. SA3, pp. 79–107, 1970.*)

of processes than those found in the tables. However, the processes shown include the most important ones, and the inclusion of additional processes only serves to extend the lengths of the matrix columns by offering a greater selection of alternative processes to accomplish a given result. It should be emphasized that the schematics found in the tables are meant to depict, in a very general way only, the configuration of the process they identify. No additional significance should be attached to them.

Table 10.1.1 lists the sources and products of water treatment and renovation systems. In the absence of water renovation, groundwaters and surface waters constitute primary sources. Where renovation is practiced, these waters are used as initial and makeup sources. Also, when water renovation is not practiced, surface water becomes the sink, or product, of the treatment system.

With the use of Tables 10.1.1 to 10.1.5, a shorthand notation can be developed to represent the composition of a treatment or renovation system along with its primary source and product. This notation uses the element P_{ij} to represent a treatment process, with the subscript i denoting the train in which the process is located and j denoting the process within the train. The individual processes are identified in the last four tables as to element notation. The primary source and product of a system are represented by S_{kl}, with the subscripts k and l denoting the primary source and product, respectively.

Using this notation, a treatment plant converting water that had been used for domestic purposes to a product suitable for discharge into a surface stream can be represented by the set

$$\{S_{32}: P_{11}, P_{12}, P_{14}, P_{21}, P_{36}; P_{42}, P_{45}, G\}$$

Table 10.1.1 SOURCES AND PRODUCTS OF WATER TREATMENT AND RENOVATION SYSTEMS

1	Groundwater
2	Surface water
3	Domestic use
4	Irrigation water
5	Industrial water
6	Cooling water
7	Boiler feed water

Table 10.1.2 TREATMENT PROCESSES USED IN GROSS-PARTICULATE REMOVAL TRAINS

	Gross particulates			
Treatment process	Trash	Inorganics	Flotable organics	Settleable organics
P_{11} Coarse screening	✓			
P_{12} Grit removal		✓		
P_{13} Flotation			✓	
P_{14} Sedimentation				✓
P_{15} Fine screening				✓

The elements in the set represent coarse screening, grit removal, sedimentation, trickling filtration (including sedimentation), disinfection of the water phase prior to discharge, the anaerobic digestion of the sludge phase, and the drying of the sludge phase. The element G indicates the ultimate disposal of the sludge (or solids residual) to the ground. The element W would identify a surface water as the ultimate sink for the sludge. It is to be noted that the convention to be followed in using the notation requires the source and product element S_{kl} to be separated from the other elements in the set by a colon and the elements of the sludge treatment train to be separated from the other elements by a semicolon.

Reuse can take the form of recycle (S_{kk}) or sequential reuse (S_{km}, S_{ml}). When recycle is practiced, treatment processes should be used to remove the *recycle removal increment*. This term identifies the types and quantities of the materials added to the water each time it is used. Failure to remove the recycle removal increment will result in the deterioration of the water with each cycle.

When sequential reuse is practiced, treatment before each use will depend upon (1) the quality requirements of the water for that use and (2) the quality of the water produced by its preceding use. The quality requirements for different uses have been discussed in Chap. 6. The quality changes wrought by each possible use are highly variable. Table 10.1.6 provides a qualitative general-

Table 10.1.3 TREATMENT PROCESSES USED IN SUSPENDED-PARTICULATE REMOVAL TRAINS

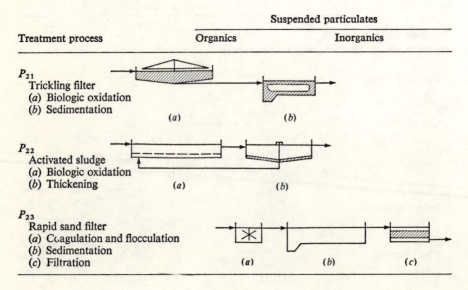

ization of those materials that should be removed from water prior to domestic use. Three water sources are selected for the generalization. It is to be noted that groundwater and surface waters are presently the most common sources of domestic supply, although recycle will become more common practice in the future.

When sequential reuse is practiced, the water is used in such a way that it is used first to satisfy that use having the most demanding quality requirements, followed by that having next to the most demanding requirements, and so forth.

Table 10.1.4 TREATMENT PROCESSES USED IN DISSOLVED-MATERIALS REMOVAL TRAINS

		Dissolved material		
Treatment process	Gases	Organics	Ions	Bacteria and viruses
P_{31} Aeration				
P_{32} Carbon adsorption				
P_{33} Chemical precipitation (a) Coagulation and sedimentation (b) Filtration				
P_{34} Ion exchange				
P_{35} Membrane separation				
P_{36} Disinfection				

The last use generally has the least demanding requirements of all uses. However, even sequential reuse may require some treatment between reuses.

The effectiveness of a treatment process or a treatment train may be expressed in terms of removal efficiency:

$$E_i = \frac{c_{i-1} - c_i}{c_{i-1}} \quad (10.1.2)$$

where E_i = removal efficiency of ith process or train

c_{i-1} = influent concentration of material being removed

c_i = effluent concentration of material being removed

For a series of processes or treatment trains, the concentration of the material being removed in the effluent of the nth process can be computed from

$$c_n = c_0 \prod_{i=1}^{n} (1 - E_i) \quad (10.1.3)$$

where c_0 = concentration in influent to first process or treatment train

Table 10.1.5 PROCESSES USED IN SLUDGE TREATMENT TRAINS

Treatment processes	Treatment result					
	Volume reduction	Biologic destruction of organics	Ion removal	Bulk water removal	Moisture reduction	Combustion of organics
P_{41} Thickening						
P_{42} Anaerobic digestion						
P_{43} Conditioning						
P_{44} Dewatering						
P_{45} Drying						
P_{46} Incineration						

Process and removal-train efficiencies can be used to express removal as measured by several parameters. Such parameters include settleable solids, suspended solids, biochemical oxygen demand (BOD), chemical oxygen demand (COD), and bacteria. In spite of the diversity of parameters employed, all measure, in one way or the other, certain characteristics displayed by one or more of the particulate and dissolved-material fractions in the water being treated.

The internal kinetics of a treatment process limit the rate at which water can be treated by the process. This rate, when expressed for the purpose of determining treatment efficiency, is called the *process loading intensity* (PLI):

$$E_i = f(L_i) \quad \text{or} \quad f(q_i) \quad (10.1.4)$$

where L_i = process loading intensity applied to ith process expressed in terms of material application

q_i = process loading intensity applied to ith process expressed in terms of hydraulic application

The process loading intensity is generally expressed either in terms of the hydraulic application rate or in terms of the rate at which materials carried in the water are applied to the process.

Table 10.1.6 MATERIALS THAT SHOULD BE REMOVED FROM WATER PRIOR TO DOMESTIC USE

	Sources of water		
Materials that should be removed	Ground S_{13}	Surface S_{23}	Domestic S_{33} (Waste water)
Gross particulates			
Trash		×	×
Inorganics			×
Flotable organics			
Settleable organics			×
Suspended particulates			
Organics			×
Inorganics		×	×
Dissolved materials			
Gases	×		
Selected ions	×		
Organics			×
All ions			×
Heat			
Bacteria and viruses	×	×	×
Sludge produced			
Organic			×
Inorganic	×	×	×

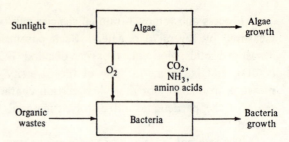

FIGURE 10.2.1
Diagram indicating the symbiotic relationship between algae and bacteria in an oxidation pond.

10.2 LAGOON SYSTEMS

Treatment trains discussed in Sec. 10.1 reflect a relatively high level of treatment technology. Lagoon systems represent a lower level of such technology.

The organics in waste discharges (the entire size spectrum of particulates) are often stabilized by retaining the discharges in large, relatively shallow ponds. Stabilization (conversion of organics to inorganics) in these ponds results from the combined metabolic activity of bacteria and algae. When ponds are designed and operated in such a manner that stabilization is completely aerobic, they are referred to as *oxidation ponds*. When conditions in a pond are anaerobic, or alternately aerobic and anaerobic, the pond is commonly called a *waste stabilization lagoon*.

Oxidation Ponds

The process in which organics are transformed in oxidation ponds involves a symbiotic relationship between two major groups of microorganisms—the bacteria and the algae. As depicted in Fig. 10.2.1 bacteria utilize organic waste materials for growth and energy, the latter being provided through an oxidation of a portion of organic carbon to carbon dioxide. The carbon dioxide, along with ammonia and other nitrogenous decomposition products that are released by hydrolysis, is utilized in algal growth (photosynthesis). Oxygen is produced in the process, which makes possible further bacterial oxidation. A predator population consisting primarily of zooplankton is supported by the bacterial and algal populations and contributes to the production of carbon dioxide. In short, an ecosystem exists in the pond that has all the features of the aquatic ecosystem discussed in Sec. 4.6.

Photosynthesis is discussed in detail in Sec. 3.6. Figure 3.7.3 depicts the energy flow through the biotic component of an oxidation pond. From this figure it is apparent that under certain circumstances the quantity of energy locked up in the living protoplasm discharged in the effluent may exceed that entering the system in waste organics. If energy degradation were the sole criterion for waste treatment, oxidation ponds would hardly qualify as treatment devices.

The chemical and biologic relationships existing in oxidation ponds are so complex that no completely rational method has been devised for the design of the ponds. However, a semirational method has been proposed in which pond surface area is related to the efficiency with which solar energy is utilized in photosynthesis.[1] The method presupposes that no deposition occurs and that the content of the pond remains homogeneous throughout.[2]

Waste Stabilization Lagoons

Waste stabilization lagoons are generally utilized in the treatment of wastes containing settling solids. The solids upon settling to the bottom of the lagoon decompose anaerobically. The nonsettling portion of the waste undergoes either aerobic or anaerobic decomposition depending upon the particular waste loading on the lagoon.

Attention is directed to the schematic diagram presented in Fig. 10.2.2. Depicted therein is a section of a lagoon receiving a waste with settleable solids. The settleable solids accumulate in proximity to the influent pipes. That fraction of the liquid that does not seep through the bottom of the pond or evaporate to the atmosphere ultimately finds its way to the surface and over the weir of the effluent structure. In some lagoons the organic loading is sufficiently high to deplete large regions of the lagoon of oxygen. Because there is little mixing taking place in the lagoon during daylight hours, aerobic and anaerobic zones result.

Most of the photosynthetic activity takes place in the aerobic zone. The depth of this zone may reach 1 ft or more during periods of high sunlight intensity. At night the zone may disappear entirely. A major portion of the bacterial decomposition may occur in the sludge deposits and the anaerobic zone. The products of anaerobic decomposition either are decomposed further as they pass up through the aerobic zone, or escape in bubbles to the atmosphere.

[1] W. J. Oswald and H. B. Gotaas, Photosynthesis in Sewage Treatment, *Trans. ASCE*, vol. 122, pp. 73–105, 1957.

[2] Since oxidation ponds designed by such a method are highly efficient, they are commonly referred to as *high-rate ponds*.

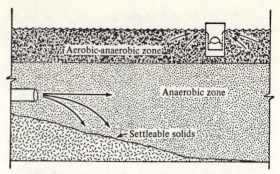

FIGURE 10.2.2
Schematic diagram of a waste stabilization lagoon.

Lagoon effluents are collected at the surface over weirs. The effluent characteristics, therefore, are essentially the same as those of the top layers of the pond. The diurnal variations in pH and dissolved oxygen of a lagoon effluent are illustrated in Fig. 10.2.3. The curves presented therein are idealized to emphasize the relationships between these properties and solar radiation. The high photosynthetic activity during the daylight hours is reflected by the dissolved-oxygen curve. Although there may not be any measurable dissolved oxygen for several hours during the night, supersaturation often will occur during many of the daylight hours. Periods of high pH coincide with high photosynthetic activity. Such pH values reflect the high uptake and, hence, the low concentration of carbon dioxide by the photosynthesizing algae. The lag of dissolved oxygen and pH behind solar radiation suggests a period in which carbon dioxide rather than solar radiation is limiting to the photosynthetic process. Spot checks of dissolved oxygen and pH at the 2-ft depth (in a 5-ft lagoon) during periods of high photosynthetic activity have yielded values of 0 mg/l and 7, respectively.

Based on the BOD parameter, efficiencies of waste stabilization lagoons may reach as high as 90 percent. However, because of the high algal content of the effluents the COD parameter will indicate only a 50 to 60 percent treatment efficiency.

From the foregoing discussion, it appears that the photosynthetic process is not necessarily a major factor in the reduction of organics in a waste stabilization lagoon. Although the oxygen concentration builds up to high values during the daylight hours, such a buildup may occur in a region limited in magnitude and with a pH too high to permit active bacterial decomposition.

The popularity of waste stabilization lagoons is due primarily to low cost both in initial outlay and operation. Their greatest disadvantage is an erratic behavior resulting from empirical design practices. Such practices include

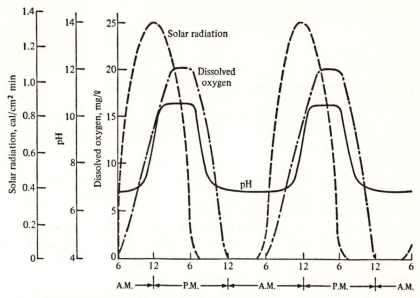

FIGURE 10.2.3
Idealized curves derived from continuous measurements of pH and dissolved oxygen in the effluent of a waste stabilization lagoon.

arbitrarily establishing the depth at 3 to 5 ft and basing the pond area on daily loadings varying from 15 to 50 lb of 5-d 20°C BOD per acre. Although some attention is paid to climatic conditions, little consideration is given to the scientific principles involved.

A semirational design relationship has been developed[1] based on the assumptions that (1) over the time frame of significance the lagoon is completely mixed and (2) the removal mechanism demonstrates first-order kinetics. Using Eq. (4.1.1) to express a weight balance across the lagoon, one has

$$V\frac{dc_1}{dt} = Qc_0 - Qc_1 - Vkc_1 \qquad (10.2.1)$$

where c_0, c_1 = concentrations of pollutant in influent and effluent, respectively. $[FL^{-3}]$

Q = flow into and out of lagoon. $[L^3 t^{-1}]$

V = volume of lagoon. $[L^3]$

k = first-order rate constant. $[t^{-1}]$

---

[1] G. v. R. Marais and V. A. Shaw, A Rational Theory for the Design of Sewage Stabilization Ponds in Central and South Africa, *Civ. Eng. South Afr.*, vol. 3, no. 11, pp. 1–23, 1961.

At steady state, $dc_1/dt = 0$, and Eq. (10.2.1) can be written as

$$c_1 = \frac{c_0}{1 + kV/Q} \quad (10.2.2)$$

For the ith lagoon in a series of lagoons of the same size

$$c_i = \frac{c_{i-1}}{1 + k_i V/Q} \quad (10.2.3)$$

and if $k_1 = k_2 = \cdots = k_n$,

$$c_n = \frac{c_0}{(1 + kV/Q)^n} \quad (10.2.4)$$

where n = number of lagoons

The first lagoon in a series is generally called the *primary lagoon*. The major portion, if not all, of the settleable solids will settle out in the primary lagoon. According to Eq. (10.2.2), long retention times V/Q favor increased removals. In practice, however, it has been found that nuisance conditions (mostly odors) develop in the primary lagoon when the retention time is too long. Such conditions apparently result from the complete elimination of the aerobic layer at the surface of the lagoon. At pH values normally occurring in this layer during the day (pH > 8), the odor-producing hydrogen sulfide H_2S is almost completely dissociated into the ionic constituents H^+ and HS^-. See Fig. 2.1.1.

Equation (10.2.2) does not take into account the temperature effect. Such an effect can be expressed in terms of the variability of the rate constant:

$$k = k_0 \Theta^{T-T_0} \quad (10.2.5)$$

where k, k_0 = rate constants at temperatures T and T_0, respectively

Θ = temperature coefficient

Solution procedure 10.2.1 From data collected in South Africa, an empirical relationship has been developed between the maximum allowable BOD in a primary lagoon and the lagoon depth:

$$c_m = \frac{750}{0.6z + 8} \quad (10.2.6)$$

where c_m = maximum allowable BOD in lagoon, mg/l

z = lagoon depth, ft

For lagoons following the primary lagoon (secondary lagoons) a minimum retention time of 7 d is recommended.

For South African lagoons, it has been found that the rate constant for BOD removal is 0.17 d^{-1}, whereas for coliform removal the rate constant is 2 d^{-1}. Although there is no evidence that the value of the rate constant for bacteria decreases with increasing retention time, evidence does exist that the rate constant for BOD removal decreases with increasing retention time. ////

10.3 INDIVIDUAL HOUSEHOLD SYSTEMS

Whenever possible, the use of municipal waste-water treatment facilities should be given preference over individual household systems. However, where municipal facilities are unavailable, the individual system—the septic-tank system—can often be used with satisfactory results, provided that the soil conditions are favorable and that the system is designed properly.

The septic-tank system consists of two components, each with its own treatment objective and special requirements. The waste water from the originating household discharges first into a septic tank, a concrete box of rectangular shape with an average hydraulic retention time of approximately 2 d. See Fig. 10.3.1. The gross particulates settle to the bottom, while at the same time the flotage (grease, oil, etc.) separates and rises to the surface where it is kept from the discharge of the tank by baffles. The organics in the solids at the bottom of the tank and the scum at the top undergo anaerobic digestion, a process which converts part of the organics to carbon dioxide and methane. Periodically, accumulated solids have to be removed.

The effluent from the septic tank flows to a tile system by which the waste water is spread uniformly at a subsurface level over a soil absorption field. The organics remaining in the waste water are thus introduced to an environment in which they decompose aerobically. See Fig. 10.3.2.

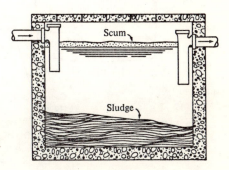

FIGURE 10.3.1
Septic tank.

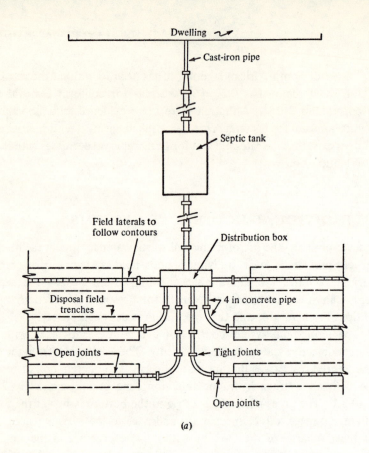
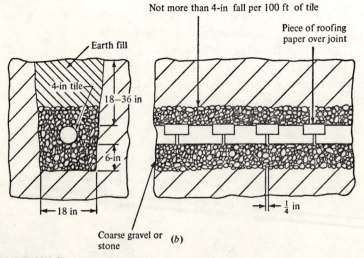

FIGURE 10.3.2
Septic-tank system. (*a*) Layout for disposal field; (*b*) details of absorption-field tile line.

The volume of the septic tank is generally based on the number of persons being served by the system or by the number of bedrooms. The area required for the absorption field is generally established on the basis of a soil percolation test. This test involves the determination of the rate at which water subsides in a test hole dug in the absorption-field area. The relationship between the rate of subsidence, as determined by the soil percolation test, and the maximum allowable rate of waste-water application to the absorption field has been found to be

$$q = \frac{195}{t^{1/2}} \quad (10.3.1)$$

where q = maximum rate of application, $m^3/(m^2)(d)$
t = rate of subsidence, min/cm

Careful attention should be paid to the topography and water supplies when locating a septic-tank system.

EXERCISES

10.1.1 Correlate each type of treatment train with the size spectrum of particles removed.

10.1.2 Why is order generally important in the sequence of treatment trains?

10.1.3 Using the notation discussed in Sec. 10.1.1, identify treatment systems to prepare water for domestic use from (*a*) groundwater, (*b*) surface water, and (*c*) domestic waste water.

10.1.4 Construct treatment systems for water reused in the sequence S_{23}, S_{34}.

10.1.5 If the BOD removal efficiencies of the GPR, SPR, and DMR trains are 0.40, 0.80, and 0.90, respectively, what will be the BOD effluent concentration if the influent to the treatment system has a concentration of 300 mg/l?

10.1.6 What is the significance of the process loading intensity?

10.2.1 Construct a sketch, similar to the one in Fig. 4.6.1, of the type of ecosystem that is normally found in an oxidation pond. Identify the more important subsystems and interactions.

10.2.2 Construct a sketch of the type of ecosystem that is normally found in a waste stabilization lagoon. Identify the more important subsystems and interactions.

10.2.3 What is the rationale for basing the process loading intensity of waste stabilization lagoons on loading per unit of surface area? Is such a rationale valid? Why?

10.2.4 What type of effluent does one expect from the surface of a waste stabilization lagoon? From middepth? From the bottom?

10.2.5 What is the role of photosynthesis in waste stabilization lagoons?

10.2.6 Identify the function of a waste stabilization lagoon with respect to particle size removed.

10.2.7 Design a two-lagoon system to accommodate 200 people assuming the per capita waste-water flow to be 0.38 m^3/d and the 5-d 20°C BOD to be 77 g/d. Assume further that the removal rate constant will be equal to 0.17 d^{-1} and an 80 percent removal efficiency is the goal.

10.2.8 Assuming the removal rate constant for coliform bacteria is 2 d^{-1}, compare the coliform removal efficiency of the two-lagoon system in Exercise 10.2.7 with a single lagoon having a volume equal to the sum of the volumes in the two-lagoon system.

10.3.1 Identify the function of the components of a septic-tank system with respect to particle size removed.

10.3.2 Sketch a layout of a septic-tank system, and indicate the critical design specifications.

11

PROCESSES USED IN GROSS-PARTICULATE REMOVAL TRAINS

11.1 SCREENING PROCESSES

Gross-particulate removal (GPR) trains are constructed from those processes that are effective in removing the gross-particulate fraction from process streams. The processes include screening, gravity sedimentation, and flotation. GPR trains when used in waste treatment systems are commonly referred to as *primary* treatment units.

Coarse screens, sometimes called *bar screens*, are generally employed as the first unit in a gross-particulate removal train. These devices are used to remove large materials which would damage equipment, interfere with the satisfactory operation of a process or equipment, or be the cause of some other objectionable condition. Coarse screens are composed of vertical or inclined bars spaced with openings of from $\frac{3}{4}$ to 6 in. They may be cleaned manually or mechanically. Materials removed by the coarse screens are usually buried or incinerated.

The velocity at which water flows through a screen is a critical factor in the screening process. The lower the velocity through the screen, the greater will be the removal efficiency. However, a velocity too low will result in deposition of solids in the channel containing the screen. As a rule, the cross-sectional area of a screen channel is designed so that the velocity of flow through the screen is not less than 2 ft/s nor greater than 3 ft/s. The cross-sectional area can be calculated from

$$S = \frac{Q}{e\overline{V}} \qquad (11.1.1)$$

where S = cross-sectional area of screen channel. $[L^2]$

Q = volumetric rate of flow. $[L^3 t^{-1}]$

$\overline{V}$ = average velocity through screen. $[Lt^{-1}]$

e = ratio of area of screen openings to cross-sectional area of channel

In those situations where the gross particulates include relatively large organic particles, a grinding mechanism, called a *comminuting device*, is used in place of, or in addition to, coarse screens. Instead of removing the large organic particles, the latter are ground up into smaller particles and left in the water for subsequent removal in the primary sedimentation basin. When used, comminuting devices are generally located between the grit chamber and primary sedimentation unit.

Fine screens are sometimes used in place of sedimentation tanks. Such screens are mechanically cleaned devices of perforated plate, woven-wire cloth, or closely spaced bars. Screen openings are usually $\frac{3}{16}$ in or less.

11.2 SEDIMENTATION PROCESSES

Sedimentation plays a key role in GPR trains. *Grit chambers*, or tanks, are used to remove the heavy inorganic particulates. Other tanks, called *primary sedimentation basins* or *primary clarifiers*, are used to remove lighter organic materials having a specific gravity slightly greater than unity. Still other tanks, called *flotation basins*, are used to remove through gravity separation materials with specific gravities less than unity. Such materials are generally liquids, or semiliquids, that are transported in water in emulsified form but which separate out under quiescent conditions.

Particles settle out of a suspension in one of four different ways, depending upon the concentration of the suspension and the flocculating properties of the particles. The effect of these factors on the settling regime is indicated in Fig. 11.2.1. When particles having little tendency to flocculate (coalesce upon contact with each other) settle out of a dilute suspension, the settling regime is identified as *class-1 sedimentation*. If the particles flocculate, the settling regime in a dilute suspension is labeled *class-2 sedimentation*. In flocculent suspensions of intermediate concentration, particles are close enough together to permit interparticle forces to hold the particles in a fixed position relative to each other.

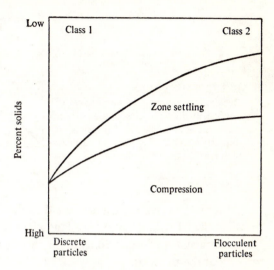

FIGURE 11.2.1
Sedimentation regimes.

As a result, the mass of particulates subsides as a whole in a regime described as *zone settling*. *Compression* occurs when the concentration becomes high enough so that the particles come into actual contact with each other and the weight of particles is supported, in part, by the structure of the compacting mass. The classical laws of sedimentation discussed in Sec. 1.4 apply only to the settling of discrete, nonflocculating particles in dilute suspensions (class-1 sedimentation).

The process loading intensity for sedimentation processes, which is expressed in terms of hydraulic application, takes its significance from the following development: Consider the sketch in Fig. 11.2.2. Here a dilute suspension of nonflocculating particles having a uniform size and shape occupies a rectangular volume. Under quiescent conditions, the particles subside at their terminal settling velocity, and the water at any depth becomes clarified as soon as those particles initially at the top surface pass through that depth. The rate of clarification, therefore, can be computed from

$$Q = \frac{z}{t} A = uA \qquad (11.2.1)$$

where Q = volumetric rate of clarification. $[L^3 t^{-1}]$

z = distance through which particles settle in time t. $[L]$

A = cross-sectional area of rectangular volume in plane perpendicular to direction of subsidence. $[L^2]$

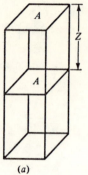
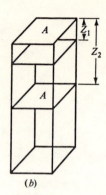

FIGURE 11.2.2
Rectangular volumes of suspension of nonflocculating particles settling under quiescent conditions. (*a*) Particles with uniform settling velocities; (*b*) particles with two settling velocities.

From Eq. (11.2.1), it is evident that the clarification capacity of a basin in which class-1 sedimentation is taking place theoretically is independent of the basin depth and is a function only of the surface area of the basin and the settling velocity of the particles.

Consider now a situation in which water is withdrawn from the top of the rectangular volume at a rate Q_2 computed from Eq. (11.2.1) for a particle size having a settling velocity of u_2. The water will be free of particles having a settling velocity equal to or higher than u_2. From Fig. 11.2.2 it is apparent that the weight-fraction removal of particles having some lower settling velocity u_1 is

$$x_1 = 1 - c_r = \frac{z_1}{z_2} = \frac{u_1}{u_2} \qquad (11.2.2)$$

where c_r = ratio of concentration of particles remaining in water c to initial concentration c_0

For a suspension containing a distribution of particles with different sizes and, hence, different settling velocities, the removal efficiency is a function of the hydraulic application to the surface area, commonly called the *surface overflow rate*:

$$E = f(q) \qquad (11.2.3)$$

where $q = Q/A$ = surface overflow rate. $[Lt^{-1}]$

In class-2 regimes clarification is a function not only of the settling properties of the particles but also of the flocculating characteristics of the suspension. Particles subsiding from such suspensions overtake and coalesce with smaller particles to form particles which settle at rates greater than did the parent particles. The greater the liquid depth, the greater will be the opportunity for contact. Therefore, class-2 sedimentation differs from class-1 in that removal is dependent both on the surface overflow rate and on the depth.

The rate of flocculation is known to be proportional to the mean velocity gradient in the system, the concentration of particles, and particle size. However, there is no satisfactory formulation available for evaluating the flocculation effect on sedimentation, and it is necessary to perform a settling-column analysis in order to measure this effect.

Solution procedure 11.2.1 The overall removal from a dilute suspension of nonflocculating particles (class-1 regime) having different settling velocities can be predicted on the basis of a settling analysis.[1] In such an analysis, the suspension is placed in a settling column constructed so that samples of the suspension may be drawn off at a certain depth. At varying time intervals, samples are drawn off, and the concentration in each sample is determined. Each sample contains no particles with settling velocities great enough to transport them from the surface past the sampling depth during the settling period. The maximum settling velocity of particles in each sample, therefore, is approximately

$$u_j = \frac{z}{t_j} \quad (11.2.4)$$

where subscript notation provides an index for the particular time interval at which the sample is collected.

If the initial concentration is uniform throughout the column, the concentration of all the particles in a sample whose settling velocities are less than u_j is the same as in the initial suspension. Hence, a settling-velocity curve such as is presented in Fig. 11.2.3 can be established as a characteristic of the suspension.

For a given surface overflow rate q_k, where

$$q_k = u_k \quad (11.2.5)$$

those particles having settling velocities $u \geq u_k$ will be completely removed. Such particles constitute $1 - c_{rk}$ of the total originally in the clarified water. According to Eq. (11.2.2) the weight fraction of particles having velocities $u < u_k$ will be

$$\int_0^{c_{rk}} \frac{u}{u_k} dc_r$$

The overall removal from the clarified water, therefore, will be

$$x_T = 1 - c_{rk} + \frac{1}{u_k} \int_0^{c_{rk}} u \, dc_r \quad (11.2.6)$$

---

[1] T. R. Camp, Sedimentation and the Design of Settling Tanks, *Trans. ASCE*, vol. 111, pp. 895–936, 1946.

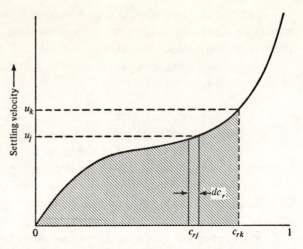

FIGURE 11.2.3
Settling-velocity analysis curve for suspension of nonflocculating particles.

The last term in Eq. (11.2.6) can be evaluated either by graphic integration or by numerical integration of the settling-analysis curve between the limits of $c_{rj} = 0$ and c_{rk}. ////

Solution procedure 11.2.2 A settling-column analysis of a class-2 suspension consists of withdrawing samples (under quiescent conditions) from several depths at different time intervals. The relationship between the weight fraction of particles removed in each sample and the depth and time that the sample was collected can be illustrated by a coordinate plot such as is shown in Fig. 11.2.4. Here the index values given to the weight fractions removed x_{ij} correspond to the particular depth z_i and time t_j at which each sample was collected. The curves illustrate typical isoconcentration lines, that is, lines connecting points of equal removal.

The isoconcentration lines describe a depth-time ratio equal to the minimum average settling velocity of the fraction of particles indicated. For example, x_D of the particles in the suspension have had an average velocity greater than z_2/t_2 by the time the depth of z_2 is reached. By the time the depth z_4 is reached, the same fraction of particles have had an average velocity of z_4/t_3 or greater. The curvature of the lines reflects the flocculating nature of the suspension. Isoconcentration lines for a settling suspension of nonflocculating particles would be linear.

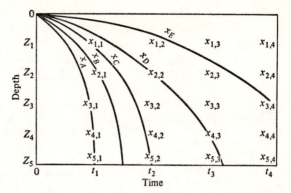

FIGURE 11.2.4
Fractional removal of flocculating particles as a function of time and depth.

The overall removal in a basin of depth z_5 at a surface overflow rate $q_{5,2}$, where

$$q_{5,2} = \frac{z_5}{t_2} \quad (11.2.7)$$

is computed as follows: From Fig. 11.2.4 it is seen that during the time interval t_2, $x_{5,2}$ of the particles have had an average settling velocity of z_5/t_2 or greater. Of the remainder, $x_{5,2} - x_{4,2}$ have had an average velocity of $(z_5 + z_4)/2t_2$; $x_{4,2} - x_{3,2}$ have had an average velocity of $(z_4 + z_3)/2t_2$; etc. Thus the overall removal is

$$x_{T5,2} = x_{5,2} + \frac{(z_5 + z_4)/2t_2}{z_5/t_2}(x_{5,2} - x_{4,2})$$

$$+ \frac{(z_4 + z_3)/2t_2}{z_5/t_2}(x_{4,2} - x_{3,2}) + \cdots$$

$$+ \frac{(z_1 + 0)/2t_2}{z_5/t_2}(x_{1,2} - 0) \quad (11.2.8)$$

or more generally for any depth z_i and surface overflow rate q_{ij}

$$x_{Ti,j} = x_{i,j} + \sum_{k=1}^{i} \frac{z_k + z_{k-1}}{2z_i}(x_{k,j} - x_{k-1,j}) \quad (11.2.9)$$

The solution of Eq. (11.2.9) is facilitated with the use of Algorithm 11.2.1. ////

Overall removal from a class 2 settling regime, Eq. (11.2.9).

$$x_{Tij} = x_{ij} + \sum_{k=1}^{i} \frac{z_{k-1} + z_k}{2z_i} (x_{k-1,j} - x_{k,j})$$

Note: Index value i and j correspond to the depth and time, respectively, at which the value of x was measured.

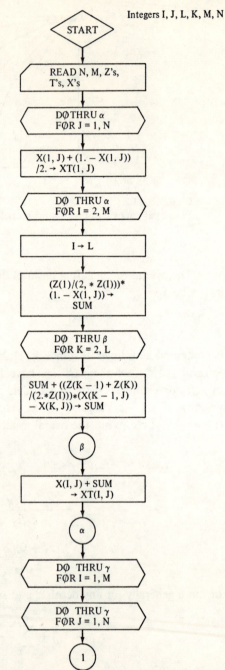

ALGORITHM 11.2.1

11.3 GRIT CHAMBERS

Inorganic particles (grit) are generally removed from a process stream first so as to (1) protect the moving mechanical equipment from abrasion and excessive wear, (2) reduce the incidence of pipe clogging, and (3) reduce the frequency at which tanks must be cleaned to remove accumulations of inorganics at the bottom of the tanks. Removal is accomplished by differential sedimentation through velocity control. An attempt is made to keep the flow velocity through the grit chamber low enough so that the heavier inorganics will settle out but high enough to keep the organics in suspension. Such a velocity is in the neighborhood of 1 ft/s. Velocity control is maintained by a variety of devices, most of which can also be used to measure flow through the system. One type of grit chamber is illustrated in Fig. 11.3.1a. Mechanical cleaning mechanisms may or may not be included.

Most inorganic suspensions dealt with in treatment systems exhibit class-1 settling behavior. Consequently, Eq. (11.2.1) is applicable to the rational design of grit chambers. Often such a design has been based on the removal of particles of 0.20 mm in size and larger. Since the size distribution of inorganics found in waste waters varies considerably, it is advisable to perform a settling-column analysis on the suspension to be treated prior to making a design. A settling-column analysis is described in Solution Procedure 11.2.1.

Traditionally, grit chambers have been designed on an empirical basis. At a 1 ft/s velocity, a retention time of 20 s to 1 min has been used for dimensioning the basin.

Solution procedure 11.3.1 The sizing of a grit chamber in which velocity control is critical may be approached in the following manner: The surface area required to remove all particles having a settling velocity u or greater is

$$A = xy = \frac{Q}{u} \quad (11.3.1)$$

where x = length of basin

y = width of basin

The cross-sectional area of the basin can be expressed as

$$S = yz = \frac{Q}{\bar{V}} \quad (11.3.2)$$

By solving the equations for Q and combining one has

$$x = \frac{\bar{V}}{u} z \quad (11.3.3)$$

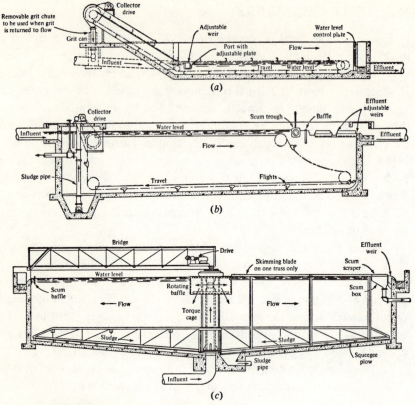

FIGURE 11.3.1
Typical horizontal-flow sedimentation basins. (*a*) Mechanically cleaned grit-removal basin; (*b*) rectangular basin, mechanically cleaned; (*c*) circular basin, mechanically cleaned. (*From Link-Belt Book* 2617, FMC Corporation, Environmental Equipment Div.)

Equation (11.3.3) is solved for the minimum depth to accommodate a scraper mechanism in the basin and for the upper limit of velocity control. For the length of basin thus determined, the depth of flow is then investigated for the lower limit of velocity control. ////

11.4 PRIMARY SEDIMENTATION BASINS

Particles that are primarily organic in nature are flocculent, and those suspended in influents to GPR trains tend to settle out in a class-2 regime. Particles subsiding from such suspensions overtake and coalesce with smaller particles to form particles which settle at rates higher than did the parent particles. The greater the water depth, the greater will be the opportunity for contact. Therefore, in sedimentation basins designed for the removal of organic particles in dilute suspensions, removal efficiency is dependent both on the surface overflow rate and on depth. In practice, primary sedimentation basins are used with depths ranging from 7 to 12 ft and with surface overflow rates which vary from 600 to 1,000 gal/(d)(ft^2). A settling-column analysis of the suspension to be removed provides a more rational basis for design.

Sedimentation processes as part of GPR trains are carried out in rectangular, circular, and square basins. The basin structure consists of four functional components, each with its own special problems of hydraulic and process design: Feed is introduced through an *influent structure*. Sedimentation takes place in the *settling zone*. Clarified water is collected by an *effluent structure*, and sludge accumulates and is withdrawn from a *sludge zone*. When sludge volumes are large, mechanical equipment is used for the continuous removal of sludge.

Typical examples of sedimentation basins are shown in Fig. 11.3.1*b* and *c*. In rectangular basins, feed is introduced at one end along the width of the basin and collected at the surface, either across the other end or at different points along the length of the basin. When employed, mechanical sludge-removal equipment pushes the settled solids down a gently sloping bottom (about 1 percent) toward a hopper at one end of the basin, where they are removed from the system. The length-to-width ratio will vary from 3:1 to 5:1, with 4:1 being the most common. Most basins are 7 to 10 ft in depth. Some, however, are as deep as 15 ft.

Circular basins from 35 to 150 ft in diameter and 7 to 12 ft deep are currently used in treatment and renovation processes. Feed is brought into the basin through a center well. The clarified effluent is collected at weirs along the periphery of the basin. A scraper mechanism forces the settled sludge down a sloping bottom (generally 6 to 8 percent) to a central sludge hopper.

Multiple, Inclined-Surface Devices

According to Eq. (11.2.1) the rate of clarification is directly proportional to the cross-sectional area in a plane perpendicular to the direction of subsidence. However, problems of flow distribution and sludge removal impose practical limitations to the utilization of shallow basins. These problems are avoided by using other methods of increasing the cross-sectional area. Two proprietary devices are available for such a purpose.

One device consists of modules of inclined tubes with small hydraulic radii which are suspended in a basin. The suspension, generally of a type that settles in a class-2 regime, enters the tube system from a distribution chamber, flows up through the tubes at an angle of 45 to 60° with the horizontal, and is discharged into a collection gallery. Inside the tubes, the particles settle to the tube bottoms and become trapped by a stream of settled particles flowing downward, countercurrent to the suspension. At the lower end of the tubes, the settled particles discharge into the distribution chamber from which they are removed by scraper mechanisms. The tubes may be of circular, square, or rectangular cross section.

Another device consists of parallel, inclined plates suspended in a basin designed so that the flow of the suspension is from top to bottom. As a result of the cocurrent flow arrangement, frictional drag of the suspension augments the force of gravity in transporting the settled sludge stream to the lower end of the plates.

Theoretically, the flow capacity of a multiple, inclined-surface device is increased according to the relationship

$$Q = \frac{uA}{\sin \theta} \quad (11.4.1)$$

where θ = angle of inclination of surfaces from horizontal

The lower limit to the magnitude of the angle of inclination θ is fixed by the flow characteristics of the settled particles along the surfaces of the device. Sedimentation efficiencies are improved over those of conventional sedimentation basins not only by additional surface area but also by such factors as reduced convection currents and short circuiting. The efficiencies of overloaded sedimentation basins can be improved through installation of these devices in the existing basin structure.

11.5 FLOTATION PROCESSES

Flotation is a unit operation employed in the separation of solid and liquid particles from a liquid phase. Separation is facilitated by the presence of fine bubbles resulting from the introduction of a gas phase, usually air, to the system. The rising bubbles either adhere to or are trapped in the particle's structure, thereby imparting buoyancy to the particles or increasing the buoyancy of the particles. Particles having a density even greater than that of the liquid phase can be separated by flotation. Separation by flotation does not depend so much on the size and relative density of particles being removed as it does on the surface properties.

Two methods of flotation are currently in use: In *dispersed-air flotation*, gas bubbles are generated by introducing the gas phase through a revolving impeller or through porous media. Bubble size is on the order of 1,000 μm in diameter.[1] Dispersed-air flotation is used to a wide extent in the metallurgical industry. In *dissolved-air flotation*, bubbles are produced as a result of the precipitation of a gas from a solution supersaturated with the gas. The average bubble size ranges from 70 to 90 μm. This method finds wide application in the treatment of industrial wastes. Although the same principles apply in both methods, the following discussion is oriented toward dissolved-air flotation.

Gas-Particle Contact

The mechanism of contact between the gas bubbles and floated particles may be one or both of two types. The first type is predominant in the flotation of flocculent materials and involves the trapping of growing and rising gas bubbles in the growing structure of the floc particle. Here the bond between the bubble and particle is one of physical capture only.

The second type of contact is one of adhesion. Adhesion results from the intermolecular attractions that are exerted at an interface between two phases and which give rise to interfacial tension. The extent to which adhesion will take place can be predicted from a consideration of the interfacial tensions involved in a gas-liquid-solid system.

Figure 11.5.1a shows a gas bubble in contact with a solid particle in a liquid constituting the continuous phase. The angle formed between the gas-liquid interface and the solid-liquid interface at the point where the three phases make contact is called the *angle of contact* α. The equilibrium existing at the point of contact between the interfacial tensions is illustrated by the force diagram in Fig. 11.5.1b. Algebraically the equilibrium can be expressed as

$$\sigma_{gs} = \sigma_{sl} + \sigma_{gl} \cos \alpha \qquad (11.5.1)$$

where σ_{gs} = interfacial tension between gas and solid phases

σ_{sl} = interfacial tension between solid and liquid phases

σ_{gl} = interfacial tension between gas and liquid phases

If σ_{gs} is equal to or less than σ_{sl}, the contact angle α is 0 or cannot exist, in which case a liquid film prohibits contact between the solid and gas phases.

---

[1] E. R. Vrablik, Fundamental Principles of Dissolved-air Flotation of Industrial Wastes, *14th Annu. Purdue Ind. Waste Conf*, 1959.

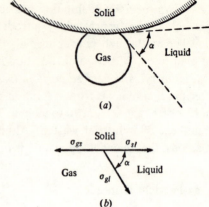

FIGURE 11.5.1
Interfacial tensions existing in a three-phase system.

When σ_{gs} is greater than σ_{sl}, the angle α is greater than 0, and the gas bubble can adhere to the solid particle. The tendency for the gas bubble to adhere to the solid increases with the size of the contact angle.

Although the adhesion mechanism has been discussed above in terms of a solid dispersed phase, the same principles apply in the case of a dispersed phase that is liquid.

Use of Chemical Agents

The entrapment of gas bubbles often is promoted by the use of coagulating chemicals. These chemicals, which include alum, ferric chloride, activated silica, etc., impart a flocculent structure to or increase the flocculent structure of the floated particles, thereby facilitating the capture of gas bubbles.

Several groups of chemicals are used that aid flotation by altering the surface properties of the phases involved. *Frothers*, such as the higher alcohols, pine oil, and cresylic acid, serve to lower the interfacial tension between the gas and liquid σ_{gl}. This promotes bubble formation as well as a more stable froth at the surface of the liquid. *Collectors* either reduce the interfacial tension between the solid and liquid σ_{sl} or increase the tension between the gas and solid σ_{gs}. Both effects serve to increase the angle of contact α. Soap, fatty acids, and amines are in common use as collectors. *Activators* are a group of chemicals that improve the effect of collectors. *Depressants* prevent the flotation of certain phases without preventing the desired phase from becoming floated. Activators and depressants often are referred to as *promoters*.

Dissolved-Air Flotation

Two methods of dissolved-air flotation have been successfully used in waste treatment processes. *Vacuum flotation* is accomplished by subjecting the waste first to a short period of aeration at atmospheric pressure. This step is followed by the application of a vacuum (approximately 9 in Hg), which induces the precipitation of dissolved air from the saturated solution. The fact that only a limited quantity of air can be precipitated out of a solution saturated at 1 atm limits the application of vacuum flotation.

The second method, commonly referred to as *pressure flotation*, involves the solution of air under an elevated pressure followed by its release at atmospheric pressure. A typical arrangement for pressure flotation is shown in Fig. 11.5.2. Equipment components commonly include a pressurizing pump, air injection equipment, a back-pressure regulating device, and a flotation unit. Operational steps are as follows: Air is introduced to the influent usually at the suction end of a centrifugal pump discharging into a retention tank at a pressure ranging from 25 to 50 lb/in^2 (gauge). In passing through the pump, the air is sheared into fine bubbles, thereby facilitating their solution in the retention tank. The latter has a retention time of from 30 to 60 s and is provided with bleed lines for removing any undissolved air that separates in the unit. From the retention tank the waste flows through a back-pressure regulating device to the flotation unit.

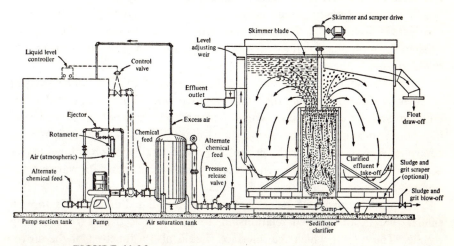

FIGURE 11.5.2
Pressure-flotation system. (*From Sediflotor Clarifier, Infilco Bull.* 6051. *Infilco Division, Westinghouse Electric Company.*)

At the flotation unit, the pressure is reduced to approximately 1 atm. Air bubbles released from solution float the suspended particles to the surface from where they are removed by mechanical scrapers. Clarified liquid is withdrawn at some depth below the surface, while any heavier solids that do not float but settle to the bottom may be moved to the center or to one end of the tank for discharge by a conventional raking mechanism.

The flotation unit may be either rectangular or circular. Units of the latter type may be divided into two compartments: the inner compartment to be used for flotation and the outer as a conventional clarifier.

When high solids-to-gas ratios are permitted, only a portion of the feed or a portion of the flotation-unit effluent may be aerated under pressure. The aerated portion is then mixed with the unaerated waste in the flotation unit.

Design of Flotation Operations

The separation of suspended particles by flotation does not depend as much on the size and relative density of the particles as it does on their structure and surface properties and the quantity of air used in their flotation. For this reason, flotation operations cannot be designed on the basis of mathematical equations, and laboratory tests must be resorted to for preliminary design criteria. Final criteria should be derived from pilot-plant studies.

The factors of greatest importance in the design of a pressure-flotation operation are the feed solids concentration, the quantity of air used, and the overflow rate. Retention time is important when thickening of the float is considered. Laboratory investigations are directed toward evaluating the effect of these variables on the float solids concentration and the suspended solids concentration in the effluent.

The feed solids concentration and the quantity of air used can be grouped together in a dimensionless ratio and correlated with effluent suspended solids and float solids concentration. No particular advantage is obtained by increasing the air-to-solids ratio beyond certain values. The quantity of air used is controlled by varying either the proportion of the total flotation volume that is aerated or the pressure employed to accomplish the aeration. Pressure-saturation relationships are discussed in Sec. 1.2.

In most suspensions being floated, a sludge-liquid interface is formed similar, but in an inverted relationship, to that produced in a zone settling regime. The rate at which the interface rises is also a function of the air-to-solids ratio. A laboratory study of the rise rate is basic to an estimate of the overflow rate. In practice, overflow rates vary from 1 to 4 gal/(ft^2)(min).

As a rule, a 6-ft depth is required in the flotation unit to minimize turbulence and short circuiting. For this depth and for the overflow rates indicated above, retention time will vary from 10 to 40 min.

EXERCISES

11.1.1 Determine the overall dimensions of a bar screen to service a domestic waste water with a flow range of 7 to 10 ft^3/s. The bar screen is to consist of 2 × $\frac{5}{16}$-in bars, inclined at an angle of 60° with the horizontal. Determine also the area of the approach channel.

11.1.2 What will be the head losses during maximum and minimum flow if the head-loss relationship is $h(\text{ft}) = 0.0222(\overline{V}_1^2 - \overline{V}_2^2)$, where $\overline{V}_1$ is the average velocity through the screen and $\overline{V}_2$ is the average velocity in the approach channel (both velocity terms are in feet per second)?

11.2.1 Identify the characteristics of each of the following:
(*a*) Class-1 sedimentation
(*b*) Class-2 sedimentation
(*c*) Zone settling

11.2.2 In Sec. 11.2, sedimentation is discussed within the context of clarified water being collected from the top of a rectangular volume of a suspension. How might this concept be applied to a sedimentation basin in which the suspension is introduced at one end and clarified water collected at the other end?

11.2.3 A settling analysis was performed on a dilute suspension of nonflocculating particles. The following data were recorded for samples collected at the 1.20-m depth:

Settling time, min	0.5	1.0	2.0	4.0	6.0	8.0
Weight fraction remaining	0.56	0.48	0.37	0.19	0.05	0.02

For a clarification rate equal to 0.0244 m^3/(s)(m^2), what is the overall removal?

11.2.4 A settling analysis is performed on a dilute suspension of flocculent particles. The following data, expressed in terms of weight fraction removed, were recorded for samples at three different depths:

Depth, cm	Settling time, min					
	10	20	30	45	60	120
61	0.31	0.59	0.63	0.70	0.73	0.78
122	0.18	0.49	0.61	0.65	0.69	0.75
183	0.18	0.40	0.61	0.65	0.68	0.71

Compute overall removals in basins with depths equal to 1.83 m for overflow rates corresponding to the range of data collected.

11.3.1 Determine the overall dimensions of a rectangular grit chamber designed to remove particles 0.20 mm in diameter and larger from a waste water with a flow of 7 to 10 ft^3/s. A scraper mechanism is to be used to clean the bottom of the tank and a proportional weir will maintain the velocity between 1.0 and 1.2 ft/s.

11.4.1 Make a sketch of a rectangular sedimentation basin, noting average dimensions. What important specifications would you apply to the four functional components of such basins?

11.4.2 Explain how multiple, inclined-surface devices improve sedimentation. What types of such devices are available?

11.5.1 Describe the nature of the mechanism of contact between gas bubbles in dissolved-air flotation and the floated particles.

11.5.2 What types of chemical agents are used to improve flotation, and how are they effective?

11.5.3 A waste containing 400 mg/l of suspended solids and with a flow of 600 gal/min is to be treated by pressure flotation using aerated effluent recycle. Laboratory investigations reveal that an air-to-solids ratio of 0.03 is an optimum value relative to the quality of effluent and float solids concentration and that the initial rise rate of the float interface at this ratio is 0.4 ft/min. Additional investigations showed that after the clarified effluent is saturated with air at 25°C under a pressure of 50 lb/in^2 (gauge), 5.6×10^{-4} lb of air will be released from 1 gal of effluent when the pressure is reduced to 1 atm. On the basis of this information, estimate the effluent recycle and dimensions of a flotation unit to treat the waste when the latter is at 25°C.

12

PROCESSES USED IN SUSPENDED-PARTICULATE REMOVAL TRAINS

12.1 ACTIVATED-SLUDGE PROCESSES

Suspended-particulate removal (SPR) trains may consist of one or both of two types of treatment processes. If the particulates to be removed are primarily organic in composition, biologic oxidation processes are generally employed. On the other hand, if the particulates are inorganic or if they are organic but present in relatively low concentrations, the rapid-sand-filter process is used. The effluents of biologic oxidation processes will always contain a residual concentration of suspended particles which are primarily organic. For this reason, a water renovation system may include a biologic oxidation process followed by a rapid-sand-filter process. Two types of biologic oxidation processes will be discussed in connection with SPR trains—activated-sludge processes and trickling-filter processes. The principles of biologic oxidation are discussed in Chap. 3.

Process Fundamentals

Several types of processes are collectively called activated-sludge processes. Although varying in detail, these processes have in common the contacting of the process stream with preformed biologic floc in an aerated tank. The end

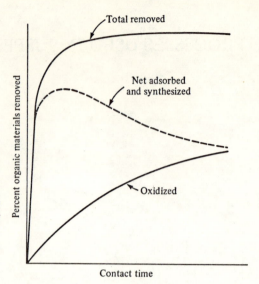

FIGURE 12.1.1
Removal of organic materials upon contact with activated sludge in an aerated batch system as a function of time. (*Adapted from C. C. Ruchhoft et al., Studies of Sewage Purification IX Total Purification, Oxidation, Adsorption, and Synthesis of Nutrient Substrates by Activated Sludge, fig.* 10, *U.S. Public Health Serv. Rep., vol.* 54, *no.* 12, *pp.* 468–496, *March* 1939.)

result is that a portion of the organics amenable to biologic degradation is converted to inorganics while the remainder is converted to additional activated sludge.

Activated sludge consists of a flocculent assemblage of microorganisms, nonliving organic matter, and inorganic materials. The microorganisms, which include bacteria, molds, protozoans, and metazoans, such as rotifers, insect larvae, and worms, are related to each other in a food chain. Bacteria and molds decompose complex organic compounds to produce, through growth, cellular material on which protozoans feed. The protozoans, in turn, are consumed by metazoans. The latter may also feed directly upon the decomposers.

The term "activated" stems from a unique property of activated sludge. The surfaces of the floc are highly active in adsorbing colloidal and suspended materials from solution. This property is illustrated by the curves in Fig. 12.1.1, showing the removal of organic materials from solution upon contact with activated sludge in an aerated batch system. The top curve defines total removal as a function of contact time. The progress of biologic oxidation is followed in the lower curve. The dotted curve is drawn through points derived from the differences between corresponding points along the other two curves and represents that portion of the organics adsorbed and synthesized to protoplasm. It is to be noted that initially removal is due almost entirely to adsorption. Synthesis is proportional to biologic oxidation and, hence, would contribute little initially to the total of the quantities adsorbed and synthesized.

That portion of the organics adsorbed initially that is not immediately oxidized or used in synthesis is stored in the activated-sludge floc. The composition of activated sludge during this period can be expressed as

$$C_a H_b O_c N_d P_e S_f \cdot C_x H_y O_z$$

where $C_x H_y O_z$ represents the stored organics. When the full storage capacity of the sludge has been utilized, the sludge floc is no longer active in the adsorptive sense. Activity can be restored only after a period of aeration during which the stored material is utilized in oxidation and synthesis. The aeration process through which activity is restored is called *sludge stabilization*.

The concentration of activated-sludge suspensions is generally expressed in terms of *volatile solids*. The latter is a measure of the organic fraction of the activated-sludge floc.

Many types of activated-sludge processes are currently employed in treatment technology. These include the conventional activated-sludge process, step-aeration process, contact-stabilization process, extended-aeration processes, and high-rate processes. The process loading intensity, PLI, in current use in the design and control of activated-sludge processes is expressed in terms of weight of organics applied per unit of time per unit weight of the volatile suspended solids fraction of the activated-sludge suspension in the aeration basin. The PLI and normal efficiencies for several types of activated-sludge processes are listed in Table 12.1.1.

Conventional Activated-Sludge Process

A diagram of the conventional activated-sludge process is shown in Fig. 12.1.2. It is to be noted that the process is carried out in two treatment units—an aeration basin and a sedimentation unit, called a *thickening tank*.

Table 12.1.1 PROCESS LOADING INTENSITIES AND EFFICIENCIES FOR ACTIVATED-SLUDGE PROCESSES

Type of activated-sludge process	Process loading intensity L*	Process efficiency $E \times 100$†
Extended-aeration process	0.05–0.20	75–85
Conventional process	0.20–0.50	90–95
Step-aeration process	0.20–0.50	90–95
Contact-stabilization process	0.20–0.50	85–90
High-rate process	0.50–5.00	60–85

* Pounds of 5-d 20°C BOD per day per pound of volatile suspended solids [(lb BOD_5)/(d)(lb VSS)].
† In terms of BOD_5 removal.

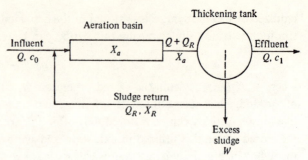

FIGURE 12.1.2
Flow diagram of a conventional activated-sludge process.

The process stream is mixed with recycled sludge and aerated for a period of 4 to 8 h in long, narrow aeration basins. From the aeration basins, the mixture of activated sludge and process stream, called *mixed liquor*, flows to a thickening tank from which clarified liquid is discharged at the top and the sludge solids are removed in the underflow. A portion of the underflow usually equal in volume to from 10 to 30 percent of the process-stream flow is returned to be mixed with the influent. The remainder of the underflow (a quantity minor compared with that returned) is wasted from the process. The concentration of activated sludge maintained in the mixed liquor varies from 1,500 to 2,500 mg/l, while the concentration in the returned sludge underflow is on the order of 4 to 5 times that in the mixed liquor.

As the process is normally utilized, the ratio of the concentration of organics in the process stream to the concentration of activated sludge is low. As a result, the growth increment through the process is small in comparison with the returned activated sludge. Sludge is wasted from the process in a quantity equal to the growth increment.

Two distinct stages in the biologic process can be followed as the mixed liquor flows through the aeration basin: The first is a *clarification stage* in which most of the colloidal and suspended organics are adsorbed to the surfaces of the floc. This stage takes place immediately and extends only a short distance from the influent end of the basin. The second stage, or *stabilization stage*, occupies the major portion of the aeration basin. During this period, the organics which are stored during the clarification stage are utilized in growth and oxidation. Since the organic concentration is highest at the influent end, synthesis and, hence, oxygen demand are greatest initially.

The PLI expressed in terms of weight of organics applied per unit of time per unit weight of the volatile suspended solids concentration in the mixed liquor can be computed from

$$L = \frac{Qc_0}{X_a V} \quad (12.1.1)$$

where L = process loading intensity. $[t^{-1}]$
Q = volumetric rate of flow of influent. $[L^3 t^{-1}]$
c_0 = concentration of organics in influent. $[FL^{-3}]$
X_a = concentration of volatile suspended solids in mixed liquor. $[FL^{-3}]$
V = volume of aeration basin. $[L^3]$

Rearranging Eq. (12.1.1) to solve for the required capacity of the aeration basin gives

$$V = \frac{Qc_0}{LX_a} \quad (12.1.2)$$

The retention time in the basin, and hence, the aeration time, is

$$t = \frac{V}{Q + Q_R} \quad (12.1.3)$$

where Q_R = volumetric flow of sludge return. $[L^3 t^{-1}]$

The concentration of the volatile suspended solids in the mixed liquor X_a is determined by the thickening properties of the activated sludge and the rate at which concentrated sludge is returned to the aeration basin. This relationship can be developed from a weight balance of volatile suspended solids around the thickening tank. Assuming no solids loss in the effluent and by ignoring the growth increment withdrawn as excess sludge, one has

$$X_a(Q + Q_R) = X_R Q_R \quad (12.1.4)$$

where X_R = concentration of volatile suspended solids in sludge return. $[FL^{-3}]$

Rearranging to solve for the concentration of volatile suspended solids in the mixed liquor gives

$$X_a = X_R \frac{Q_R/Q}{1 + Q_R/Q} \quad (12.1.5)$$

In conventional activated-sludge processes the sludge recycle rate Q_R will normally range between 10 and 30 percent of the influent flow Q. When considering the value of the concentration of volatile suspended solids in the return sludge X_R for design purposes, it is desirable to perform thickening studies on the activated sludge developed from the particular process stream to be treated.

This is done to determine the maximum concentration to which the sludge can be thickened.

When process control is the objective, the rate of sludge return becomes the parameter of concern. Equation (12.1.4) can be rearranged to give

$$Q_R = \frac{X_a Q}{X_R - X_a} \quad (12.1.6)$$

In such cases the value of the concentration of volatile suspended solids in the return sludge in milligrams per liter can be estimated from

$$X_R = \frac{10^6}{I} \quad (12.1.7)$$

where I = sludge-volume index, ml/g

The sludge-volume index is the volume in milliliters occupied per gram of activated sludge after a 1-l sample of the mixed liquor collected at the outlet of the aeration tank has been allowed to settle in a graduated cylinder for 30 min. Its value is determined from

$$I = \frac{1{,}000 P_V}{X_a} \quad (12.1.8)$$

where P_V = volume concentration of suspended solids in mixed liquor, ml/l

The inconsistency, here, of using a ratio of the volume concentration of the total suspended solids to the weight concentration of the *volatile* suspended solids is ignored.

The specific relationship between the removal efficiency and the PLI for any particular process stream can be established with data derived from a batch process. Often it is found that the relationship approximates

$$E = 10^{-mL} \quad (12.1.9)$$

where m = constant

The performance of the thickening tank is an important factor in determining the overall efficiency of the activated-sludge process. The surface area required in such tanks is based primarily on two factors: the clarification and thickening capacities. The clarification capacity is stated in terms of the surface overflow rate, whereas the thickening capacity is stated in terms of the weight of solids applied per unit of time per unit of thickener surface area. Each capacity is determined by a particular characteristic of the suspension being thickened. The surface area of the thickening tank is made large enough to accommodate either capacity.

In practice, thickening tanks used in activated-sludge processes have surface overflow rates of 800 to 1,000 gal/(d)(ft^2). Solids loading capacities of these units range from 15 to 30 lb/(d)(ft^2). Both capacities are best determined from batch settling tests on the activated-sludge suspension to be thickened.

The net production of activated sludge resulting from the conversion of organics to cellular growth can be estimated from

$$W = Y_N Q(c_0 - c_1) \quad (12.1.10)$$

where W = net sludge production. $[Ft^{-1}]$

Y_N = net yield

c_0, c_1 = concentration of organics in process influent and effluent, respectively. $[FL^3]$

Q = volumetric flow of process stream. $[L^3 t^{-1}]$

The net production of sludge is wasted from the process and must be processed by some sludge treatment train prior to disposal.

Two types of aeration devices are used in activated-sludge processes—mechanical aerators and air diffusers. Plate and tube air diffusers are generally used in the conventional process. These diffusers, which are located along the bottom of the aeration tank, supply air at pressures of from $5\frac{1}{2}$ to 8 lb/in^2, depending on the tank depth. For domestic waste waters, 0.5 to 1.5 ft^3 of air is employed per gallon of waste water treated. Aeration not only supplies the oxygen required for biologic oxidation but also provides the agitation necessary to keep the activated sludge in suspension. Aeration tanks used in the conventional process have depths of from 10 to 15 ft. The length of such tanks varies from 100 to 400 ft, and the ratio of length to width generally exceeds 5:1. Thickening tanks employed in activated-sludge processes are generally round and similar to the type of sedimentation tank shown in Fig. 11.3.1c.

Step-Aeration Processes

The step-aeration process is a modification of the conventional activated-sludge process. (See Fig. 12.1.3.) The process stream in three or four equal increments is introduced to the system at different points, equal distances apart, along the aeration basin. The organic load is distributed over the length of the basin. As a result, accelerated growth and oxidation are not confined to one end of the basin as in the conventional process but, instead, take place over most of the basin. The difference between the oxygen-demand patterns of the conventional process and step aeration is illustrated in Fig. 12.1.4.

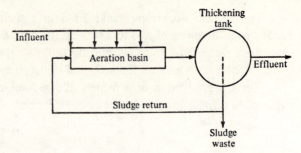

FIGURE 12.1.3
Flow diagram of step-aeration process.

By distributing the organic load over the length of the basin, shorter retention times and lower activated-sludge concentrations in the mixed liquor are possible.

Contact-Stabilization Process

The contact-stabilization process was developed to take advantage of the adsorption characteristics of activated sludge. A flow diagram of the process is presented in Fig. 12.1.5.

Settled waste is mixed with returned activated sludge in an aerated contact basin having a nominal retention time of 20 to 40 min. During this period, organics are adsorbed by the sludge floc in what constitutes the clarification stage of the conventional process. Following the contact period, the activated sludge is separated from the mixed liquor in a thickening tank. A small portion of the sludge is wasted, while the remainder flows to a stabilization basin. Here

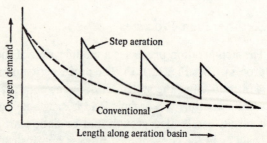

FIGURE 12.1.4
Comparison between oxygen demands exerted in conventional and step-aeration activated-sludge processes.

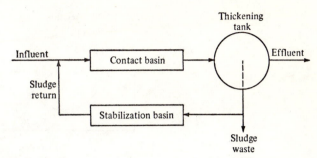

FIGURE 12.1.5
Flow diagram of contact-stabilization process.

the sludge is aerated for 1.5 to 5 h depending upon the strength of the waste. During this period, the stored organics are utilized in growth and respiration and, as a result, become *stabilized* or *activated*. The stabilized sludge is returned to the contact basin to be mixed with the incoming waste.

Use of the process permits a reduction in aeration-basin capacity. Whereas in the conventional process the entire volume of the mixed liquor is aerated through the stabilization stage, in contact stabilization, aeration through this stage is confined to the return sludge volume.

Extended-Aeration Process

The extended-aeration process is carried out in aeration basins designed for relatively long retention times (30 to 36 h). All the underflow from the thickening tank is returned to the aeration basin. As a result, the activated-sludge solids are recycled to a point where they are highly oxidized. The only solids that are wasted from the system are those that are discharged in the overflow from the thickening tank. The solids in the overflow account for the somewhat lower efficiency of the process. The extended-aeration process is used most often in small installations where minimal supervision is anticipated.

High-Rate Process

The conventional, step-aeration, and contact-stabilization processes accomplish organic removals on the order of 85 to 95 percent. High-rate processes are utilized where removals of a lower degree are acceptable.

The high-rate process is similar to the conventional process except that the return sludge is equal in volume to about 5 to 10 percent of the waste flow and

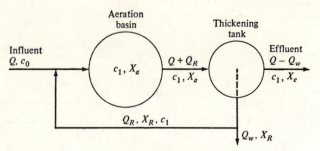

FIGURE 12.1.6
Flow diagram of a completely mixed activated-sludge process.

the aeration time is limited to 2 to 3 h. The high ratio of organics to activated sludge results in a high growth rate but a poorly clarified effluent.

Completely Mixed Process

The completely mixed activated-sludge process is finding application in the treatment of industrial wastes containing high concentrations of organic materials. The influent is distributed uniformly throughout the aeration basin, thereby protecting the activated sludge from shock loadings commonly encountered in the treatment of such wastes. Furthermore, the design and operation of the completely mixed process can be established on a more rational basis than is possible with other modifications of the activated-sludge process.[1]

The basic models for the completely mixed process are discussed in Sec. 4.4. However, the models discussed therein do not apply to systems which include sludge recycle. Attention is directed to Fig. 12.1.6. The latter is a flow diagram of a completely mixed process with sludge recycle. In order to develop the following design and operational relationships, two assumptions are made: It is assumed, first, that all waste utilization by the microorganisms takes place in the aeration basin and, second, that the total biomass in the system is equal to the biomass in the aeration basin; i.e., the biologic solids in the thickening tank are ignored.

The process loading intensity as applied to the activated-sludge process has one major deficiency: Normally, this parameter, as a matter of convenience, is expressed in terms of the weight of organics applied per unit of time per unit weight of the volatile suspended solids concentration in the mixed liquor. As a

---

[1] A. W. Lawrence and P. L. McCarty, Unified Basis for Biological Treatment Design and Operation, *J. Sanit. Eng. Div., ASCE,* vol. 96, no. SA3, pp. 757–778, June 1970.

biologic parameter, the PLI should be expressed in terms of the weight of organics applied per unit of time per unit weight of biologically active solids, or viable cell mass. For a completely soluble waste the two types of PLI may be approximately the same. For normal domestic waste waters, however, the volatile solids concentration may be 2 to 5 times that of the biologically active solids.

The biologic solids retention time θ_s can be used in place of the PLI as a design and control parameter. Furthermore, it is more readily measurable and easier to control. The solids retention time is defined as the total active solids in the system divided by the total quantity of active solids wasted per unit of time:

$$\theta_s = \frac{VX_a}{(Q - Q_w)X_e + Q_w X_R} \quad (12.1.11)$$

where X_a = concentration of biologically active solids in aeration basin. $[FL^{-3}]$

V = aeration-basin volume. $[L^3]$

X_e, X_R = concentrations of biologically active solids in effluent and recycle, respectively. $[FL^{-3}]$

Q_w = flow rate of fraction of underflow from thickening tank which is wasted from system. $[L^3 t^{-1}]$

It is to be noted that when no recycle is practiced the solids retention time is equal to the hydraulic retention time V/Q or θ.

A weight balance of the active solids across the system in Fig. 12.1.6 yields

$$V\frac{dX_a}{dt} = \hat{\mu} \frac{c_1}{K_s + c_1} X_a V - k_d X_a V - (Q - Q_w)X_e - Q_w X_R \quad (12.1.12)$$

The biologic coefficients $\hat{\mu}$, K_s, and k_d are defined in Sec. 3.3.

By multiplying Eq. (12.1.12) by the ratio $1/VX_a$, setting $dX_a/dt = 0$, and rearranging, a steady-state form of Eq. (12.1.12) is

$$\frac{1}{\theta_s} = \hat{\mu} \frac{c_1}{K_s + c_1} - k_d \quad (12.1.13)$$

Solving for the concentration of organic substrate in the effluent, Eq. (12.1.13) becomes

$$c_1 = \frac{K_s(1 + k_d \theta_s)}{\theta_s(\hat{\mu} - k_d) - 1} \quad (12.1.14)$$

At minimum retention time, i.e., that retention time below which the biologically active solids are washed out of the system, $c_1 = c_0$, and the steady-state form of Eq. (12.1.13) can be expressed as

$$\frac{1}{\theta_{sc}} = \hat{\mu} \frac{c_0}{K_s + c_0} - k_d \quad (12.1.15)$$

where θ_{sc} = critical solids retention time. $[t]$

Where $c_0 \gg K_s$, Eq. (12.1.15) becomes

$$\theta_{sc} = \frac{1}{\hat{\mu} - k_d} \quad (12.1.16)$$

The total weight of the biologically active solids present in the aeration basin can be determined from an expression developed from a weight balance of the organic substrate across the aeration basin:

$$V \frac{dc_1}{dt} = Qc_0 + Q_R c_1 - (Q + Q_R)c_1 - \frac{1}{Y} \hat{\mu} \frac{c_1}{K_s + c_1} X_a V \quad (12.1.17)$$

With rearrangement, the steady-state form of Eq. (12.1.17) can be solved for the total weight of biologically active solids in the basin:

$$X_a V = \frac{YQ\theta_s(c_0 - c_1)}{1 + k_d \theta_s} \quad (12.1.18)$$

Furthermore, a relationship between the active solids retention time θ_s and the hydraulic retention $V/Q = \theta$, the hydraulic recycle ratio Q_R/Q, and the solids concentration ratio X_R/X_a can be developed from a solids balance across the aeration basin:

$$V \frac{dX_a}{dt} = Q_R X_R - (Q + Q_R)X_a + \hat{\mu} \frac{c_1}{K_s + c_1} X_a V - k_d X_a V \quad (12.1.19)$$

At the steady state and with the substitution of Eq. (12.1.13), Eq. (12.1.19) can be written

$$\frac{1}{\theta_s} = \frac{Q}{V}\left(1 + \frac{Q_R}{Q} - \frac{Q_R}{Q}\frac{X_R}{X_a}\right) \quad (12.1.20)$$

Equations (12.1.13), (12.1.18), and (12.1.20) are of key importance in the design and operation of the completely mixed activated-sludge process. The use of these equations is dependent upon prior knowledge of the values of the biologic coefficients Y, $\hat{\mu}$, K_s, and k_d for the particular organic substrate concerned. Such values vary from substrate to substrate and must be determined for each individual substrate. A procedure for determining the values of these coefficients is described in Solution Procedure 4.4.1. The biologic coefficients calculated for several organic substrates in aerobic, mixed culture systems are listed in Table 12.1.2. In waste treatment, biologic coefficients are generally calculated on the basis of BOD_5 or COD removal.

The procedure to be followed in the preliminary design of the completely mixed activated-sludge process is as follows: The solids retention time required to achieve an effluent with a given concentration of organics remaining c_1 is determined from Eq. (12.1.13). Using this value of θ_s along with the design flow rate Q, the weight of active solids in the aeration basin can be computed with Eq. (12.1.18).

The maximum possible concentration of active solids in the recycle X_R is a function of the thickening characteristics of the solids suspension. The evaluation of these characteristics is discussed in Sec. 14.1. Once X_R has been determined experimentally for the particular waste of concern, the required volume of the aeration basin is calculated for different values of the hydraulic recycle ratio Q_R/Q. The solids retention time θ_s and weight of active solids in the aeration basin $X_a V$, determined earlier, are used in the calculations. The optimal choice of the design volume and hydraulic recycle ratio is made on the basis of economic considerations of capital costs, recycle pumping costs, and aeration costs.

When attention is focused on the operational mode, X_R is determined from the sludge-volume index I. [See Eq. (12.1.7).] A constant solids retention time θ_s can be maintained for fluctuating values of X_R and Q simply by varying Q_R.

Solution procedure 12.1.1 Calculate the volume of the aeration basin in a conventional activated-sludge process treating a waste with a 5-d 20°C BOD of 150 mg/l and a flow of 2 Mgal/d.

Table 12.1.2 BIOLOGIC COEFFICIENTS CALCULATED FOR SEVERAL ORGANIC SUBSTRATES IN AEROBIC, MIXED CULTURE SYSTEMS*

Organic substrate	Y	$\hat{\mu}$, d^{-1}	K_s, mg/l	k_d, d^{-1}	Coefficient basis	Temperature, °C
Domestic waste	0.5			0.055	BOD_5	
Domestic waste	0.67			0.048	BOD_5	20
Skim milk	0.48	2.4	100	0.045	BOD_5	
Glucose	0.42	1.2	355	0.087	BOD_5	
Peptone	0.43	6.2	65		BOD_5	30
Domestic waste	0.67	3.7	22	0.07	COD	

* Taken from A. W. Lawrence and P. L. McCarty, Unified Basis for Biological Treatment Design and Operation, table 2, *J. Sanit. Eng. Div., ASCE*, vol. 96, no. SA3, p. 768, June 1970.

ASSUMPTIONS
1. Process loading intensity $L = 0.35$ d^{-1}.
2. Recycle ratio $Q_R/Q = 0.25$.
3. Sludge-volume index $I = 100$ ml/g.

SOLUTION
1. From Eq. (12.1.7)

$$X_R = \frac{10^6}{100} = 10,000 \text{ mg/l}$$

2. From Eq. (12.1.5)

$$X_a = 10,000 \frac{0.25}{1 + 0.25} = 2,000 \text{ mg/l}$$

3. From Eq. (12.1.2)

$$V = \frac{2 \times 10^6 (3.785)(150)}{0.35(2,000)} = 1.62 \times 10^6 \text{ l} \qquad ////$$

Solution procedure 12.1.2 Calculate the area of a thickening tank for the conventional process described in Solution procedure 12.1.1.

ASSUMPTIONS
1. Surface overflow rate $q = 1,000$ gal/(d)(ft^2).
2. Solids loading capacity $w = 20$ lb/(d)(ft^2).

SOLUTION
1. Required area based on clarification capacity:

$$A = \frac{2 \times 10^6}{10^3(10.8)} = 185 \text{ m}^2$$

2. Required area based on solids loading capacity:

$$A = \frac{10,000(2 \times 10^6)(0.25)(3.785)}{20(454 \times 10^3)(10.8)} = 193 \text{ m}^2$$

3. Since $193 > 185$, use $A = 193$ m^2. $\qquad ////$

12.2 TRICKLING-FILTER PROCESS

Trickling filters consist of shallow, circular tanks of large diameter filled with crushed stone or similar media. The process stream is applied intermittently or continuously over the top surface of the filter by means of a rotating distributor and is collected and discharged at the bottom. The use of coarse media provides

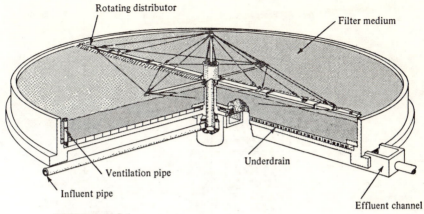

FIGURE 12.2.1
Cutaway view of a trickling filter. (*Dorr-Oliver, Inc.*)

interconnecting void spaces large enough to accommodate simultaneously the process stream passing through the filter and circulating air. For the same reason the collection channels at the bottom of the filter are sized generously. A cutaway view of a trickling filter is shown in Fig. 12.2.1.

The term "filter" is a misnomer because the removal of organic material is not accomplished with a filtering or straining operation. Removal is the result of an adsorption process which occurs at the surfaces of biologic slimes covering the filter media. Subsequent to their adsorption, the organics are utilized by the slimes for growth and energy.

The composition of the biologic slimes is very similar to that of activated sludge. A large community of various species of heterotrophic bacteria supported by the adsorbed organics is fed upon by a predator population consisting of protozoans and small metazoans. Included, also, are inorganic and non-metabolizable organic materials. The zoological nature of the bacteria imparts a slimy consistency to the mass.

The exchanges taking place along the surfaces of the trickling-filter slimes are depicted in Fig. 12.2.2. As it passes through the filter, the process stream contacts the slimes in thin sheets. Concurrent with the adsorption of organics from the liquid onto the slimes, there is discharged to the liquid inorganics resulting from the oxidation of previously adsorbed organics. Oxygen diffuses from the atmosphere to support aerobic oxidation in the outer layers of the slime. Resulting carbon dioxide is discharged to the air.

Growth is most active at the surface of contact where the concentration of organics is the highest. Consequently, the oxygen demand is the greatest at

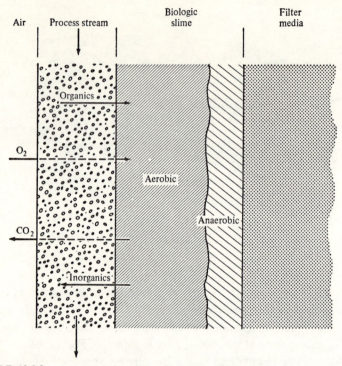

FIGURE 12.2.2
Schematic representation of exchanges taking place along surfaces of biologic slimes in a trickling filter.

this surface, and the amount of oxygen remaining to diffuse to greater depths in the slime layer is limited. After the slime layer builds up to a certain thickness, anaerobiosis becomes established in the portion of the layer farthest from the contacting surface.

The growing slime layer constantly is being scoured by the flow of the process stream through the filter. After the layer reaches a critical thickness, outer portions of the layer slough off and are discharged in the effluent. Sloughing will occur either intermittently or more or less continuously depending upon the hydraulic loading on the filter.

The trickling-filter process embodies both a trickling-filter unit and a sedimentation basin, commonly referred to as a *secondary clarifier*. (See Fig. 12.2.3.) The clarifier is used for the removal of the biologic slimes that are discharged from the filter. Since such slimes are contributed by the process itself and are not a part of the organics in the influent, filter efficiencies are computed on the basis of the quality of effluent from the clarifier.

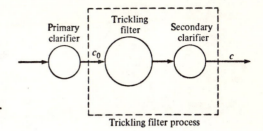

FIGURE 12.2.3
Flow diagram of a single-stage trickling-filter process with no recirculation.

Many trickling-filter processes incorporate recirculation to improve the quality of the final effluent. Although there is no theoretical basis for quality improvement, such factors as more equalized hydraulic loads, better distribution over the filter, and less clogging contribute to increase treatment efficiency when recirculation is utilized.

A wide variety of recirculation schemes are in current use; many are protected by patents. Three popular schemes are presented in Fig. 12.2.4. Despite the wide variation in arrangement, few differences in performance have been observed. Generally, the ratio of recirculated flow to influent flow falls within the range from 0.5 to 3.0, although greater ratios have been used. Some schemes involve recirculation during low-flow periods only, at a constant rate at all times, or at rates proportional to the influent flow.

Two-stage filtration schemes are sometimes used in which two filters are in series either next to each other or separated by an intermediate sedimentation tank. Two-stage processes are accompanied by a recirculation arrangement.

For trickling-filter processes, PLI is expressed in terms of weight of organics applied per unit of time per unit volume of filter medium:

$$L = \frac{Qc_0}{VF} \quad (12.2.1)$$

where L = process loading intensity. $[Ft^{-1}L^{-3}]$

Q = volumetric rate of flow of influent. $[L^3 t^{-1}]$

c_0 = concentration of organics in influent. $[FL^{-3}]$

F = recirculation factor

For process streams composed of a mixture of complex organics, the amount of organics removed when recirculation is practiced decreases with each pass through the filter. The retardant effect is the result of a more rapid removal of those components more readily assimilated by the biologic slimes. The re-

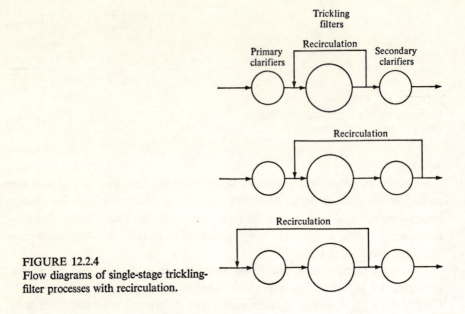

FIGURE 12.2.4
Flow diagrams of single-stage trickling-filter processes with recirculation.

circulation factor takes this retardant effect into account. The recirculation factor has been expressed as

$$F = \frac{1 + Q_R/Q}{(1 + 0.1 Q_R/Q)^2} \quad (12.2.2)$$

where Q_R = volumetric rate of recirculated flow. $[L^3 t^{-1}]$

The relationship existing between the removal efficiency and the PLI for any particular process stream can be derived experimentally from pilot-plant data. Such a relationship often takes the form of that given in Eq. (12.1.9).

Table 12.2.1 COMPARISON OF LOW-RATE AND HIGH-RATE TRICKLING FILTERS

Feature	Low-rate	High-rate
Hydraulic loading, gal/(d)(ft^2)	25–100	200–1,000
Organic loading, (lb BOD$_5$)/(1,000 ft^3)(d)	5–25	25–300
Depth, ft		
Single-stage	5–8	3–6
Multistage	2.5–4	1.5–4
Dosing interval	Intermittent	Continuous
Recirculation	Generally not included	Always included
Effluent	Highly nitrified, 20 mg/l of BOD$_5$	Not fully nitrified, 30 mg/l or more of BOD$_5$

General practice has been to use an empirical relationship for process design. A bewildering variety of such relationships have been developed, mostly from data collected at domestic waste treatment installations. One of the most popular relationships of this type was developed by the National Research Council. The NRC formula for a single-stage unit can be written as

$$E = (1 + 0.0085\sqrt{L})^{-1} \qquad (12.2.3)$$

where L is expressed in terms of pounds per day of 5-d 20°C BOD of settled waste water per acre-foot of filter medium.

As a result of empirical development, trickling filters are designed and operated either as low-rate filters or as high-rate filters. A comparison of the two types is outlined in Table 12.2.1.

12.3 RAPID-SAND-FILTER PROCESS

As indicated in Sec. 12.1, the rapid-sand-filter process is used in SPR trains when the particulates to be removed are inorganic or exist as organics in relatively low concentrations. The process is carried out in three basic units: Chemicals, or a solution of chemicals, are added to the process stream prior to its introduction to a *flocculation basin*. The chemicals react with the natural alkalinity in the process stream or with added alkalinity to form a precipitate in a process called *coagulation*. This process is rapid and may take place even before the process stream enters the flocculation basin. While flowing through the latter, the process stream is agitated gently, a process referred to as *flocculation*. Flocculation induces the growth of precipitate particles, called *floc*, to a size that facilitates sedimentation. As the floc particles grow, suspended particulate materials in the process stream collide with the floc, they are caught and enmeshed, and they become part of the floc.

From the flocculation basin the process stream flows to a *sedimentation basin*, where most of the floc particles with the entrapped suspended particulate materials settle out. The stream is then passed through a *filter* consisting of a bed of sand, or similar materials, to remove the residual of floc particles that fail to settle out in the sedimentation basin. Although it is convenient to think of the three units—flocculation basin, sedimentation basin, and filter—as being separate processes (they must be dealt with as such when designed), they are all part of a single process. All three are required to accomplish the process objective, which is to completely remove the suspended particulates initially in the process stream. Sometimes a single basin will contain both the flocculation and sedimentation processes. In such cases, however, the processes are separated physically in compartments.

Coagulation

Several types of chemicals are used in the coagulation process. Alum (aluminum sulfate) is of common use in such processes. The stoichiometric equation between alum and the natural alkalinity in water can be written as

$$Al_2(SO_4)_3 \cdot xH_2O + 6HCO_3^- \rightarrow$$
$$2Al(OH)_3\downarrow + 3SO_4^{--} + 6CO_2 + xH_2O \quad (12.3.1)$$

Aluminum hydroxide constitutes the precipitated product of the reaction. It is important to note that for every mole of alum reacting six equivalents of bicarbonate alkalinity are required and six moles of carbon dioxide are produced. The decrease in bicarbonate alkalinity and the increase in carbon dioxide results in a reduction in pH (see equilibrium 2 in Fig. 2.3.1).

When insufficient bicarbonate is present in the process stream, alkalinity must be added for the alum reaction to occur. When the added alkalinity is hydroxyl,

$$Al_2(SO_4)_3 \cdot xH_2O + 6OH^- \rightarrow 2Al(OH)_3\downarrow + 3SO_4^{--} + xH_2O \quad (12.3.2)$$

Copperas (ferrous sulfate) is also used as a coagulant. When copperas is added to water having a pH falling in the neutral range,

$$FeSO_4 \cdot 7H_2O + 2OH^- \rightleftharpoons Fe(OH)_2\downarrow + SO_4^{--} + 7H_2O \quad (12.3.3)$$

The formation of ferrous hydroxide (a precipitate) is promoted by increasing the pH. For copperas to be useful as a coagulant in systems devoid of oxygen, the pH must be greater than 9.5. Such high pH values are normally attained by adding lime $Ca(OH)_2$.

When oxygen is present in a system, the ferrous hydroxide is oxidized to ferric hydroxide:

$$4Fe(OH)_2 + O_2 + 2H_2O \rightleftharpoons 4Fe(OH)_3\downarrow \quad (12.3.4)$$

At normal pH values, ferric hydroxide is more insoluble than ferrous hydroxide. Therefore, copperas has greater utility as a coagulant when oxygen is available. Occasionally chlorine is added with copperas to oxidize the iron to the ferric form:

$$6FeSO_4 \cdot 7H_2O + 3Cl_2 \rightleftharpoons 6Fe^{3+} + 6SO_4^{--} + 6Cl^- + 42H_2O \quad (12.3.5)$$

The ferric ion hydrolyzes to ferric hydroxide:

$$Fe^{3+} + 3OH^- \rightleftharpoons Fe(OH)_3\downarrow \quad (12.3.6)$$

Equation (12.3.6) also expresses the hydrolysis of ferric sulfate and ferric chloride, two other coagulants that are used in treatment technology.

Aluminum hydroxide is the most insoluble between the pH values of 5 and 7. At values less than 5, aluminum hydroxide dissociates to form the aluminum ion, whereas at values greater than 7 the hydroxide dissociates producing the aluminate ion:

$$Al^{3+} + 3OH^- \rightleftharpoons Al(OH)_3 \rightleftharpoons AlO_2^- + H^+ + H_2O \quad (12.3.7)$$

Ferric hydroxide, on the other hand, does not exhibit an amphoteric nature.[1] The solubilities of the hydrous oxides of aluminum and iron as a function of pH are presented in Fig. 12.3.1.

The type of coagulant needed will vary with the characteristics of the process stream being treated. Consequently the dosage required to effect optimum coagulation of a specific process stream must be determined in the laboratory.

Coagulation can be facilitated through the addition of polyelectrolytes. These materials are high-molecular-weight long-chain organic polymers with a multitude of ionizable sites along the chain length. When a polyelectrolyte is added to a colloid, the ionizable sites dissociate to form charges on the polymer, the signs of which depend upon the nature of the polyelectrolyte. The charged sites attract colloid particles of opposite charge, and the polymer structure provides a link between numerous particles, thus forming larger particles more easily separated. Since colloid particles in natural waters normally have a negative charge, cationic polyelectrolytes are popular in water treatment, either as a primary coagulant or as a coagulant aid.

In many waters low in natural silica, coagulation is improved through the addition of activated silica. Activated silica upon addition to water forms long-chain inorganic polymers. Although the polymers have no ionized sites, they function as a coagulant aid by forming bridges through adsorption between the hydrous oxide precipitates produced by the solution of primary coagulants.

The coagulant (and coagulant aid, if used) is applied in an operation called a *flash mix* in which the water is agitated violently for a period of 1 to 3 min to disperse the chemicals evenly throughout the bulk of the water. The flash mix can be induced at points of high-velocity flow, such as at the suction end of a pump or a hydraulic jump. More commonly, the flash mix is induced in a small tank provided with a high-speed mixer.

---

[1] An *amphoteric compound* is a compound capable of dissolving both in acid and alkaline solutions.

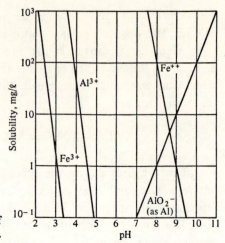

FIGURE 12.3.1
Solubilities of the hydrous oxides of aluminum and iron as a function of pH.

Flocculation

The subsidence velocity of a particle which is spherical or nearly spherical in shape will increase as the particle increases in size. Thus, when the stability of a suspension is such that coalescence of the suspended particles occurs upon contact with each other, any operation that promotes contact between particles will facilitate sedimentation.

The total number of contacts made by particles in a suspension per unit of time is a function of the mean velocity gradient existing in suspensions:

$$N = \Psi(G) \qquad (12.3.8)$$

where G = mean velocity gradient. $[t^{-1}]$

The velocity gradient is induced by agitation and, thus, is a function of power input to the suspension, or friction loss in the case where agitation is the result of the suspension flowing through a baffled basin.

For different values of the velocity gradient, corresponding shear stresses are established in the suspension. Floc particles become more fragile as they grow, and for a given value of G there is a maximum size produced before the floc particle is destroyed by shear stress. The practical limit of the velocity gradient and, hence, turbulence will depend upon the size of particle desired.

Flocculation basins are commonly fitted with paddle devices which rotate around either vertical or horizontal shafts. Peripheral speeds of paddles vary from 0.5 to 3 ft/s. Retention time in the basin ranges from 10 to 30 min.

Sedimentation

After flocculation, the process stream is passed through a sedimentation basin. There under quiescent conditions the floc particles are allowed to settle out. Basins used for such purposes are similar to other types of sedimentation basins except that in many instances mechanical sludge-removal equipment is not used. Mechanical removal equipment becomes necessary only when the organic fraction of the suspended particulates removed from the process stream becomes significant and failure to remove the sludge promptly results in putrescible conditions becoming established in the basin.

The removal efficiency in the sedimentation basin can be expressed by Eq. (11.2.3). Normally, surface overflow rates of 0.25 to 0.50 gal/(min)(ft^2) are used for settling basins designed for the removal of alum floc.

Rapid Sand Filter

The term "rapid-sand-filter process" is used in this text in a generic sense. In addition to coagulation, flocculation, and sedimentation, the term includes filtration through some type of porous medium. The porous medium can be sand, anthracite, or diatomaceous earth. Flow through such filters can be by gravity or under pressure. The term "rapid" distinguishes the filter from the slow sand filter which at one time was used in a process that did not include coagulation and flocculation and which operated at much slower rates. The slow sand filter is primarily of historical interest now.

The object of filtration in the rapid-sand-filter process is to remove the residual floc remaining in the process stream after sedimentation. Several mechanisms are involved in floc removal, the most important of which is thought to be adsorption of the floc on the surface of the filter grains.

The typical rapid sand filter of the gravity type consists of a bed of sand 24 to 30 in deep overlying 18 to 24 in of graded gravel. (See Fig. 12.3.2.) At the bottom of the gravel on the filter floor is a perforated pipe or some other type of underdrainage system. The sand ordinarily used has an effective size of from 0.4 to 0.55 mm and a uniformity coefficient ranging from 1.35 to 1.75.[1] Gravel size will vary from 1½ in at the bottom to ⅛ in at the top of the gravel layer. Filter basins are normally constructed in some even multiple or fraction of 1 Mgal/d and with a length-to-width ratio of approximately 1.25:1. Data relating to the behavior of rapid sand filters under different operating conditions

---

[1] *Effective size* is equal to the sieve size in millimeters that will pass 10 percent (by weight) of media. The *uniformity coefficient* is equal to the sieve size passing 60 percent of the media divided by that size passing 10 percent.

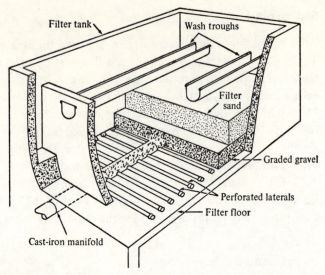

FIGURE 12.3.2
Cutaway view of typical rapid sand filter. (*Reprinted from "Water Treatment Plant Design" book by permission of the Association. Copyright 1969 by the American Water Works Assoc., Inc., 2 Park Avenue, New York, N.Y. 10016.*)

are limited. For this reason, design is still based largely on empirical values.

The process stream—coagulated, flocculated, and settled—is applied to the filter at a rate of 2 to 4 gal/(min)(ft^2) of filter surface. Filtered water is collected by the underdrains and piped to a clear well. When the filter becomes clogged, it is washed by reversing the flow. Filtered water, introduced by the underdrainage system, is forced up through the bed at a rate of approximately 2 ft^3/(s)(ft^2) of filter surface. Backwashing expands the sand bed from 20 to 60 percent above its normal depth. The suspended materials trapped in the top layers of the sand are released to the rising water. The wash water flows over the sides of the wash-water gutter and is carried to a gullet approximately 2 ft wide and extending the length of the filter. From there the wash water discharges into the drain. Normally, a wash period of 4 to 5 min is sufficient to remove the materials clogging the filter. Surface sprays are often used to supplement backwashing. Perforated pipes applying water immediately above the sand surface at a rate of 4 to 6 gal/(min)(ft^2) and at a pressure of 10 lb/in^2 have been found satisfactory for this purpose. Total water used for washing amounts to 1 to 4 percent of the water filtered.

Wash-water gutters of concrete or metal and of different cross sections can be used in a variety of arrangements. As a rule, the gutters are spaced 4 to

6 ft from edge to edge. The edge of the gutter is located slightly above the surface of the sand bed at maximum expansion. The bottom should clear the top of the unexpanded bed by at least 2 in.

The length of filter run usually is fixed by the head loss through the sand bed. As a rule, 8 to 10 ft of head are available for loss through the sand and rate controller. Initially, with a clean filter the loss through the sand will be on the order of 6 to 12 in, with the remainder being taken up by the controller. As the operation proceeds, losses build up in the sand until the controller is wide open. The filter must be backwashed at this point since further increase in head loss through the sand would result in a reduction of flow through the controller. Total operating head (difference between the elevation of the water level over the filter and the maximum water level in the filtered-water clear well) must be at least 10 ft.

Mixed-Media Filter

In simple media filters such as the rapid sand filter, backwashing grades the media hydraulically so that the smallest particles are at the top of the bed. As a result, most of the suspended floc is removed during filtration at or near the surface of the filter bed. Floc which manages to penetrate the upper few inches of the bed is likely to pass completely through the filter. Consequently, only the upper part of the bed is effective in accomplishing the objective of the filter.

The mixed-media bed (for which patents are pending) employs two or more materials of different specific gravities. These materials form a bed with particles uniformly distributed so that the coarse, low-specific-gravity particles are at the top and the fine, high-specific-gravity particles are at the bottom. The result is a filter in which the entire bed is utilized for floc removal. Both the efficiency and length of filter runs are significantly greater than those for rapid sand filters. Furthermore, with the use of filter aids, such as polyelectrolytes, mixed-media beds can be operated at rates as high as 5 gal/(min)(ft^2) or more.

EXERCISES

12.1.1 What factors will determine the choice between a biologic oxidation process and a rapid-sand-filter process for inclusion in a suspended-particulate removal train? Under what circumstances might both be included?

12.1.2 What is "activated sludge," and what are its unique properties? What characteristics do all activated-sludge processes have in common? Explain. What is meant by "sludge stabilization"?

12.1.3 Make a flow diagram of the conventional activated-sludge process, list the major process specifications, and explain the operation.

12.1.4 A waste water with a 5-d 20°C BOD of 400 mg/l and a flow of 1.5 Mgal/d is to be treated with the conventional activated-sludge process. Pilot-plant studies have demonstrated that it is practical to thicken the sludge from the process to 12,000 mg/l (volatile suspended solids). Determine the volume of the aeration tank.

12.1.5 Thickening studies performed on the waste water in Exercise 12.1.4 indicate that the maximum surface overflow rate in the thickening tank will be 800 gal/(d)(ft^2) whereas the maximum solids loading capacity will be 30 lb/(d)(ft^2). Calculate the required area of the thickening tank.

12.1.6 If the net sludge yield to be expected in Exercise 12.1.4 is 0.5, estimate the maximum net sludge production.

12.1.7 After the treatment system in Exercise 12.1.4 has been constructed and is on-stream, the sludge-volume index is found to be 120 ml/g. Determine the sludge return rate required to control the process.

12.1.8 Describe in general detail each of the following modifications of the conventional activated-sludge process:
(a) Step-aeration process   (b) Contact-stabilization process
(c) Extended-aeration process (d) High-rate process

12.1.9 The waste water described in Exercise 12.1.4 is to be treated by the completely mixed activated-sludge process. An effluent with a BOD$_5$ equal to 20 mg/l is desired. Establish the relationship between the sludge recycle rate and the aeration-tank volume. Assume that $Y = 0.5$, $K_s = 50$ mg/l, $\hat{\mu} = 4$ d^{-1}, $k_d = 0.05$ d^{-1}, and $X_R = 12{,}000$ mg/l.

12.1.10 After the treatment system in Exercise 12.1.9 has been constructed and is on-stream, the sludge-volume index is found to be 120 ml/g. Determine the sludge return rate required to control the process.

12.2.1 Explain the removal mechanism in trickling filters.

12.2.2 Sketch and label a trickling-filter system with recirculation.

12.2.3 What basic differences are there between low- and high-rate trickling filters?

12.2.4 Explain the "retardant effect" and the limitation it imposes on recirculation.

12.2.5 Design a high-rate single-stage trickling-filter system to treat the waste described in Exercise 12.1.4. What efficiency do you expect?

12.3.1 Make a flow diagram of the units of the rapid-sand-filter process. Describe the function of each component, the empirical design details, and the relationships between components.

12.3.2 List the coagulants commonly used in coagulation processes. Give the stoichiometric equations for typical reactions involving each of these coagulants.

12.3.3 A water containing 10 mg/l of natural alkalinity (as CaCO$_3$) is to be coagulated using 50 mg/l of alum [as $Al_2(SO_4)_3 \cdot 18 H_2O$]. How much lime must be added to the water to provide sufficient alkalinity for the coagulation process?

12.3.4 How much carbon dioxide will be introduced as the result of the coagulation process in Exercise 12.3.3?

12.3.5 In what pH range is aluminum hydroxide the most insoluble? When oxygen is not present in a system being coagulated with copperas, what is the optimal pH range for coagulation? How can such a pH be induced? What advantage is incurred when oxygen is present in a system being coagulated with copperas?

12.3.6 What materials are used as coagulant aids, and how do they function?

12.3.7 What factors determine the optimal mean velocity gradient in flocculation?

12.3.8 Make a sketch of the cross section of a typical rapid sand filter giving the dimensions of the significant parts. Describe the pertinent operating procedures, and give the normal rates at which the filter is operated.

13
PROCESSES USED IN DISSOLVED-MATERIALS REMOVAL TRAINS

13.1 AERATION PROCESSES

Aeration processes employed in dissolved-materials removal (DMR) trains are used for the purpose of transferring volatile substances to and from water. Such applications include the removal and replacement of carbon dioxide, the addition of oxygen, the removal of hydrogen sulfide, and the removal of various organic compounds responsible for tastes and odors. Aeration processes are carried out in *spray aerators, multiple-tray aerators, cascade aerators,* and *diffusion aerators*. The principles of gas transfer are discussed in Sec. 1.2.

The time rate change in concentration of a gas in solution can be expressed by

$$\frac{dc}{dt} = K_L a(c^* - c) \quad (1.2.6)$$

where c = concentration of gas in solution. $[FL^{-3}]$

c^* = saturation concentration of gas. $[FL^{-3}]$

$K_L a$ = overall volumetric mass-transfer coefficient, liquid-phase basis. $[t^{-1}]$

Upon integration, Eq. (1.2.6) becomes

$$\frac{c^* - c}{c^* - c_0} = \frac{D_2}{D_1} = e^{-K_L at} \quad (13.1.1)$$

where D_1, D_2 = saturation deficit initially and at time t

The efficiency of the process can be written as

$$E = \frac{D_1 - D_2}{D_1} = 1 - e^{-K_L at} \quad (13.1.2)$$

Thus $K_L at$ becomes a key term in the determination of the efficiency of an aeration process.

Spray aerators are arranged so that the water is sprayed upward, vertically or at an inclined angle, in such a manner that the water is broken into small drops. Spray-aeration systems commonly consist of fixed nozzles on a pipe grid. Spray aerators are fairly efficient, particularly for carbon dioxide removal and for the addition of oxygen. However, they require considerable area and are vulnerable to freezing during freezing weather.

The term $K_L at$ in Eq. (13.1.2) for any particular spray-aerator system is a function of the volumetric liquid flow rate and the hydraulic head:

$$K_L at = f(Q, h) \quad (13.1.3)$$

As a rule, from 50 to 150 ft^2 of ground area are required to accommodate a system aerating 1 million gal/d of liquid.

Multiple-tray aerators consist of a vertical stack of trays equipped with slots, perforations, or wire-mesh bottoms over which water is distributed and allowed to fall to a collection basin at the bottom of the stack. In many such aerators, coarse media such as coke, stone, or ceramic balls are placed in the trays to facilitate gas transfer and to take advantage of certain catalytic effects in some applications.

The value of the term $K_L at$ for a specific tray-aerator installation is a function of the volumetric flow rate of the liquid and the number of trays:

$$K_L at = f(Q, n) \quad (13.1.4)$$

As a rule, a tray aerator will have from three to ten trays, 1 to 2 ft apart, and a horizontal area requirement of 25 to 50 ft^2/(Mgal)(d) capacity. When tray aerators are housed in an enclosure, forced ventilation must be provided.

Cascade aerators have only limited application in DMR trains. Generally, these aerators consist of a series of steps down which the water flows in thin layers. Efficiencies are a function of the number of steps. Total area requirements for most applications are on the order of 40 to 50 ft^2/(Mgal of water processed).

Diffusion aerators usually are tanks 10 to 15 ft deep in which the process stream is aerated for 10 to 30 min. Air is diffused through perforated pipes, porous tubes, or porous plates located along one side of the tank at about middepth. As a rule, from 0.01 to 0.15 ft^3 of air is used per gallon of water treated. The advantages offered by diffusion aerators are several: Area requirements are moderate, practically no head is lost through the process, and cold-weather operating problems are minimal.

13.2 CARBON-ADSORPTION PROCESSES

The activated-carbon processes are adsorptive processes by which organic materials of very small size can be removed from a process stream. Removal is accomplished by physical adsorption at the highly active surfaces of the carbon. Such surfaces can be renewed through a destructive-distillation process commonly referred to as *regeneration*.

Activated carbon will remove most organic materials from solution. However, the degree to which any organic material can be removed by a particular carbon varies and must be determined by laboratory tests.

Adsorption of organic materials from solution by activated carbon has been shown to conform to the empirical relationship

$$\frac{c_0 - c}{m} = kc^{1/n} \qquad (13.2.1)$$

where c_0, c = concentration of organic materials in solution, initially and after contact with activated carbon. [FL^{-3}]

m = concentration of activated carbon. [FL^{-3}]

k, n = constants, values of which vary with organic solute and temperature

Adsorption data from batch tests can be evaluated by plotting on logarithmic scales the organics removed per unit of carbon dosage $(c_0 - c)/m$ versus the residual concentration of organics in solution c. The straight line of best fit through the plot is commonly referred to as an *adsorption isotherm*. Extrapolation of the adsorption isotherm through the initial-concentration intercept c_0 yields a measure of the *adsorptive capacity* of the carbon. Tests performed at different temperatures yield other isotherms.

Process streams are contacted with activated carbon in two ways. Powdered carbon can be added to the process stream in the coagulation step of the filtration process in SPR trains. In such applications, dosages of 2 to 200 mg/l are used, and no attempt is made to recover the carbon for reuse.

When the concentrations of dissolved organics are high, removal with granular carbon in columns or counterflow fluidized beds is more economic. In such applications, spent carbon is regenerated and reused. Several processes consist of multiple columns in series through which the process stream is passed in an upflow arrangement.

Actual design details will vary with the size of carbon particle used. However, the rate of application of the process stream to a carbon column is on the order of 7 to 10 gal/(min)(ft^2), with a retention time of 40 to 50 min provided for in the process.

13.3 CHEMICAL-PRECIPITATION PROCESSES

Chemical-precipitation processes involve the removal of undesirable ions from solution through chemical precipitation. Precipitation is induced by other ions added to the solution in the form of chemical compounds. The processes are dependent upon the solubility relationships discussed in Sec. 2.1. Precipitates formed by the processes create fine suspensions which then must be removed by an SPR train. Two chemical-precipitation processes are of considerable importance in water treatment—*water softening* (removal of calcium and magnesium ions) and iron and manganese removal. In waste treatment, chemical precipitation is used to remove phosphorus.

Water Softening

Hardness in water is caused by the ions of calcium, magnesium, iron, manganese, strontium, and aluminum. These ions combine with fatty-acid radicals introduced by soap to produce undesirable precipitates. Practically all the hardness-producing elements can be found in natural waters in some concentration. However, in most waters only the ions of calcium and magnesium are present in sizable concentrations and, hence, are of major concern in the majority of softening processes.

The anions generally present with calcium and magnesium are bicarbonate and sulfate. Occasionally, significant quantities of nitrates and chlorides are present also. The calcium ions are removed by introducing, directly or indirectly, the normal carbonate ion. Since the equilibrium constant for calcium carbonate at 25°C is

$$[Ca^{++}][CO_3^{--}] = K_s = 4.82 \times 10^{-9} \quad (13.3.1)$$

the solubility is quite low, and calcium carbonate is precipitated from solution.

The normal carbonate ion is produced by adding hydroxyl ions to the solution. The latter react with the bicarbonate ions:

$$HCO_3^- + OH^- \rightleftharpoons CO_3^{--} + H_2O \quad (13.3.2)$$

If the concentration of bicarbonate in solution is insufficient to produce the number of normal carbonate ions required to precipitate out with the calcium ions present, then normal carbonate ions are introduced directly to the solution through the addition of a soluble carbonate.

Since the solubility of $Mg(OH)_2$ is quite low,

$$[Mg^{++}][OH^-]^2 = K_s = 5.5 \times 10^{-12} \quad (13.3.3)$$

the magnesium ion too can be removed by adding hydroxyl ions to the solution.

$$Mg^{++} + 2OH^- \rightleftharpoons Mg(OH)_2 \quad (13.3.4)$$

The removal of calcium and magnesium ions from solution must be compensated for by the addition of other cations in order to maintain an ionic balance in the system. This is generally performed by adding a non-hardness-producing cation, such as sodium. The change in the ionic composition of a particular water brought about by softening might be represented as follows:

Before softening		After softening	
Cations	Anions	Cations	Anions
Ca^{++}	HCO_3^-	$3Na^+$	SO_4^{--}
Mg^{++}	SO_4^{--}		Cl^-
	Cl^-		

Softening reactions are not complete, so there will still be some bicarbonate, calcium, and magnesium ions remaining in solution after softening.

Carbon dioxide in solution H_2CO_3 dissociates to form hydrogen and bicarbonate ions. (See Fig. 2.3.1.) Both species of ions react with hydroxyl ions. The overall reaction can be expressed as

$$H_2CO_3 + 2OH^- \rightleftharpoons CO_3^{--} + 2H_2O \quad (13.3.5)$$

When the hydroxyl ion is added to soften water, the demand exerted by dissolved carbon dioxide must be taken into account when computing quantities to be added.

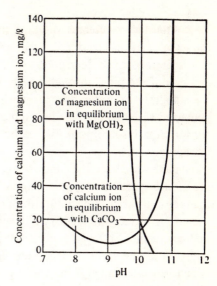

FIGURE 13.3.1
Equilibrium concentrations of calcium and magnesium ions as a function of pH. (*After S. T. Powell, "Water Conditioning for Industry," fig. 6, p. 94, McGraw-Hill, New York, 1954.*)

The hydroxyl ion is conveniently and most economically provided by adding hydrated lime $Ca(OH)_2$ to water. The normal carbonate ion is generally introduced by adding soda ash Na_2CO_3. At normal temperatures the softening reactions are slow and incomplete. However, reactions are considered to go to completion when computing quantities of lime and soda ash required for the process.

Precipitation of calcium carbonate and magnesium hydroxide is facilitated by contacting the water solution with slurries of previously precipitated calcium carbonate and magnesium hydroxide. In some softening processes, alum or some other coagulating agent is used to hasten the sedimentation of the precipitates resulting from the process.

The precipitation of magnesium hydroxide is aided by adding lime in excess of the quantity computed for the process. The excess hydroxyl radical exerts a common ion effect on the solubility of magnesium hydroxide. The equilibrium concentrations of both magnesium and calcium ions as a function of pH are illustrated in Fig. 13.3.1. It is to be noted that maximum insolubility of magnesium hydroxide is reached at a pH value of approximately 10.4. As a rule, approximately 1 (g)(meq)/l of lime is required to elevate the pH to this value.

The softening reactions are accelerated and driven further to completion by increasing the temperature of the water to approximately 212°F. The latter practice, called the *hot process*, is used in the softening of industrial water supplies.

After being softened with lime and soda ash, water is generally unstable in the sense that it will tend to deposit the calcium carbonate precipitates remaining in suspension. Such depositions will, among other things, reduce the carrying capacity of the pipes through which the water is transported to the consumer. To stabilize the water and avoid this problem, softened water is generally contacted with carbon dioxide in a process called *recarbonation*. Recarbonation converts the insoluble calcium carbonate precipitates into calcium and bicarbonate ions.

Plants used for the lime–soda-ash softening of municipal water supplies are of two types: One type consists of a sludge contact unit in which a sludge blanket is utilized to facilitate the precipitation reaction. Water flowing into the unit is mixed with a slurry of previously precipitated solids, lime, soda ash, and coagulant. Softened water separates from the sludge blanket, rises, and is collected at the surface. Units are designed on the basis of surface loadings which generally range from 1 to 3 gal/(min)(ft^2). Equipment for such units is almost all proprietary.

The other type of plant consists of a flash-mix unit, a flocculation basin, and a sedimentation basin. Softening chemicals and coagulant dosage are added to the water in the flash-mix unit. Following a short period (1 to 3 min) of violent agitation, the water is subjected for from 30 to 60 min to gentle stirring to promote the growth of precipitate and coagulant flocs. From the flocculation basin, the water flows to a sedimentation basin having a retention time of from 2 to 6 h. The settled water is then filtered through a rapid sand filter.

Iron and Manganese Removal

Iron and manganese impart the same characteristics to water, have approximately the same chemical behavior, and are removed from water by the same processes. For this reason, the discussion which follows deals primarily with iron, the most commonly occurring of the two elements.

At normal pH values, iron in the ferrous state Fe(II) is quite soluble. When oxygen is added to a system containing the ferrous ion, the latter is oxidized to the ferric state Fe(III):

$$2Fe^{++} + 1\tfrac{1}{2}O_2 + 3H_2O \rightarrow 2Fe(OH)_3 \quad (13.3.6)$$

Since

$$[Fe^{3+}][OH^-]^3 = K_s = 4 \times 10^{-38} \quad (13.3.7)$$

iron precipitates out as ferric hydroxide. However, even with an abundant supply of oxygen the precipitation of ferric hydroxide is impeded when the pH

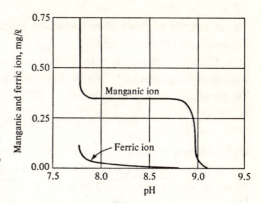

FIGURE 13.3.2
Removal of manganic and ferric ion in treated water at various pH values. (*Based on data presented in R. S. Weston, Manganese in Water, Its Occurrence and Removal, J. Am. Water Works Assoc., vol. 24, p. 1272, 1932.*)

of the water is less than 7.8. The pH effect on the solubility of iron and manganese is illustrated in Fig. 13.3.2.

Iron removal generally results from softening water with lime or soda ash. Removal is also accomplished by aeration. Aeration provides oxygen for the oxidation of ferrous iron to ferric iron. Furthermore, when water contains carbon dioxide in excess of the amount required for the carbonate equilibrium, aeration removes some of the excess, thereby elevating the pH of the water. If aeration fails to elevate the pH above 7.8, lime or soda ash must be added to induce maximum precipitation of ferric hydroxide.

Aeration for the purpose of iron removal is performed with dispersed aerators, sprays, cascading devices, and coke trays. The latter consists of a series of trays, one over the other, filled with coke. Water applied to the top tray trickles over the coke from the top tray to the bottom. The coke, besides providing a maximum of contact surface between the air and water, serves to catalyze the oxidation process. Following aeration, provision must be made for the separation of the iron precipitate from the water.

Phosphorus Removal

Normally, phosphorus is present in untreated waste waters in concentrations of approximately 10 mg/l as P. Such phosphorus exists in three forms—orthophosphate, polyphosphate (dehydrated orthophosphate), and organic phosphorus. Biologic treatment converts most of the polyphosphates and organic phosphorus to orthophosphates. The latter are the most easily removed by precipitation. Orthophosphates exist as three species PO_4^{3-}, HPO_4^{--}, and $H_2PO_4^-$ in equilibrium with each other. The equilibrium is sensitive to pH, and at the pH of most domestic waste waters HPO_4^{--} and $H_2PO_4^-$ are the predominant species. (See Fig. 2.3.5.)

Orthophosphates can be precipitated with lime. Calcium ions combine with phosphate ions and hydroxyl ions to form hydroxylapatite. Assuming that the phosphate is present in the form of HPO_4^{--}

$$3HPO_4^{--} + 5Ca^{++} + 4OH^- \rightleftharpoons Ca_5(OH)(PO_4)_3\downarrow + 3H_2O \qquad (13.3.8)$$

Hydroxylapatite, which has a variable composition, is only slightly soluble. The reaction is pH-sensitive. However, a large portion of the orthophosphates in waste waters can be precipitated at pH values as low as 9.0. As long as $[Ca^{++}] \gg [P]$ in the waste stream being processed, any base that will elevate the pH to 10.0 or above will provide for substantial phosphorus removal.

Alum also has been used to remove phosphorus. In such reactions, the aluminum ion reacts with the orthophosphate ion to form aluminum phosphate:

$$HPO_4^{--} + Al^{3+} \rightleftharpoons AlPO_4\downarrow + H^+ \qquad (13.3.9)$$

Minimum solubility of $AlPO_4$ occurs at approximately pH = 6.0.

Both ferrous and ferric ions react with orthophosphates in much the same manner as does the aluminum ion. See Solution procedure 2.1.3. Minimum solubility of $FePO_4$ is achieved at pH ≈ 5, a pH level not usually attained in waste waters. On the other hand, minimum solubility of $Fe_3(PO_4)_2$ is obtained at pH ≈ 8. Ferrous salts lower the pH of waste waters. For this reason, lime is usually added to raise the pH to a level more favorable to the precipitation of $Fe_3(PO_4)_2$.

Although focus here has been placed on orthophosphate, other forms of phosphorus—polyphosphates and organic phosphorus—will also be removed to varying degrees by the precipitation processes discussed. Where lime is used as the precipitating chemical, the precipitation process is generally included in the DMR train following biologic treatment. In such cases, the process stream may be flocculated, settled, and then filtered prior to pH stabilization by recarbonation. Where alum or iron is used, they may be added either before the primary settling basin or, in those instances where activated-sludge processes are included in the system, in the aeration basin.

Solution procedure 13.3.1 An analysis of a water yielded the following information:

Carbon dioxide	0.4 (g)(meq)/l
Total alkalinity	2.0 (g)(meq)/l
Total hardness	4.0 (g)(meq)/l
Magnesium hardness	1.0 (g)(meq)/l

Calculate the necessary quantities of hydrated lime and soda ash required to soften the water.

SOLUTION

Assuming that the bicarbonate alkalinity is equal to the total alkalinity (an assumption that is approximately true for most natural waters that are hard), the quantity of lime required is:

CO_2	0.4 (g)(meq)/l
HCO_3^-	2.0 (g)(meq)/l
Mg^{++}	1.0 (g)(meq)/l
Excess to elevate pH	1.0 (g)(meq)/l
	4.4 (g)(meq)/l

For each 1 million gal of water, one needs 4.4(74.1/2)(8.34) = 1,360 lb.

The soda ash required to furnish the additional normal carbonate ion required to precipitate out all original and applied calcium ions in solution is

Initial Ca^{++} in water	3.0 (g)(meq)/l
Ca^{++} added to provide OH^- for Mg^{++}	1.0 (g)(meq)/l
Ca^{++} added to provide OH^- for elevating pH	1.0 (g)(meq)/l
	5.0 (g)(meq)/l
CO_3^{--} formed from HCO_3^-	−2.0 (g)(meq)/l
	3.0 (g)(meq)/l

For each 1 million gal of water one needs $3.0(\frac{106}{2})(8.34)$ = 1,330 lb.

No soda ash is added to remove the calcium ions resulting from the neutralization of the carbon dioxide by lime. These calcium ions are required in the subsequent stabilization of the softened water by recarbonation. ////

13.4 ION-EXCHANGE PROCESSES

Ion exchange involves the displacement of ions of given species from insoluble exchange materials by ions of different species when solutions of the latter are brought into contact with the exchange materials. The exchange can be expressed by the general equilibriums

$$B_1^+ + R^- B_2^+ \rightleftharpoons B_2^+ + R^- B_1^+ \quad (13.4.1)$$

$$A_1^- + R^+ A_2^- \rightleftharpoons A_2^- + R^+ A_1^- \quad (13.4.2)$$

where B_1^+, B_2^+ = cations of two different species

A_1^-, A_2^- = anions of two different species

R^-, R^+ = cationic and anionic exchange materials, respectively

The exchange behaves as a chemically reversible interaction between a fixed, ionized exchange site on the exchange material and ions in solution. The equilibrium expression for Eq. (13.4.1) is

$$K = \frac{[B_2^+][R^-B_1^+]}{[B_1^+][R^-B_2^+]} \quad (13.4.3)$$

where K = equilibrium constant sometimes referred to as *equilibrium selectivity coefficient*

A similar expression can be written for Eq. (13.4.2).

When the equilibrium selectivity coefficient is greater than unity, the exchange process is said to involve a favorable equilibrium, and in the case of Eq. (13.4.1) the exchange material will preferentially take up cations of species B_1^+. If a solution containing equal concentrations of cation species B_1^+ and B_3^+ is brought into contact with exchange material $R^-B_2^+$, that species of cation having the greater equilibrium selectivity coefficient with respect to the exchange material will be preferentially picked up over the other species. The relative order in which ions generally displace other ions held by an exchange material is indicated in Table 13.4.1. Ion species appearing low in the series are displaced from the exchange material by species listed higher in the series.

Table 13.4.1 DISPLACEMENT SERIES FOR AN ION-EXCHANGE MATERIAL

Cations	Anions
La^{3+}	SO_4^{--}
Y^{3+}	CrO_4^{--}
Ba^{++}	NO_3^-
Sr^{++}	AsO_4^{3-}
Ca^{++}	PO_4^{3-}
Mg^{++}	MoO_4^{--}
Cs^+	I^-
Rb^+	Cl^-
K^+	F^-
Na^+	OH^-
Li^+	
H^+	

However, it must be emphasized that since equilibriums are involved, displacement is also a function of concentration (or activity in the case of concentrated solutions).

Originally, natural and synthetic aluminosilicates were the only materials used in ion exchange. These materials, commonly called *zeolites*, are now being displaced by synthetic resins and by sulfonated carbonaceous compounds such as coal. The former are becoming increasingly more popular and in the future probably will be used in virtually all new applications. Most of these resins are polymeric materials containing ion-active groups such as SO_3H^- and NH_4^+. Physically, the synthetic resins are spherical and have an average diameter of 0.5 to 0.75 mm.

The capacity of ion-exchange materials can be expressed in a number of ways. In water treatment, exchange capacity has long been expressed in terms of kilograins as $CaCO_3$ per cubic foot of exchange material. However, since ion exchange is now used in applications other than water softening, a more general expression for exchange capacity is needed. A growing preference is noted for the usage of gram equivalents or milliequivalents per some unit of volume.

The capacity of an ion-exchange material will vary with the nature and the concentration of ions in the solution coming into contact with the exchange material, and with the concentration of the regenerating solution. The term *ultimate capacity* is used to describe the capacity of an exchange material in terms of the regeneration method used. The application of ion-exchange materials in most processes is economic only if the materials are regenerated for use.

Although ion exchange has been used to concentrate valuable ionized materials from dilute solutions and for the selective fractionation of certain ionized solutes in solution, its widest application is found in the removal of objectionable ions from water used for domestic and industrial purposes.

When the objective in water treatment is the removal of the hardness ions Ca^{++} and Mg^{++} only, exchange materials are used which operate on a sodium-ion cycle, that is, which exchange Na^+ for Ca^{++} and Mg^{++}. The equilibrium equation for the removal of calcium with such exchange materials is

$$Ca^{++} + R^{--}(Na^+)_2 \rightleftharpoons 2Na^+ + R^{--}Ca^{++} \qquad (13.4.4)$$

When the exchange material is *spent*, that is, when equilibrium is reached, the exchange material in the form of $R^{--}Ca^{++}$ is regenerated to $R^{--}(Na^+)_2$ so that it can be reused. Regeneration is accomplished by contacting the exchange material with a solution of sodium ions. Since the equilibrium selectivity coefficients for equilibriums such as Eq. (13.4.4) have large values, at equilibrium almost all the exchange material is in the $R^{--}Ca^{++}$ form, and solutions with high concentrations of Na^+ are required for regeneration. Normally, regeneration is accomplished using solutions of 5 to 10 percent NaCl. The approximate exchange capacities of several ion-exchange materials along with regeneration requirements are presented in Table 13.4.2. Occasionally seawater and water from brine wells are used for this purpose.

Demineralization (essentially the total removal of all ions) is accomplished by contacting the water being treated with both cationic and anionic exchange materials. Cationic exchange materials used in demineralization operate on a hydrogen cycle and are regenerated with acid. For example, the equilibrium equation for the removal of sodium ions can be written as

$$Na^+ + R^-H^+ \rightleftharpoons H^+ + R^-Na^+ \quad (13.4.5)$$

Anionic exchange materials utilized in demineralization operate on a hydroxyl cycle; a strong base is used as a regenerant. Using the removal of the sulfate ion for an example, one has

$$SO_4^{--} + R^{++}(OH^-)_2 \rightleftharpoons 2OH^- + R^{++}SO_4^{--} \quad (13.4.6)$$

The hydrogen and hydroxyl ions combine in conformance with the equilibrium expression for water (see Fig. 2.3.1), thereby effecting an almost complete deionization.

Although mixed beds are sometimes used in demineralization, more often the water is contacted with the exchange materials in separate units.

Table 13.4.2 APPROXIMATE EXCHANGE CAPACITIES AND REGENERATION REQUIREMENTS OF ION EXCHANGERS*

Exchanger and cycle	Operating capacity, (g)(eq)/ft³†	Regenerator	Regeneration requirements, (g)(eq)/ft³‡
Cation exchangers			
Natural zeolite, Na	4–8	NaCl	3–6
Synthetic zeolite, Na	8–20	NaCl	2–3
Carbonaceous and resin, Na	7–50	NaCl	1.8–3.6
Carbonaceous and resin, H	7–50	H_2SO_4	2–4
Anion exchanger			
Resin, OH	16–33	NaOH	5–8

* After G. M. Fair and J. C. Geyer, "Water Supply and Waste-Water Disposal," table 23-6, p. 642, Wiley, New York, 1954.
† To convert to kilograins as $CaCO_3$ per cubic foot, multiply by 0.77.
‡ To convert to pounds of NaCl per kilograin as $CaCO_3$, multiply by 0.167.

The use of ion-exchange materials in water treatment is limited to waters containing little turbidity. Turbidity serves to coat the surfaces of the ion-exchange material, thereby reducing the exchange capacity and regeneration efficiencies. A similar condition results if ferrous ions are oxidized by dissolved oxygen and precipitate out as ferric hydroxide.

Nitrogen in the form of the ammonium ion NH_4^+ and nitrate ion NO_3^- can be removed with ion-exchange resins. The cationic exchange resins used for removing NH_4^+ are regenerated with lime. To remove the ammonium ion, the pH of the process stream must be reduced to a value of approximately 7.0. (See Solution procedure 2.1.2.)

Nitrate ions can be removed with anionic exchange resins regenerated with brine. These resins will remove orthophosphate ions as well.

13.5 MEMBRANE-SEPARATION PROCESSES

Membrane-separation processes are capable of removing a large fraction of the total ions in a process stream. Two membrane processes are currently being employed in treatment technology—*electrodialysis* and *reverse osmosis*. Both processes produce, in addition to the processed stream, a concentrate stream for which disposal must be found. The concentrate may have a volume of more than 10 percent of the influent stream.

Electrodialysis

Electrodialysis membranes are sheetlike barriers made out of highly cross-linked high-capacity ion-exchange resins that pass ions but prevent the passage of water. These membranes are of two types: cation membranes, which transmit only cations, and the anion membranes, which will only permit anions to pass. The selective permeability of the membranes is due to the ion-exchange nature of the materials out of which the membranes are manufactured.

The mechanism of the electrodialytic process is illustrated in Fig. 13.5.1. When a potential difference is established across a solution of electrolytes, current will flow as a result of ion migration toward the electrode of opposite charge. If cation- and anion-permeable membranes (*A* and *C* in Fig. 13.5.1) are imposed in alternate sequence across the electric field, the anion-permeable membranes will obstruct the movement of cations toward the cathode, while the cation-permeable membranes will obstruct the migration of anions toward the anode. As a result, cells are formed in alternate sequence in which the water becomes either more dilute or more concentrated with respect to the electrolytes.

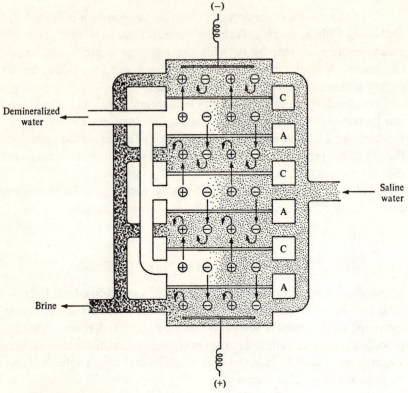

FIGURE 13.5.1
Mechanism of the electrodialytic process.

Electrodialysis of brackish and salt waters is carried out in units consisting of as many as 500 membranes (250 cells) arranged in a stack. The cross-sectional area of the membranes in these units ranges from 400 to 800 in^2. Stack height varies from 2 to 4 ft. Membrane sheets are separated by hydraulic spacers having a thickness of as little as 0.04 in. The spacers are designed so that the water is forced to follow a tortuous path over the surface of the membrane. The cathodes used are generally made of stainless steel or some steel alloy, whereas anodes are made of some metal coated with platinum. Gaseous products are formed at the electrodes. Pressure losses through electrodialysis units range up to 60 lb/in^2.

Reverse Osmosis

Reverse osmosis is a membrane process in which the process stream is forced to flow from a solution of high salt concentration to one of lower concentration. In natural osmosis, water flows in the opposite direction, and pressures of 600

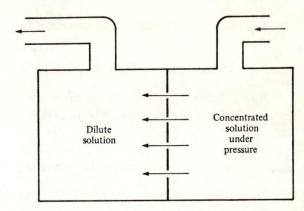

FIGURE 13.5.2
Schematic representation of reverse osmosis.

to 800 lb/in² are required to reverse this flow. A schematic representation of the process is presented in Fig. 13.5.2.

Membranes are constructed of cellulose acetate and can be made to reject almost 99 percent of most ion species. The transport mechanism in the reverse-osmosis process is explained in terms of the bound-water content of the cellulose-acetate membrane. Water from the salt solution will permeate the membrane by interacting with the tightly bound water in the membrane. The solute ions, however, are partially prevented from permeating because of the relative inability of the bound water to solvate them.

The process-stream flow rate through the reverse-osmosis membrane has been expressed as

$$q = K_w(\Delta p - \Delta \pi) \qquad (13.5.1)$$

where q = flow rate per unit area of membrane. $[Lt^{-1}]$

Δp = applied pressure difference across membrane. $[FL^{-2}]$

$\Delta \pi$ = osmotic pressure difference across membrane. $[FL^{-2}]$

K_w = membrane constant. $[L^3 F^{-1} t^{-1}]$

The flux of salt through the membrane can be expressed as

$$J = K_s \Delta c \qquad (13.5.2)$$

where J = salt flux. $[FL^{-2} t^{-1}]$

Δc = difference in salt concentration across membrane. $[FL^{-3}]$

K_s = salt permeation constant. $[Lt^{-1}]$

Equations (13.5.1) and (13.5.2) are related by

$$J = qc_f \quad (13.5.3)$$

where c_f = salt concentration in feed stream. $[FL^{-3}]$

The salt-rejection performance of a reverse-osmosis process can be described in terms of the *desalinization ratio* (DR):

$$DR = \frac{c_f}{c_p} \quad (13.5.4)$$

where c_p = salt concentration in product stream. $[FL^{-3}]$

By assuming that $\Delta c \approx c_f$, Eqs. (13.5.1) to (13.5.3) can be substituted into Eq. (13.5.4) to yield

$$DR = \frac{K_w}{K_s}(\Delta p - \Delta \pi) \quad (13.5.5)$$

Equation (13.5.5) relates the performance stated in terms of the desalinization ratio to the characteristics of the membrane and to the process variables.

Commercial reverse-osmosis equipment is designed to have one of three basic configurations: In one configuration, the membrane is wound into a spiral with a porous separator. In another, the membrane lines a porous supporting tube. In the third type, the membrane is fitted in a plate and frame assembly. All configurations are designed so that polarizing-salt concentration regions on the feed side of the membrane are minimized by the flowing process stream.

Flow rates of approximately 10 gal/(d)(ft^2) are normally achieved in process applications.

13.6 DISINFECTION PROCESSES

Strictly defined, *sterilization* refers to the total destruction or total removal of all microorganisms from the treated medium, whereas *disinfection* is a term applied to those processes in which pathogenic microorganisms, but not their spores, are destroyed. However, disinfection is used here in a broader sense to describe the destruction of a particular species of microorganism at some stage of its development. Although disinfection can be brought about by a variety of processes, only chlorination will be discussed here.

Chlorine and chlorine compounds are popular as disinfectants. Chlorine gas, in particular, finds wide application in the treatment of municipal water supplies. The popularity of chlorine gas in such a use stems from the fact that in concentrations that are tasteless and nonpoisonous to humans, chlorine has a specific high toxicity for microorganisms responsible for waterborne diseases.

Furthermore, chlorine gas is relatively inexpensive, easy to apply, and can be used to establish temporary residuals.

Atomic chlorine can exist in any one of several oxidation states. Several compounds in which chlorine exists in different oxidation states are listed below in descending order of oxidation level.

Compound	Chlorine valence
ClO_2	+4
$NaClO_2$	+3
$HOCl$	+1
Cl_2	0
$NaCl$	−1

The higher the oxidation level, the more powerful is the oxidizing power of the chlorine compound. Compounds containing chlorine at higher levels of oxidation may be more effective toward minimizing tastes and odors that often develop when chlorine compounds are used as disinfectants. However, with the exception of chlorides, little difference in bactericidal efficiency is noted between the oxidation levels. Chlorides in small concentrations do not behave as disinfectants.

Chlorine gas is quite soluble in water (7,160 mg/l at 20°C and 1 atm). When dissolved in water, chlorine hydrolyzes rapidly to form hypochlorous acid:[1]

$$Cl_2 + H_2O \rightleftharpoons HOCl + H^+ + Cl^- \qquad (13.6.1)$$

For chlorine concentrations of less than 1,000 mg/l and pH values greater than 3, the hydrolysis goes virtually to completion.

Hypochlorous acid in water ionizes to form the hypochlorite ion:

$$HOCl \rightleftharpoons OCl^- + H^+ \qquad (13.6.2)$$

The dissociation of hypochlorous acid in water is strongly dependent upon the hydrogen-ion concentration. The dissociation constant varies linearly from a value of 2.0×10^{-8} at 0°C to approximately 3.7×10^{-8} at 25°C. The effect of pH on the distribution of hypochlorous acid and hypochlorite ion in water at two different temperatures is shown in Fig. 13.6.1. The chlorine species HOCl and OCl⁻ in water constitute what is known as *free available chlorine*.

---

[1] The following discussion is taken largely from G. M. Fair et al., The Behavior of Chlorine as a Water Disinfectant, *J. Am. Water Works Assoc.*, vol. 40, pp. 1051–1061, 1948.

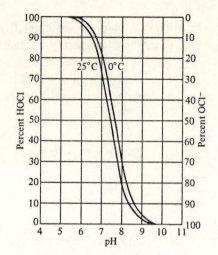

FIGURE 13.6.1
Percent distribution of HOCl and OCl⁻ in water as a function of pH.

Hypochlorites, such as calcium hypochlorite, when dissolved in water ionize to form the hypochlorite ion:

$$Ca(OCl)_2 \rightarrow Ca^{++} + 2OCl^- \quad (13.6.3)$$

The hypochlorite ions thus produced along with the hydrogen ions in solution establish an equilibrium with hypochlorous acid in accordance with Eq. (13.6.2).

Hypochlorous acid in water reacts with ammonia to produce monochloramine, dichloramine, and trichloramine:

$$NH_3 + HOCl \rightarrow NH_2Cl + H_2O \quad (13.6.4)$$

$$NH_3 + 2HOCl \rightarrow NHCl_2 + 2H_2O \quad (13.6.5)$$

$$NH_3 + 3HOCl \rightarrow NCl_3 + 3H_2O \quad (13.6.6)$$

The amount of each species of chloramine produced is dependent upon the relative quantities of hypochlorous acid and ammonia present, pH, and temperature. Chloramines are also formed by the reaction of hypochlorous acid with any organic amines present in solution. Chlorine present in water as chloramines is called *combined available chlorine*.

When chlorine is added to water containing reducing agents, ammonia, and organic amines, residuals develop; a plot of the residuals versus the dosages required to attain the residuals yields a curve similar to the one shown in Fig. 13.6.2. Chlorine added initially reacts with the reducing agents present, is reduced to chlorides, and develops no measurable residual. The reaction with

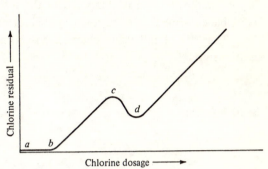

FIGURE 13.6.2
Chlorine residual curve.

reducing agents is represented in Fig. 13.6.2 by the portion of the curve extending from *a* to *b*. The chlorine dosage at *b* represents the quantity of chlorine required to meet the demand exerted by the reducing agents. Reducing agents often found in water and sewage include hydrogen sulfide, nitrites, and ferrous ions.

After the chlorine demand exerted by reducing agents has been met, further addition of chlorine results in the formation of chloramines. Chloramines thus formed impart a combined available chlorine residual. When the applied chlorine has reacted with all the ammonia and organic amines in solution, a free available chlorine residual begins to develop. At a certain critical dosage (corresponding to point *c* on the curve in Fig. 13.6.2) the concentration of free available chlorine becomes large enough to initiate and propagate the oxidation of those chloramines formed by the ammonia initially present in the water. The destruction of this chloramine fraction, which decreases the chlorine residual (from *c* to *d* on the curve), is accompanied by the formation of oxidized nitrogen compounds, such as nitrous oxide, nitrogen, and nitrogen trichloride. The decrease in chlorine residual is the result of the reduction of atomic chlorine to its lowest oxidation state, that is, to chloride.

Upon completion of the oxidation, additional chlorine added to the system results in the development of a free available chlorine residual. The point at which oxidation of the ammonia-chloramine products is complete (point *d* on the curve) is commonly referred to as the *breakpoint*.

In summary, from *b* to *c* the chlorine residual is predominantly in the combined available state with some free available residual developing as *c* is approached. From *c* to *d* the chloramines resulting from the chlorination of ammonia are destroyed. At *d* the residual remaining consists of the resistant chloramines produced by the chlorination of organic amino compounds. Beyond d the residual is composed of the resistant chloramines, and the free available chlorine formed as additional chlorine is added to the system.

The rate of kill using chlorine has been found to be expressed by the relationship

$$\frac{dN}{dt} = -kNt \quad (13.6.7)$$

or its integral

$$t^2 = \frac{4.6}{k} \log \frac{N_1}{N_2} \quad (13.6.8)$$

where N_1, N_2 = number of microorganisms living initially and at time t, respectively

k = rate constant

The rate constant k can be determined from a semilogarithmic plot of the percent surviving versus the square of the contact time.

As a practical guide to the chlorination of water for domestic purposes, concentrations of chlorine residuals required to ensure effective disinfection are presented in Table 13.6.1. These values are purported to include a factor of safety.

The use of ammonia in conjunction with chlorine has been practiced in water treatment for the purpose of inducing the formation of combined available

Table 13.6.1 RECOMMENDED MINIMUM CONCENTRATIONS OF FREE AVAILABLE AND COMBINED AVAILABLE CHLORINE RESIDUALS TO ENSURE EFFECTIVE DISINFECTION*

pH	Minimum concentration of free available chlorine residual, disinfecting period at least 10 min, mg/l	Minimum concentration of combined available chlorine residual, disinfecting period at least 60 min, mg/l
6.0	0.2	1.0
7.0	0.2	1.5
8.0	0.4	1.8
9.0	0.8	1.8
10.0	0.8	

* From Bur. of Environ. Sanit. and Off. of Prof. Train., New York State Dep. of Health, "Manual of Instruction for Water Treatment Operators."

residuals. The reasons cited for such a practice include the minimization of tastes and odors and a more persistent chlorine residual.

Disinfection of domestic waste-treatment-plant effluents with chlorine is generally deemed to be adequate when sufficient chlorine has been added to obtain a combined available residual of 0.5 mg/l after a 15-min contact period. The quantity of chlorine added to obtain a measurable residual is called the *chlorine demand* and is represented by the dosage at point *b* along the residual curve in Fig. 13.6.2.

EXERCISES

13.1.1 For what purpose are aeration processes used in water and waste-water treatment systems? Identify and describe the devices used in aeration processes.

13.1.2 Referring to Solution procedure 2.1.2, explain the pH adjustment that might have to be made in water prior to removal of hydrogen sulfide and ammonia by aeration.

13.2.1 For what purposes is carbon adsorption used in treatment technology? How is the process stream contacted with the carbon?

13.2.2 What is an "adsorption isotherm," and how is it used to measure the adsorptive capacity of carbon?

13.3.1 What causes hardness in water? How does hardness manifest itself?

13.3.2 What anions are commonly associated with the cations causing hardness?

13.3.3 Using the solubility relationships discussed in Sec. 2.1, explain the removal of hardness through the addition of lime. What is the name of this effect?

13.3.4 What cation is normally introduced to water being softened to replace the hardness-producing cations? In what form is this cation generally introduced?

13.3.5 Explain the effect of excess lime treatment on the removal of magnesium ions.

13.3.6 Describe how chemical softening processes are carried out in a treatment system.

13.3.7 For what purpose is water recarbonated?

13.3.8 An analysis of a water yields the following information:

Carbon dioxide	10 mg/l as CO_2
Total alkalinity	250 mg/l as $CaCO_3$
Total hardness	400 mg/l as $CaCO_3$
Magnesium hardness	40 mg/l as Mg

Calculate the necessary quantities of hydrated lime and soda ash required to soften the water.

13.3.9 How may the removal of iron and manganese from water be accomplished? What effect does carbon dioxide removal have on the removal of iron and manganese? What effect does the addition of lime and soda ash have?

13.3.10 In what form is phosphorus generally present in effluents from biologic treatment processes? What chemical agents can be used to remove phosphorus from such effluents? What are the pH optima for the use of these agents?

13.4.1 What is the "equilibrium selectivity coefficient" in ion exchange, and what is its significance?

13.4.2 When ion exchange is used to soften water, what ions are exchanged, and what type of solution is generally employed in regeneration?

13.4.3 To what level can water be softened with ion exchange? Is such a level desirable?

13.4.4 A small municipal treatment plant is to soften water with ion exchange at a rate of 0.2 Mgal/d. The total hardness of the untreated water is 200 mg/l (as $CaCO_3$), and it is desired to reduce the hardness to 75 mg/l. The ion-exchange material to be used has an exchange capacity of 20 $(g)(eq)/ft^3$ and a regeneration requirement of 3 $(g)(eq)/ft^3$. The ion-exchange material is to be regenerated 2 times per day.
(*a*) What proportion of untreated water should be mixed with treated water to obtain water with 75 mg/l hardness?
(*b*) What volume of ion-exchange material must be used?
(*c*) How many gallons of 10 percent brine will be required for each regeneration?

13.5.1 What types of materials are removed from water by the membrane-separation processes? How large a concentrate volume results from these processes? What is done with the concentrates?

13.5.2 Describe the equipment and explain the separation mechanism in electrodialysis processes.

13.5.3 Explain the separation mechanism in reverse osmosis. Describe the types of equipment and the operating conditions used in the process. What flow rates can normally be obtained through reverse-osmosis membranes?

13.5.4 What is the significance of the desalinization ratio?

13.6.1 What is the difference between sterilization and disinfection? Which does chlorine provide.

13.6.2 How does the bactericidal efficiency and oxidizing power vary with the valence level of the chlorine compound used? Of what importance is the oxidizing power of chlorine compounds when the latter are used in water treatment?

13.6.3 Write the equations illustrating the hydrolysis of chlorine in water. How do the hydrolysis products vary with the pH of the water?

13.6.4 What is meant by "free available chlorine"? What is meant by "combined available chlorine"? Which is the most bactericidal? For what purpose will a combined available chlorine residual be sought?

13.6.5 Construct a curve showing the relationship between chlorine residual and chlorine dosage. Explain the nature of the reactions that are reflected by the various regions along the curve. Where is the "breakpoint," and what is its significance?

13.6.6 What four factors determine the disinfection efficiency of the chlorination process?

13.6.7 What is meant by "chlorine demand"? What is its significance?

14

PROCESSES USED IN SLUDGE TREATMENT TRAINS

14.1 THICKENING PROCESSES

Thickening processes are used to reduce the volume of sludge that is to be treated by subsequent processes in the sludge treatment train. Two types of processes can be used to concentrate sludges—flotation thickening and gravity thickening. Thickening by flotation is carried out with the dissolved-air flotation process discussed in Sec. 11.5. Sludge solids are floated to the surface by bubbles formed by the release of dissolved air from the depressurized sludge suspension. Concentrated sludge at the surface is removed by mechanical scrapers. Flotation processes are most effective for sludges of a gelatinous nature.

Gravity thickening processes are widely employed in waste-treatment technology. One such use was discussed in Sec. 12.1. There, gravity thickening was used to concentrate activated sludge prior to its recycle and waste to other sludge treatment processes. Although the thickening tank is an integral unit of the activated-sludge process, its concentration function is considered a part of the sludge treatment train.

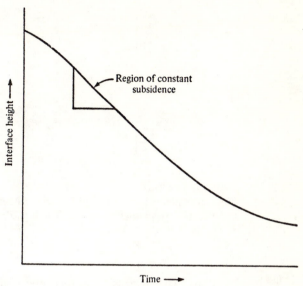

FIGURE 14.1.1
Position of subsiding particle-liquid interface as a function of time.

Sludges are concentrated suspensions that exhibit zone settling characteristics (see Fig. 11.2.1). These characteristics are somewhat different from those of dilute suspensions.

In dilute systems, particles settle unhindered at their own individual velocities of subsidence. However, as the concentration of a suspension is increased, a concentration is reached in which the fastest settling particles form a zone and settle from then on collectively and at a reduced rate. As the concentration is increased further, this zone forms at progressively earlier periods until a point is reached where initial subsidence is collective and no individual particle movement is observed. For concentrations as large as or greater than this concentration, a distinct interface is formed between the subsiding particles and the clarified liquid. The position of such an interface will vary with time in a manner similar to that described by the curve in Fig. 14.1.1.

All batch settling processes produce four zones, each with its own particular characteristics. Figure 14.1.2 is a sketch showing the relative positions of the different zones as they are formed at different time intervals in a thick flocculent suspension. Initially, the concentration is uniform throughout the suspension. The suspension having the initial concentration is labeled b. Immediately, a solids-liquid interface develops, and a zone of clarified liquid, a, is formed. In zone b, the particles settle at a uniform velocity under conditions

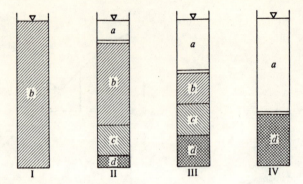

FIGURE 14.1.2
Zone formation in a concentrated flocculent suspension after four different time intervals. [*By permission from A. Anable, in J. H. Perry (ed.), "Chemical Engineers' Handbook," copyright, McGraw-Hill, New York, 1950.*]

of hindered settling. The suspension in this zone remains constant at uniform concentration, and for a given depth of settling column the magnitude of the velocity at which this part of the suspension settles is a function of the concentration:

$$u = \Psi(c) \qquad (14.1.1)$$

The velocity of subsidence for a given suspension has been found to decrease with increasing concentration and to increase as the settling depth is increased. The effect of depth is particularly significant at higher concentrations.

Concurrent with the formation of zone *a*, two other zones *c* and *d* are formed. Zone *c* is a zone of transition through which the settling velocity decreases in response to an increased concentration of solids. When the suspension becomes concentrated to the point where the solids can be considered to be supported mechanically by the solids below, they become a part of a compression zone *d*.

When a thickening tank is operated in continuous mode, the bottom portion of the tank is occupied by a blanket of solids in which zones *b*, *c*, and *d*, observed in a batch process, are present. Above the solids blanket and extending to the overflow weir, there exists a zone of clarified liquid. Actually, conditions prevail that are a static counterpart of those illustrated by sketch II or III in Fig. 14.1.2. The influent suspension is introduced through a center well at an elevation near the solids-liquid interface.

For continuous-flow sedimentation or thickening tanks, the surface area required to separate concentrated suspensions is based primarily on two factors: the clarification and thickening capacities. The clarification capacity must be sufficient to produce an overflow relatively free of settleable solids, whereas the thickening capacity must be such as to produce in the underflow all the solids which have been removed from the system. Both capacities can be estimated from batch settling tests.

The *clarification capacity* can be estimated from the initial rate at which the solids-liquid interface subsides. The rate can be determined from a settling curve such as is shown in Fig. 14.1.1. Followingا brief period of reflocculation caused by decaying turbulence in the system, the interface subsides at a more or less constant rate. The clarification capacity is determined from a measurement of the slope during this period of constant subsidence. The surface area of a continuous-flow tank must be large enough so that the rate of liquid rise (computed on the basis of liquid overflow only) is less than the subsidence velocity.

The estimation of the *thickening capacity* is more involved. In a batch sedimentation the concentration initially is the same throughout the suspension, and the solids settle at a uniform velocity. Before the solids are deposited at the bottom of the settling column, they must pass through all concentrations ranging from the initial concentration to that of the deposited solids. If the solids handling capacity at any intermediate concentration is less than that of the lower concentration just above it, the solids cannot pass through the concentration as rapidly as they are settling into it and as a result a zone of the intermediate concentration will propagate upward through the column. By the time the zone reaches the solids-liquid interface, all the solids will have passed through it.

For the purpose of developing a design relationship for a continuous-flow thickener, consider that part of the solids blanket existing between a layer i and the bottom of the blanket at the underflow. A solids balance can be written as

$$V \frac{d\bar{c}}{dt} = Q_i c_i - Q_u c_u \qquad (14.1.2)$$

where V = volume of solids blanket below layer i. $[L^3]$

$\bar{c}$ = average concentration of solids below layer i. $[FL^{-3}]$

c_i, c_u = solids concentrations in layer i and underflow, respectively. $[FL^{-3}]$

Q_i, Q_u = volume rates of flow of suspension subsiding through layer i and underflow, respectively. $[L^3 t^{-1}]$

A volumetric balance across the same region yields

$$\frac{dV}{dt} = Q_i - Q_u - Q_{ci} \qquad (14.1.3)$$

where Q_{ci} = volume rate of flow of clarified liquid rising through layer i. $[L^3 t^{-1}]$

At steady state, Eqs. (14.1.2) and (14.1.3) can be combined to give

$$Q_{ci} = Q_i c_i \left(\frac{1}{c_i} - \frac{1}{c_u}\right) \quad (14.1.4)$$

From a solids balance for that part of the solids blanket existing above the layer i and assuming all solids are removed from the effluent of the thickener, one has

$$Q_0 c_0 = Q_i c_i \quad (14.1.5)$$

where Q_0 = volume rate of flow of influent to thickener. $[L^3 t^{-1}]$

c_0 = concentration of solids in influent. $[FL^{-3}]$

By substituting Eq. (14.1.5) into (14.1.4), dividing both sides of the latter by the horizontal cross-sectional area of the thickener, and rearranging, one has[1]

$$G_i = \frac{Q_0 c_0}{A} = \frac{Q_{ci}/A}{1/c_i - 1/c_u} \quad (14.1.6)$$

where G_i = solids flux at layer i in continuous-flow thickener. $[Ft^{-1}L^{-2}]$

A = horizontal cross-sectional area. $[L^2]$

In a batch system, a layer having a concentration c_i subsides at a velocity of u_i relative to the clarified liquid flow. In a continuous-flow steady-state system, however, such a layer remains fixed, and the clarified liquid flow rises through the layer at a velocity equal to Q_{ci}/A.

Correlating the batch system with the continuous-flow system, one has

$$\frac{Q_{ci}}{A} = u_i \quad (14.1.7)$$

where u_i = settling velocity of solids from concentration layer c_i. $[Lt^{-1}]$

The relationship between u_i and c_i can be determined experimentally from interface subsidence curves constructed for suspensions with different initial concentrations of solids. The rate at which the interface subsides at the beginning of batch sedimentation is equal to the settling velocity of the suspension at its initial concentration (conditions existing in zone b in a batch process).

---

[1] Equation (14.1.6) was first published in H. S. Coe and G. H. Clevenger, Methods for Determining the Capacities of Slime Settling Tanks, *Trans. AIME*, vol. 55, pp. 356–384, 1916.

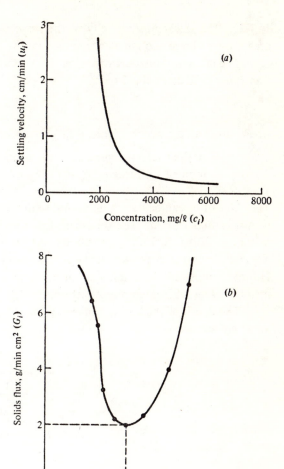

FIGURE 14.1.3
Settling velocity and solids flux curves. (a) Settling velocity-concentration curve; (b) solids flux-concentration curve.

Therefore, the slopes of this region along the subsidence curves can be used to construct a plot of the relation between settling velocity and concentration. See Fig. 14.1.3a.

At some concentration c_i Eq. (14.1.6) will be a minimum. This minimum can be identified from a plot of the equation in which corresponding values of u_i and c_i are obtained from a curve similar to the one in Fig. 14.1.3a. See Fig.

14.1.3b. The thickening capacity of the suspension is such that the thickener area must be large enough to accommodate the minimum solids flux.

The cross-sectional-area requirement from the standpoint of the clarification capacity is determined by

$$q = u_0 = \frac{Q_0 - Q_u}{A} \quad (14.1.8)$$

where q = surface overflow rate. $[Lt^{-1}]$

u_0 = settling velocity corresponding to suspended solids concentration of influent to thickener c_0. $[Lt^{-1}]$

The value of u_0 can be determined from settling-velocity curves, one of which is shown in Fig. 14.1.3a. The thickener area should be large enough to accommodate both the thickening and clarification requirements. Other methods for determining the area requirements of thickeners are to be found elsewhere.[1,2] However, most of these methods give results similar to those given by the method discussed in the present section.

The depth of thickening tanks is governed by experience. Typical allowances are as follows:

	Allowance, ft
Freeboard	1
Clear zone	1–3
Thickening zone	3
Storage capacity	0–2
Bottom pitch	1–2
Total	6–11

Solution procedure 14.1.1 A series of batch settling tests were performed on the mixed-liquor suspended solids (MLSS) from an activated-sludge process. Each test yielded a curve such as is shown in Fig. 14.1.1, which gives the relationship between the position of the solids-liquid interface and the subsidence time. The slopes of the curves in the region of constant subsidence were used to construct the settling-velocity–concentration curve shown in Fig. 14.1.3a. Using this curve determine the overall dimensions of the final clarifier if the process flow stream is 1 Mgal/d and if it is desired to maintain the MLSS in the aeration tank at 2,000 mg/l.

---

[1] R. I. Dick, Role of Activated Sludge Final Settling Tanks, *Proc. Am. Soc. Civ. Eng., J. Sanit. Eng. Div.*, paper 7231, vol. 96, April 1970.

[2] P. A. Vesiland, Design of Prototype Thickeners from Batch Settling Tests, *Water and Sewage Works*, vol. 115, pp. 302–307, 1968.

SOLUTION

1 From the curve in Fig. 14.1.3a, it can be seen that the solids can be concentrated to 6,000 mg/l. Consequently, the concentrated underflow c_u (or X_R, using the notation in Sec. 12.1) is set at 6,000 mg/l. Using Eq. (12.1.6) one finds that the underflow rate Q_u is

$$Q_R = \frac{X_a Q}{X_R - X_a} = \frac{2,000}{6,000 - 2,000} \quad (1) = 0.5 \text{ Mgal/d}$$

2 Using the curve in Fig. 14.1.3a, Eq. (14.1.6) is solved for corresponding values of c_i and u_i to obtain a set of values for G_i. See Table 14.1.1.

3 The set of values for G_i calculated in step 2 are plotted as a function of solids concentration. See Fig. 14.1.3b. The minimum flux value is found to be approximately 2 g/(min)(cm²). The minimum clarifier area to accommodate the thickening function is estimated to be

$$A = \frac{Q_0 c_0}{G} = \frac{(1.5)(2,000)(2.65 \times 10^{-4})}{2 \times 10^{-3}} = 394 \text{ m}^2$$

4 From Table 14.1.1, it is seen that the settling velocity of the suspension when the concentration is 2,000 mg/l is 1.89 cm/min. This defines the clarification capacity of the suspension, i.e., the permissible hydraulic loading intensity. The area required, therefore, to accommodate the clarification function is

$$A = \frac{Q_0 - Q_u}{u_0} = \frac{(1.5 - 0.5)(263)}{1.89} = 139 \text{ m}^2$$

Table 14.1.1 CALCULATIONS FOR STEP 2 IN SOLUTION PROCEDURE 14.1.1

c_i, mg/l	u_i, cm/min	$1/c_i$, cm³/g	$1/c_u$, cm³/g	$1/c_i - 1/c_u$, cm³/g	G_i, g/(min)(cm²)
1,700	2.74	589	167	422	6.50×10^{-3}
2,000	1.89	500		333	5.68×10^{-3}
2,200	0.915	455		288	3.18×10^{-3}
2,700	0.457	370		203	2.25×10^{-3}
3,100	0.305	323		156	1.96×10^{-3}
3,700	0.244	270		103	2.37×10^{-3}
4,700	0.183	213		46	3.98×10^{-3}
5,300	0.152	189		22	6.90×10^{-3}
5,700	0.122	175		8	15.20×10^{-3}
6,300	0.122	159			

5 Since 394 m² > 139 m², the thickening requirement controls, and the area of the clarifier will be 394 m².

6 Center depth of the clarifier will be:

Freeboard	0.31
Clear zone	0.61
Thickening zone	0.91
Storage capacity	0.31
Bottom pitch	0.61
Total	2.75 m

////

14.2 ANAEROBIC DIGESTION

The term *anaerobic digestion* is applied to a process in which organic material is decomposed biologically in an environment devoid of oxygen. Decomposition results from the activities of two major groups of bacteria. One group, called the *acid formers*, consists of *facultative bacteria*[1] which are found also in many aerobic environments and which in an anaerobic environment convert carbohydrates, fats, and proteins to organic acids and alcohols. The other group, the methane bacteria, converts the organic acids and alcohols produced by the acid formers to methane and carbon dioxide. Some organic materials such as lignin are quite resistant to the activity of both groups and, hence, pass through the process relatively unaltered. The role of predator populations in anaerobic digestion is considered to be minor compared with that in aerobic processes. Anaerobic processes are discussed in Sec. 3.5.

Ideal Process

Anaerobic digestion as applied in the past has departed considerably from a process in which all factors influencing digestion were optimum. However, for a better understanding of recent innovations and future developments in digestion, it is helpful to consider the ideal process.

The stoichiometric equation for the ideal process can be written as

$$C_aH_bO_cN_d \rightarrow nC_wH_xO_yN_z + mCH_4 + sCO_2 + rH_2O + (d - nz)NH_3 \quad (14.2.1)$$

where $s = a - nw - m$

$r = c - ny - 2s$

The terms $C_aH_bO_cN_d$ and $C_wH_xO_yN_z$ represent on an empirical-mole basis the compositions of the organic material at the start and at the conclusion of the

---

[1] Bacteria having the ability to live and grow both in the presence of and in the absence of free oxygen.

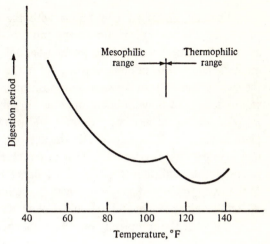

FIGURE 14.2.1
Time required for digestion.

digestion process. Methane, carbon dioxide, and ammonia are gaseous products of the process. In actual processes, these gases constitute approximately 95 to 98 percent of the gas evolved. The remaining volume is composed of hydrogen sulfide and hydrogen. The residual organic matter $C_w H_x O_y N_z$ is often similar in composition and characteristics to composted material.

The heat liberated in the process is equal to the difference between the heat of combustion of the initial material and the sum of the heats of combustion of the products. The heats of combustion can be estimated with the use of Eqs. (3.1.2) and (3.1.3).

Temperature exerts a profound effect on anaerobic digestion. Attention is directed to Fig. 14.2.1, where the time required for digestion is plotted as a function of temperature. Ordinate values are omitted since the digestion period is affected by other factors as well. The curve illustrates the existence of two temperature ranges. The *mesophilic* range extends up to 110°F and exhibits a minimum value in the vicinity of 95 to 100°F. The *thermophilic* range extends beyond 110°F and has a minimum value at approximately 130°F. It should be noted that digestion is more rapid at the thermophilic optimum than it is at the optimum value in the mesophilic range.

The ideal process would be carried out in a continuous-flow system conforming to the completely mixed model. The system would be contained in a closed vessel from which all oxygen is excluded and gaseous products are removed as they are formed. Temperature and other environmental conditions would be optimized for the process.

Feeding would be continuous with the organics being utilized by the suspended microorganisms for energy and growth. The dilution rate (reciprocal of retention time) would be kept low enough to prevent the slower-growing methane bacteria from being washed out of the system. Effluent displaced from the system would have the same composition as the contents of the system and would consist of unassimilated organics, organic and inorganic products, and microorganisms. The ideal process would provide (1) an immediate and complete dispersal of influent organics throughout the vessel, (2) an optimum contact opportunity between the organics and the microorganisms, and (3) a retention time sufficient to establish a balance between the metabolic activities of the two groups of bacteria involved in the process. The kinetics of the ideal anaerobic digestion process should follow the same relationships developed in Sec. 4.4 for the completely mixed aerobic process.

Standard-Rate Digestion

Standard-rate digestion has had wide application in sewage treatment, where it is used to digest sludge consisting of solids separated from the sewage, trickling-filter slimes, and waste-activated sludge. Normally, the process is carried out in closed tanks at temperatures ranging from 85 to 95°F. Standard-rate digestion departs considerably from the ideal process, in that little provision is made for mixing. As a result, influent organics are concentrated locally at points in the tank, and contact with the microorganism population is limited. Furthermore, discharge is not continuous, and portions of the tank volume are occupied by material not in active stages of digestion.

Attention is directed to Fig. 14.2.2, where the stratification existing in a conventional standard-rate digestion unit is illustrated. Sludge introduced at two or three points in the digester soon rises to a scum layer and becomes part of it. Here the organics undergo initial decomposition, and much of the gas produced in the process is released. As decomposition proceeds, the partially decomposed solids fall to the bottom of the tank and build up a layer of digesting and digested solids. The volume between the scum layer and the sludge layer is occupied by supernatant liquid with a high concentration of dissolved and suspended materials. The latter are removed periodically to other units for further treatment.

The retention time required to effect a given degree of volatile-solids reduction has been found to be a function of the ratio of the volatile solids to the fixed solids. The relationship is presented graphically in Fig. 14.2.3. These curves were established on the basis of operational data collected at 50 domestic waste treatment plants using standard-rate digestion. For retention time less than 15 d

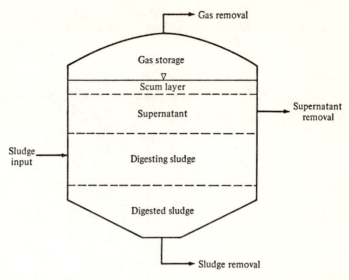

FIGURE 14.2.2
Schematic representation of the stratification existing in a conventional standard-rate digestion unit.

balanced digestion does not become established in the standard-rate process, and little reduction in volatile solids takes place.

During the digestion period, sludge becomes more concentrated as a supernatant separates from the solids. For greater economy of digestion capacity, the supernatant is withdrawn frequently during the process. The reduction in sludge volume with time appears to follow a parabolic relationship. In a steady-state process, the digester volume at any instant will be composed of incremental volumes of daily accretions having retention periods ranging from 1 d to the number of days constituting the nominal retention period:

$$V = \sum_{i=1}^{t} v_i \quad (14.2.2)$$

where V = volume of digester, ft^3

t = digestion period, d

v_i = volume occupied by daily accretion to sludge, ft^3

The volume composition is illustrated in Fig. 14.2.4. From the figure, it is clear that the capacity requirement for a standard-rate digester can be expressed as

$$V = \left[v_t + \frac{1}{3(v_1 - v_t)} \right] t \quad (14.2.3)$$

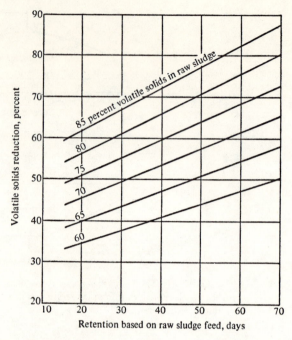

FIGURE 14.2.3
Reduction in volatile solids in raw sludge, for retentions from 15 to 70 days, $T = 85$ to $95°F$. (*After R. S. Rankin, Digester Capacity Requirements, fig. 4, Sewage Works J., vol. 20, p. 478, 1948.*)

where v_1, v_t = volume occupied by daily accretion to sludge, initially and at end of retention period

The volume of the daily accretion to the sludge can be computed with

$$v_i = v_s + v_w = \frac{w}{s\gamma} + \frac{w(1-x)/x}{\gamma} \quad (14.2.4)$$

where v_s, v_w = volumes occupied by solids and water, respectively, ft^3

w = weight of daily accretion to sludge, lb/d

s = specific gravity of solids

x = weight fraction of solids in sludge

γ = specific weight of water, lb/ft^3

The specific gravity of the solids can be estimated from

$$s = \frac{1}{p/s_v + (1-p)/s_f} \quad (14.2.5)$$

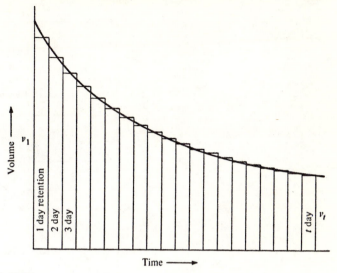

FIGURE 14.2.4
Volume composition of standard-rate digester.

where s_v, s_f = specific gravities of volatile and fixed solids, respectively

p = percent of solids that are volatile, expressed as decimal

When temperatures of 85 to 95°F are maintained in the digester, a retention time of 30 to 50 d is normally allowed for the digestion of sludges from domestic waste water. Regulatory agencies often require the capacity of standard-rate digestion units to be based on per capita allowances.

High-Rate Digestion

High-rate digestion incorporates the main features of the ideal process. Raw sludge is fed to the process continuously or at frequent intervals, and the contents of the digester are maintained in a mixed state by means of vigorous agitation. Effluent from the process consists of the mixed liquor displaced from the digester by the influent sludge. Temperature is generally maintained at the mesophilic optimum, that is, at 90 to 95°F.

Retention time for high-rate processes is on the order of 10 to 15 d. Volatile-solids reduction to be expected for these periods is indicated in Fig. 14.2.5. Capacity requirements are reduced by prethickening the influent sludge. However, when the influent sludge is concentrated beyond a solids concentration of 6 percent, the fluidity of the digester contents decreases to an

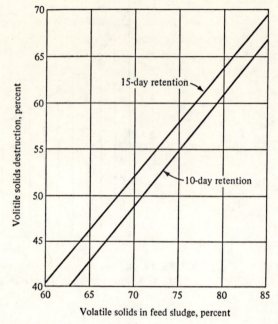

FIGURE 14.2.5
Expected volatile-solids destruction during high-rate digestion. (*After C. N. Sawyer and J. S. Grumbling, Fundamental Considerations in High-rate Digestion, fig. 8, J. Sanitary Eng. Div., Proc. ASCE, vol. 86, no. SA2, March* 1960.)

extent where it is difficult to maintain proper mixing.[1] Mixing is accomplished in several different ways: by mechanical agitators, by the diffusion of recirculated gas, and by gas-lift pumps. The effluent from a high-rate process often is discharged to a second tank in which supernatant is allowed to separate from the digested solids prior to the disposal of the latter.

14.3 CONDITIONING PROCESSES

Digested and undigested domestic waste sludges are coagulated to improve their filtering properties. Such sludges consist primarily of hydrophilic colloids and suspended particles having a wide variety of composition and size. Coagulation serves to agglomerate the particles in a structure that facilitates the separation of much of the water contained in the sludge.

[1] C. N. Sawyer and J. S. Grumbling, Fundamental Considerations in High-rate Digestion, *ASCE*, vol. 86, no. SA2, March 1960.

The actual mechanism involved in sludge conditioning through coagulation is not clearly understood. However, it appears that the effects exerted by conditioning chemicals is primarily one of pH. These chemicals are generally acid salts of heavy metals. The net charge on the hydrophilic particles is depressed as the pH of the sludge system is shifted toward the isoelectric point.

Digested sludges contain high concentrations of bicarbonate ion. These ions exert a demand upon the coagulant. In the case of ferric chloride, a common coagulant,

$$FeCl_3 + 3NH_4HCO_3 \rightleftharpoons Fe(OH)_3\downarrow + 3NH_4Cl + 3CO_2 \quad (14.3.1)$$

Bicarbonates exert a buffering effect. Consequently, the bicarbonate demand must be met before a reduction in pH can result. The quantity of coagulant needed for the proper conditioning of sludge can be reduced by decreasing the buffering capacity of the sludge through the addition of hydroxyl alkalinity:

$$OH^- + HCO_3^- \rightleftharpoons H_2O + CO_3^{--} \quad (14.3.2)$$

The bicarbonate demand can also be reduced by washing out a portion of the bicarbonate concentration in an operation referred to as *elutriation*.

Water requirements for elutriation depend upon the elutriation method used. The *single-stage method* involves one-step dilution, mixing, sedimentation, and decantation in a single tank. The alkalinity of the elutriated sludge when processed by the single-stage method is

$$c_2 = \frac{c_1 + Rc}{R+1} \quad (14.3.3)$$

where c_1, c_2 = alkalinity of sludge before and after elutriation, respectively

c = alkalinity of elutriating water

R = ratio of volume of elutriating water to volume of water in sludge

Multistage elutriation involves repeating the single-stage step on the washed sludge, using fresh water at each step. For this method

$$c_2 = \frac{c_1 + c[(R+1)^n - 1]}{(R+1)^n} \quad (14.3.4)$$

where n = number of washings

The *countercurrent method* involves carrying out the elutriation process in two tanks connected in series. Fresh water is added only during the second-stage washing, and the overflow from this stage is used as elutriating water in the first-stage washing. For the countercurrent method

$$c_2 = \frac{c_1 + (R^2 + R)c}{R^2 + R + 1} \quad (14.3.5)$$

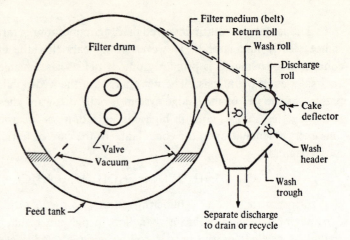

FIGURE 14.4.1
Rotary-drum filter. (*Taken from Water Pollut. Control Fed., Sludge Dewatering, fig. 15, WPCF Man. Pract. 20, Washington, 1969.*)

14.4 DEWATERING PROCESSES

The removal of bulk water from sludges and slurries can be accomplished by filtration. In process engineering, filtration is performed in four different ways: by gravity, pressure, vacuum, and centrifugal force. Filtration by gravity generally is limited to the removal of solids from relatively dilute suspensions by passage through porous beds of unconsolidated media.

Vacuum filtration finds wide application in the partial separation of liquids from concentrated suspensions, sludges, and slurries. When the liquid phase is highly viscous or when the solids are so fine that vacuum filtration is too slow, pressure filtration provides a convenient solution to the separation problem. Centrifugal filtration is used when the solids are easy to filter and a filter cake of low moisture content is desired. Discussion here will be limited to vacuum filtration.

Vacuum filtration is commonly carried out on slowly rotating drum filters in a more or less continuous operation. Such a filter consists of a cylindrical structure, the outside surface of which is covered with a filter medium. Internally, the drum is divided into shallow drainage compartments, each of which is connected by piping to an automatic valve located at one end of the filter (Fig. 14.4.1). Drum diameters vary from 3 to 12 ft, and lengths vary from as little as 1 ft to as much as 20 ft on the average.

The filter is suspended in a trough containing the sludge to be dewatered at a depth providing submergence to 20 to 40 percent of the circumferential area of the drum. As the filter is rotated, a sludge mat of ¼ in or more in thickness is formed on the filter medium as a result of a vacuum (20 to 26 in Hg) applied to the drainage compartments servicing the submerged portion of the drum. The vacuum is continued as the mat rises above the trough to dewater the sludge further. Just before the mat reaches the trough again, it is removed from the filter. Removal may be accomplished in several different ways. Filter speed, vacuum, and drum submergence are adjusted to meet operating conditions.

Filter media include wool, cotton, and synthetic fiber cloth as well as coil springs arranged in a two-layer pattern. Ancillary equipment typical of a vacuum filtration system includes vacuum receivers, filtrate pumps, moisture traps, and vacuum pumps.

Vacuum filtration is a special case of flow through beds of solid particles. As such, the movement of liquid through a filter cake can be thought of as taking place inside innumerable channels with small, indeterminate cross sections. Since the channels are so small, the flow of liquid through them is considered to be laminar. The volumetric rate at which the filtrate flows through the filter cake per unit time has been formulated as[1]

$$\frac{dV}{dt} = \frac{S\,\Delta p}{\mu L R} \quad (14.4.1)$$

where V = volume of filtrate. $[L^3]$

S = area of filter cake normal to direction of filtrate flow. $[L^2]$

Δp = pressure drop due to resistance to flow. $[FL^{-2}]$

μ = liquid viscosity. $[FL^{-1}t^{-1}]$

L = thickness of filter cake in direction of filtrate flow. $[L]$

R = specific resistance. $[t^2 L^{-3}]$

The specific resistance is a property of the filter cake. Its value depends upon such characteristics as porosity, average specific surface area of the cake particles, and compressibility. Since

$$L = \frac{vV}{S} \quad (14.4.2)$$

where v = volume of cake deposited per unit volume of filtrate

$$\frac{dV}{dt} = \frac{S^2 \Delta p}{\mu v V R} \quad (14.4.3)$$

---

[1] L. G. Rich, "Unit Operations of Sanitary Engineering," pp. 161–164, Wiley, New York, 1961.

Because of the nature of compressible cakes, it is more convenient to express sludge concentration in terms of weight of dry cake solids per unit volume of filtrate than by volume of cake per unit volume of filtrate. Therefore

$$\frac{dV}{dt} = \frac{S^2 \Delta p}{\mu w V R'} \quad (14.4.4)$$

where w = weight of dry cake solids per unit volume of filtrate. $[FL^{-3}]$

R' = specific resistance. $[t^2 F^{-1}]$

In vacuum filtration, the vacuum level is kept constant during cake formation. Therefore, Eq. (14.4.4) can be integrated to yield

$$\frac{V}{S} = \left(\frac{2 \Delta p\, t}{\mu w R'}\right)^{1/2} \quad (14.4.5)$$

Equation (14.4.5) is an expression for the volume of filtrate obtained per unit of cake area during time t. A relationship can be developed for filter yield by multiplying both sides by w/t_c:

$$\frac{V}{S}\frac{w}{t_c} = B = \left(\frac{2 \Delta p\, wt}{\mu R' t_c^2}\right)^{1/2} \quad (14.4.6)$$

where t_c = time required for one filter cycle. $[t]$

B = filter yield. $[FL^{-2} t^{-1}]$

The relation between the filter cycle time and the time required for the cake to be formed can be written as

$$t = \phi t_c \quad (14.4.7)$$

where ϕ = fraction of cycle time occupied by cake formation

With substitution, Eq. (14.4.6) becomes

$$B = \left(\frac{2 \Delta p\, w\phi}{\mu R' t_c}\right)^{1/2} \quad (14.4.8)$$

Actually, the total resistance encountered in a filtering operation is composed of two resistances in series: the resistance offered by the filter (primarily the medium) and that offered by the cake. With proper selection of the medium, however, the filter resistance is made negligible, and Eq. (14.4.8) holds.

The variables in Eq. (14.4.8) can be separated into two groups: *operational variables* and *sludge variables*. Operational variables include pressure drop due to friction Δp, fraction of cycle time occupied by cake formation ϕ, and filter cycle time t_c. An increase in yield is promoted by increasing Δp and ϕ and by decreasing t_c. As a rule, the highest vacuum is maintained, and operational control is exerted by varying drum speed t_c and drum submergence ϕ. In filtering sludges that form compressible cakes, drum submergence is usually controlled by the time required to dewater the cake to the desired moisture content. Since cake thickness decreases as cycle time is reduced, drum speed is limited by the minimum thickness that can be removed conveniently from the filter.

Sludge variables include liquid (filtrate) viscosity, solids concentration, and specific resistance. For a given sludge at a given temperature, filtrate viscosity is fixed. Solids concentration and specific resistance will vary, however, and are dependent upon sludge conditioning prior to filtration. As noted from Eq. (14.4.8) increased solids concentration and decreased specific resistance operate to give increased filter yields.

Solution procedure 14.4.1 The weight of dry cake solids per unit volume of filtrate w can be related to the solids concentration of the unfiltered sludge c by an expression derived as follows: If wV is the total weight of filterable solids both in the filter cake and unfiltered sludge, m is the weight ratio of wet cake to dry cake, and $(mwV - wV)/\gamma$ is the volume of liquid retained in wet filter cake, then

$$c = \frac{wV}{(mwV - wV)/\gamma + V} \quad (14.4.9)$$

and

$$w = \frac{c}{1 - (m-1)c/\gamma} \quad (14.4.10)$$

or, in terms of weight fractions, it can be shown that

$$w = \frac{\gamma}{(1-x)/x - (1-x_c)/x_c} \quad (14.4.11)$$

where γ = specific weight of water. $[FL^{-3}]$
$\quad\quad\quad x$ = weight fraction of filterable solids in unfiltered sludge
$\quad\quad\quad x_c$ = weight fraction of filterable solids in filter cake   ////

Solution procedure 14.4.2 The specific resistance is used not only for predicting filter yields but also as a parameter for evaluating the efficiency of sludge conditioning operations. Its value may be estimated from an analysis of laboratory data.

Equation (14.4.5) may be modified to include filter resistance to flow:

$$\frac{t}{V} = \frac{\mu w R'}{2S^2 \Delta p} V + \psi(R_f) \quad (14.4.12)$$

or

$$\frac{t}{V} = bV + a \quad (14.4.13)$$

where $\psi(R_f)$ = function of filter resistance. $[tL^{-3}]$

Thus it is seen that a plot of data relating values of t/V with V will give a straight line, the slope of which is

$$b = \frac{\mu w R'}{2S^2 \Delta p} \quad (14.4.14)$$

Briefly, the experimental apparatus used to obtain values of t/V and V consists of a *Büchner funnel* modified so that the measurements can be made on the filtrate volumes at various times during filtration. Information must be obtained concerning the filter area, solids content (w or c), filtration pressure, and viscosity of the filtrate.

Most sludges of domestic waste origin form compressible filter cakes. For these cakes, specific resistance will vary with the vacuum level. Empirically, it has been found that the relationship is expressed satisfactorily by

$$R' = C (\Delta p)^s \quad (14.4.15)$$

where C = cake constant

s = coefficient of compressibility

The cake constant and coefficient of compressibility for a given sludge cake can be determined from a logarithmic plot of data relating the specific resistance with vacuum level. The slope of the straight line of best fit is equal to s, while the intercept at $\Delta p = 1$ gives the value of C. ////

14.5 DRYING AND INCINERATION PROCESSES

Air drying beds are the most commonly used means for dewatering digested sludges. Such beds are generally 12 to 18 in deep and consist of graded layers of gravel or crushed stone overlain by 4 to 6 in of sand. The beds are underdrained with open-joint pipe. Sludge is applied to the bed to a depth of 8 to 12 in. The drying time required will depend upon meteorological conditions. Glass enclosures decrease the drying time. Sludge moisture is lost to the atmosphere through evaporation and to the drainage system by percolation. Sludge moisture can be reduced to 40 percent through air drying. For sludges produced in plants treating domestic wastes, 1 ft^2 of air drying bed is required per capita served.

Heat dryers that are used to reduce the moisture in sludge are of two types—kilns and flash dryers. Heat drying is expensive and is most often used where the dried material has some commercial value. Such dryers will reduce the sludge moisture to approximately 10 percent.

Incineration removes all sludge moisture. Furthermore, the organics remaining in the sludge are converted to inert fly ash. Heat drying may serve as the first stage of incineration. Generally, 1800 to 2500 Btu will be required for each pound of water evaporated from the sludge. For drying, this energy must be supplied by waste heat from some source or auxiliary fuel. When sludge is incinerated, the same quantity is required per pound of water in the sludge, but a considerable part of that heat is obtained from the combustion of the volatile solids in the sludge.

Most sludges which are disposed of through incineration are heterogeneous mixtures of organic materials. The heats of combustion of these sludges must be obtained from combustion analysis.

EXERCISES

14.1.1 Identify the four zones that occur in the batch sedimentation of sludge systems, and give the characteristics of each.

14.1.2 On what factors is the design of continuous-flow thickening tanks based?

14.1.3 The following data were collected from batch settling tests performed on the mixed-liquor suspended solids (MLSS) from an activated-sludge process. The settling tests were performed in columns 4 ft deep.

X, mg/l	1,700	2,000	2,200	2,700	3,700	5,300	6,300
u, ft/min	0.090	0.062	0.030	0.015	0.008	0.005	0.004

Determine the solids flux to be used in the design of a thickening tank from which sludge is to be removed continuously at a concentration of 5,000 mg/l when the flow of clarified effluent is 1 Mgal/d.

14.1.4 What will be the minimum surface area required for the thickener in Exercise 14.1.3 if it is desired to maintain the MLSS in the aeration tank at 2,000 mg/l?

14.2.1 List the characteristics of the ideal anaerobic digestion process.

14.2.2 Construct a sketch of the vertical cross section of the conventional, standard-rate digester, and label the stratified contents. In what ways does the conventional process differ from the ideal?

14.2.3 When the pH drops in a digester and the process literally ceases, what diagnosis can be made of the situation? What remedies are available?

14.2.4 List the major features of high-rate digestion.

14.2.5 A sludge consisting of 4 percent solids and 96 percent water is to be digested for 30 d in a standard-rate process. The solids initially are 70 percent volatile (specific gravity = 1.0) and 30 percent fixed (specific gravity = 2.5). If the sludge withdrawn from the digester is expected to contain 8 percent solids, estimate the required tank capacity on the basis of 500 kg of dry solids introduced to the process daily.

14.3.1 Why and how are sludges conditioned?

14.3.2 A digested sludge containing 2,500 mg/l alkalinity is to be elutriated with water containing 100 mg/l alkalinity. Compare the volume of wash water that will be used in a two-stage elutriation operation using fresh water at each step with that of a two-stage countercurrent operation if the alkalinity of the sludge is to be reduced in both cases to 400 mg/l.

14.4.1 What methods are available to the process engineer for dewatering sludges? Under what conditions is each method used?

14.4.2 Describe how a vacuum filter of the rotary-drum type operates. Identify the operational and sludge variables of significance in determining filter yield.

14.4.3 Sludge containing 5 percent filterable solids is to be dewatered at a rate of 0.467 m³/h on a drum filter under a vacuum of 635 mm Hg. Laboratory analyses reveal that at the same vacuum level the specific resistance of the filter cake is 14.5×10^{12} s²/kg and a cake with 30 percent solids can be obtained with $3\frac{1}{2}$ min of dewatering. The specific gravity is approximately 1, and the filtration temperature of the sludge approximately 25°C. Assuming a filter cycle of 5 min, estimate the filter area required.

14.4.4 A vacuum filtration test was performed on a sludge with the use of a modified Büchner funnel apparatus. Test conditions were as follows:

Temperature of sludge = 25°C
Filtration vacuum = 635 mm Hg
Filter area = 44.2 cm²
Weight of solids per unit volume of filtrate = 48 mg/ml

Filtrate volumes obtained after various time intervals are recorded below. Compute the specific resistance.

t, s	60	120	180	240	300	360	420	480
V, ml	1.3	2.4	3.4	4.3	5.2	6.0	6.7	7.5

14.5.1 Identify the most common methods employed for drying sludge, and give the sludge moisture attainable by each.

APPENDIX

A.1 FACTORS FOR CONVERSION TO THE INTERNATIONAL SYSTEM (SI) UNITS*

To convert from	To	Multiply by†	
Acceleration			
ft/s^2	m/s^2	3.048	E − 01
in/s^2	m/s^2	2.540	E − 02
Area			
acre	m^2	4.047	E + 03
ft^2	m^2	9.290	E − 02
in^2	m^2	6.452	E − 04
mi^2 (statute)	m^2	2.590	E + 06
yd^2	m^2	8.361	E − 01
Torque			
dyne-cm	N-m	1.000	E − 07
kg$_f$-m	N-m	9.807	E + 00
lb$_f$-in	N-m	1.130	E − 01
lb$_f$-ft	N-m	1.356	E + 00
oz$_f$-in	N-m	7.062	E − 03
Torque/length			
lb$_f$-ft/in	N-m/m	5.338	E + 01
lb$_f$-in/in	N-m/m	4.448	E + 00
Electricity and magnetism			
Ah	C	3.600	E + 03
Faraday (chem)	C	9.650	E + 04
G	T	1.000	E − 04
Gb	A (turn)	7.958	E − 01
Mx	Wb	1,000	E − 08
Oer	A/m	7.958	E + 01
unit pole	Wb	1.257	E − 07
Energy (includes work)			
Btu‡	J	1.054	E + 03
cal‡	J	4.184	E + 00
eV	J	1.602	E − 19
erg	J	1.000	E − 07
ft-lb$_f$	J	1.356	E + 00
kWh	J	3.600	E + 06
watt-sec	J	1.000	E + 00
Energy/(area)(time)			
Btu‡/ft^2 min	W/m^2	1.891	E + 02
Btu‡/ft^2 h	W/m^2	3.152	E + 00
cal‡/cm^2 min	W/m^2	6.973	E + 02

* Adapted from *Natl. Bur. Stand. Handb.* 102.
† E indicates the power of 10 by which the number must be multiplied; that is, 4.047 E + 03 = 4.047 × 10^3.
‡ Thermochemical.

To convert from	To	Multiply by†
Force		
dyn	N	1.000 E − 05
kg$_f$	N	9.807 E + 00
oz$_f$ (av)	N	2.780 E − 01
lb$_f$ (av)	N	4.448 E + 00
lb$_f$ (av)	kg$_f$	4.536 E − 01
Force/length		
lb$_f$/in	N/m	1.751 E + 02
lb$_f$/ft	N/m	1.459 E + 01
Heat		
Btu‡ in/s ft²-°F	W/m-°K	5.189 E + 02
Btu‡ in/h ft²-°F	W/m-°K	1.441 E − 01
Btu‡/ft²	J/m²	1.135 E + 04
Btu‡/h ft²-°F	W/m²-°K	5.674 E + 00
Btu‡/lb$_m$ °F	J/kg-°K	4.184 E + 03
Btu‡/s ft²-°F	W/m²-°K	2.043 E + 04
cal‡/cm²	J/m²	4.184 E + 04
cal‡/cm² s	W/m²	4.184 E + 04
cal‡/cm s-°C	W/m-°K	4.184 E + 02
cal‡/g	J/kg	4.184 E + 03
cal‡/g °C	J/kg-°K	4.184 E + 03
Length		
astronomical unit	m	1.496 E + 11
ft	m	3.048 E − 01
in	m	2.540 E − 02
light year	m	9.461 E + 15
mil	m	2.540 E − 05
mi (statute)	m	1.609 E + 03
yd	m	9.144 E − 01
Light		
fc	lm/m²	1.076 E + 01
fc	lx	1.076 E + 01
fL	cd/m²	3.426 E + 00
lx	lm/m²	1.000 E + 00
Mass		
oz$_m$ (av)	kg	2.835 E − 02
lb$_m$ (av)	kg	4.536 E − 01
ton (2000 lb$_m$)	kg	9.072 E + 02
Mass/volume (includes density)		
lb$_m$/ft³	kg/m³	1.602 E + 01
lb$_m$/in³	kg/m³	2.768 E + 04
oz$_m$ (av)/in³	kg/m³	1.730 E + 03
lb$_m$ (av)/gal	kg/m³	1.198 E + 02
Power		
Btu‡/s	W	1.054 E + 03
Btu‡/min	W	1.757 E + 01
Btu‡/h	W	2.929 E − 01

To convert from	To	Multiply by†	
Power			
cal‡/s	W	4.184	E + 00
cal‡/min	W	6.973	E − 02
erg/s	W	1.000	E − 07
ft-lb$_f$/h	W	3.766	E − 04
ft-lb$_f$/min	W	2.260	E − 02
ft-lb$_f$/s	W	1.356	E + 00
hp (elec)	W	7.460	E + 02
Pressure (force/area)			
atm (760 torr)	N/m^2	1.013	E + 05
bar	N/m^2	1.000	E + 05
dyn/cm^2	N/m^2	1.000	E − 01
g$_f$/cm^2	N/m^2	9.807	E + 01
in of Hg (60°F)	N/m^2	3.377	E + 03
in of water (60°F)	N/m^2	2.488	E + 02
mm of Hg (0°C)	N/m^2	1.333	E + 02
lb$_f$/ft^2	N/m^2	4.788	E + 01
lb$_f$/in^2 (psi)	N/m^2	6.895	E + 03
lb$_f$/in^2 (psi)	kg$_t$/mm^2	7.031	E − 04
torr (mm Hg, 0°C)	N/m^2	1.333	E + 02
Velocity (includes speed)			
ft/h	m/s	8.467	E − 05
ft/min	m/s	5.080	E − 03
ft/s	m/s	3.048	E − 01
in/s	m/s	2.540	E − 02
mi/h	m/s	4.470	E − 01
mi/min	m/s	2.682	E + 01
mi/s	m/s	1.609	E + 03
mi/h	km/h	1.609	E + 00
Viscosity			
ft^2/s	m^2/s	9,290	E − 02
poise	N-s/m^2	1.000	E − 01
lb$_m$/ft-s	N-s/m^2	1.488	E + 00
lb$_f$ s/ft^2	N-s/m^2	4.788	E + 01
stoke	m^2/s	1,000	E − 04
Volume (includes capacity)			
bushel (US)	m^3	3.524	E − 02
ft^3	m^3	2.832	E − 02
gal (US)	m^3	3.785	E − 03
in^3	m^3	1.639	E − 05
l (new)	m^3	1.000	E − 03
oz (US fluid)	m^3	2.957	E − 05
stere	m^3	1.000	E + 00
yd^3	m^3	7.646	E − 01
Volume/time (includes flow)			
ft^3/min	m^3/s	4.719	E − 04
ft^3/s	m^3/s	2.832	E − 02
in^3/min	m^3/s	2.731	E − 07
gal/min	m^3/s	6.309	E − 05

A.2 NOMOGRAM FOR THE SOLUTION OF HAZEN-WILLIAMS EXPRESSION WHEN $C = 100$

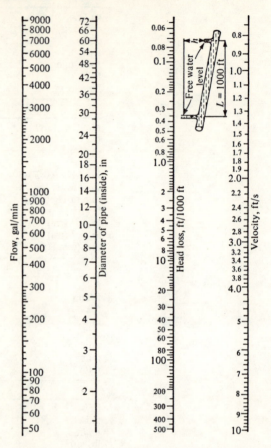

A.3 NOMOGRAM FOR THE SOLUTION OF THE MANNING FORMULA ($n = 0.013$)

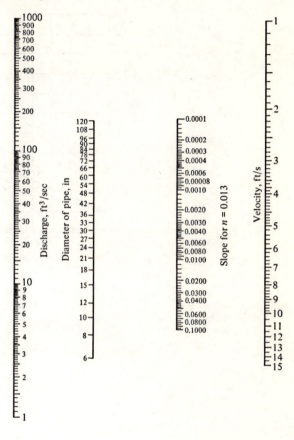

A.4 HYDRAULIC ELEMENTS OF CIRCULAR SECTIONS

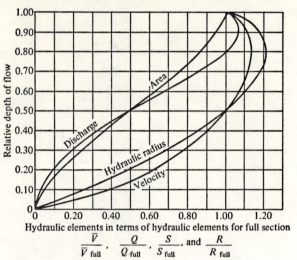

A.5 ATOMIC WEIGHTS AND VALENCES OF SELECTED CHEMICAL ELEMENTS

Element	Symbol	International atomic weight	Valence*
Aluminum	Al	26.98	3
Arsenic	As	74.92	±3, 5
Barium	Ba	137.34	2
Boron	B	10.81	3
Bromine	Br	79.91	±1, 5
Cadmium	Cd	112.40	2
Calcium	Ca	40.08	2
Carbon	C	12.01	±4, 2
Chlorine	Cl	35.45	±1, 7, 5
Chromium	Cr	52.00	3, 6, 2
Cobalt	Co	58.93	2, 3
Copper	Cu	63.54	2, 1
Fluorine	F	19.00	−1
Gold	Au	196.97	3, 1
Hydrogen	H	1.01	1
Iodine	I	126.90	−1, 5, 7
Iron	Fe	55.85	3, 2
Lead	Pb	207.19	2, 4
Magnesium	Mg	24.31	2
Manganese	Mn	54.94	2, 7, 4, 6, 3
Mercury	Hg	200.59	2, 1
Nickel	Ni	58.71	2, 3
Nitrogen	N	14.01	−3, 5, 2
Oxygen	O	16.00	−2
Phosphorus	P	30.97	5, ±3
Platinum	Pt	195.09	4, 2
Potassium	K	39.10	1
Selenium	Se	78.96	4, 6, −2
Silicon	Si	28.09	4
Silver	Ag	107.87	1
Sodium	Na	22.99	1
Strontium	Sr	87.62	2
Sulfur	S	32.06	6, 4, −2
Tin	Sn	118.69	4, 2
Zinc	Zn	65.37	2

* Most stable valences stated first.

A.6 ALGORITHM CONSTRUCTION

The computer algorithms found in the text are presented as ordered sequences of mathematical operations adapted for solution by digital computers. The type of operations used are indicated by the set of symbols shown in Fig. A.6.1. The statements written inside the symbols are abbreviated and generalized so that they can be easily coded in any one of several computer languages available to the engineer.

Three basic types of statements are used in the algorithms: input and output statements, arithmetic statements, and control statements. The READ statement introduces the input variables to the program, whereas the WRITE statement identifies the variables to be included in the output of the program. See Fig. A.6.1a and b.

The arithmetic statement calls for the specific mathematical operation to be performed. These are enclosed in the type of symbols shown in Figure A.6.1c and d. The mathematical operations performed are addition (+), subtraction (−), division (/), multiplication (∗), and exponentiation (∗∗). The operational hierarchy for these operations are exponentiation (∗∗) first, multiplication and division (∗, /) second, and addition and subtraction (+, −) last. Multiplication is not implicit but must be indicated by ∗. Results of the mathematical operations are stored in the variables indicated by arrows. For example, the statement, A + B → D, indicates that the sum of A and B are stored and labeled D. Both A and B are unchanged and available as such for other operations. However, a variable may be changed explicitly. The result of a sequence of operations contained in a subprogram, or subroutine, can be called several times in a program by statements including the name of the subprogram.

Variables are coded in names with five or less characters, each of which can be alphabetic or numeric. The first character, however, must be alphabetic. Examples of proper names are, X, A1, BB, PROD, DELT, SUM1, and K411. When a variable assumes an array of different values $(x_1, x_2, \ldots, x_n)$ the variable can be coded by another variable which expresses the integer subscript. For example, the alphabetic character I in X(I) expresses the subscript integer which takes on different values to identify different values of the x variable. Variables in a matrix can be identified by two-subscript variables, X(I, J).

Four different types of control statements are used in the algorithms. The iteration statement is enclosed in the symbol shown in Fig. A.6.1e. The statement shown in the figure calls for all the operations between the statement and the symbol containing α to be performed in an iterative manner for I = 1, I = 2,..., I = N. When all the operations have been completed for I = N, control is passed to the next statement beyond the α symbol.

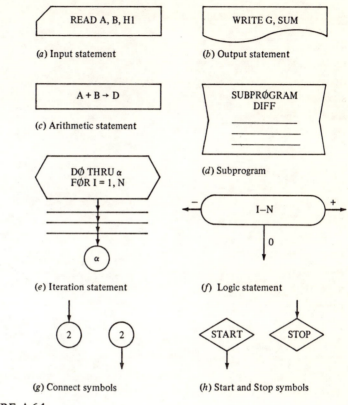

FIGURE A.6.1

The logic statement can be any arithmetic expression enclosed in the type of symbol shown in Fig. A.6.1f. If the value of the expression is negative, control passes to the statement signified by the negative branch. If it is positive, control passes to the statement signified by the positive branch. A zero value results in control being passed through the zero branch. Sometimes, the occasion arises when two of the branches can be combined.

The start and stop symbols (see Fig. A.6.1b) are the first and last symbols in the algorithm. The connect symbol (Fig. A.6.1g) connects different parts of the flow diagram that are physically separated, such as a diagram on two different pages.

For a more comprehensive explanation of algorithm construction the reader is referred elsewhere.[1]

[1] Royce Beckett and James Hurt, "Numerical Calculations and Algorithms," McGraw-Hill, 1967.

Index

Abiotic substances, 91
Absorption coefficient, 19
Absorptivity of atmosphere, 21
Acid forming bacteria, 87
Activated carbon, 382
   adsorptive capacity, 382
Activated energy, 47
Activated silica, 373
Activated-sludge processes, 353–366
   clarification stage, 356
   completely mixed, 362–366
      figure, 262
   contact-stabilization, 360
      figure, 361
   conventional, 355–359
   extended-aeration, 361
   high-rate, 361–362
   mixed liquor, 356
   process fundamentals, 353–355
   process loading intensities, table, 355
   sludge stabilization, 355

Activated-sludge processes:
   sludge volume index, 358
   stabilization stage, 356
   step-aeration, figure, 359–360
   thickening tanks for, 358–359, 366
Activators used in flotation processes, 348
Activity:
   ion products of water, 41
   solubility product, 41
   of a solution, 40
Activity coefficient, 40, 43
Adenosine diphosphate (ADP), 78
Adenosine triphosphate (ATP), 78
Adiabatic lapse rate, 178
Admissible output in matrix formation, 131
ADP (adenosine diphosphate), 78
Adsorption isotherm, 382
Advection, 1–2
   factor, 140
      value of, 157

INDEX

Advective prism, 154
Aeration processes, 380–382
Aerosols, 62
Air pollution, 176–181
   air-quality standards, table, 181
   carbon monoxide, 177–178
   control of, 179–180
   emission sources of, table, 180
   hydrocarbons, 178
   nitrogen oxides, 178
   particulate matter, 177
   sulfur oxides, 177
   from vehicles, 180
Air-quality management, 180
Air-quality standards, table, 181
Air-temperature lapse rates, figure, 179
Aldrin, 111
Algae, 89
   blooms, 114
   blue-green, 114
   cell synthesis of, figure, 90
   nutrient requirements, 114
      sources, table, 115
Alkalinity:
   definition of, 53
   effect on condition processes, 418–419
   species as a function of pH, figure, 54
Alum, use in water treatment, 372
American Insurance Association, 290, 293
Ammonia, 44
   in digester gases, 413
   equilibrium, 44
   reactions with chlorine, 398
Anabolism, 77
Anaerobic decomposition, 87–89
Anaerobic digestion, 412–418
   high-rate, 417–418
   ideal process, 412–414
   reduction of volital solids in, figure, 416
   standard-rate, 414–417
   stoichiometric equation for, 412
   thermophilic, 413
   time required for, figure, 413
Analytic solutions, 102–104
   ease of solution, table, 104
   to first-order linear differential equations, table, 102
Aquatic ecosystem, figure, 116
Arithmetic probability paper, 226

Arsenic:
   limits of, in drinking water standards, table, 172
   in pesticides, 111
Atomic Energy Commission standards, 185–186
ATP (adenosine triphosphate), 78
Atrazine, 111
Autocovariance function, 227–228
Autrotrophs, 77

Bar screens, 335
Beer's law, 19, 90
Beneficial uses of water, 170
Benthal deposits, BOD of, 151
Benthic decomposition, 88
Biochemical oxygen demand (BOD), 84–87
   determination of parameters, 86–87
   determination of rate constants for removal of, 150–151
   effect of bottom slime on, 144
   effect of temperature, 86
   first-stage, 85
   as function of time, 85
   influence of adsorption on, 144
   influence of carbonaceous materials on, 144–145
   influence of nitrogenous materials on, 144–145
   influence of sedimentation on, 144
   mathematical expression, 85–86
   stages of, 145
   as treatment efficiency parameter, 325
   water-quality differentials, table, 275
Biologic coefficients (see Biologic constants)
Biologic constants, 80
   determination of, 109–111
   effect of temperature on, 82
   growth yield, 84
   maximum specific growth rate, 80
   saturation constant, 80
   specific decay rate, 80
   specific growth rate, 80
   used in design of activated sludge processes, 363–364
      table, 365
Biologic solids retention time, 111
Biological oxidation processes:
   aerobic, 353–371

Biological oxidation processes:
  anaerobic, 412–418
  in lagoon systems, 326–331
  in septic tanks, 331
Biomass, 76
Blackbody, 20
Block diagram:
  elements of, 104
  table of: constant, 105
    divider, 105
    function generator, 105
    function switch, 105
    gain, 105
    integrator, 105
    multiplier, 105
    sign inverter, 105
    summer, 105
Blue-green algae, 114
BOD (*see* Biochemical oxygen demand)
Boolean algebra, 133
Bottom deposits, effect on BOD, 145
Boundary layer, 27, 28
Bronchial asthma, 176

C curves, 33
Carbon-adsorption processes, 382–383
Carbon cycle, figure, 71
Carbon dioxide:
  in carbonate equilibriums, 50–56
  demand in water-softening processes, 384
  free, 121
  as a function of pH, figure, 54
  hydration, reaction rates of, 55
  model of subsystem, 121–122
    table, 118
  partial pressure of, in atmosphere, 53
  use in photosynthesis, 89
Carbon monoxide, 177
Carbonate system, 121
  storage of carbon dioxide in, 121
Carbonates:
  distribution coefficients, 52
  equilibriums, 49–56
    constants of, 50
      variation with temperature, figure, 51
Carnivores, 91
Cascade aerators, 380–381
Catabolism, 77
Chemical oxygen demand (COD), 325

Chemical-precipitation processes, 383–389
Chloramines, 398
Chlorine:
  breakpoint chlorination, 398–399
  chloramines, 398–399
  demand, 400
  distribution of species, figure, 398
  gas, 397
  rate of kill by, 400
  residuals: combined available, 397
    figure, 399
    free available, 397
  valence of compounds of, 397
Chronic diseases, 176
Clarification capacity of sludges, 407
Clarification stage in activated sludge process, 356
Coagulation, 372–374
  aids in, 373
  of colloids, 67
  in sludge conditioning, 418–419
Coarse screens, 335
Coefficient of absorption, 19
Coefficient of compressibility, 424
Coefficient of drag, 26
  for spheres, disks, and cylinders, figure, 26
Collectors in flotation processes, 348
Colloid systems:
  behavior of, 61–67
  classification of, 62
  methods of destabilization, 66–67
  primary charges on, 62–64
  stability of, 65
  types of, 62
Comminuting device, 336
Common cold, 176
Common ion effect, 42
Compensation point, 90
Completely mixed process, 362–366
Composting, 267–268
Concentration equilibrium constant, 42
Conditioning processes, 418–419
  coagulation, 418–419
  elutriation, 419–424
Conduction, 14–16
  thermal conductivity, 14, 16
Conjugate depths, 297
Conservative substance, 5
Constraints in linear programming, 261

Consumers in ecosystem, 91, 116
　model of subsystem, 120
　　table, 118
Contact-stabilization process, 360–361
Continuous flow systems:
　effects of retention time and limiting nutrient, figure, 110
　microbiological, 106–111
　models of, 31–35
Continuous System Modeling Program (CSMP), 124
　coded program of, figure, 126
Continuous transport model, 138–139
　use in segmented streams, 150
Convection of thermal energy, 16–19
　flux, 22
Cooling pond, 24
Copperas, 372
Correlation coefficient, 241
Cost allocation methods, 255–264
　cost minimization method, 255–257
　uniform treatment method, 255, 258, 261
　zoned-optimization method, 256, 258
Critical dissolved oxygen deficit, 149
Critical flow, 297
Critical retention time, 109
Critical velocity, 297
CSMP (*see* Continuous System Modeling Program)

Dalton's law of partial pressures, 11
DDT, 111
Decision variables in linear programming, 263
Decomposers in ecosystems, 91, 117
Demineralization, 392
Density of water, $28n$.
Deoxygenation constants, 85
　determination from stream samples, 150–151
Depressants in flotation processes, 348
Detritus, 116, 120
　table, 118
Dewatering processes, 420–424
Diatomaceous earth, 375
Dieldrin, 111
Diffuse layer, 64
Diffusion, 1–2
　coefficients, figure, 6

Diffusion:
　Fick's law, 2
　molecular diffusion coefficient, 3
　thermal, 14
　turbulent diffusion coefficient, 5
Diffusion aerators, 380, 382
Dilution rate, 102
Dimensions, systems of, $3n$.
Disinfection processes, 396–401
Disperse medium, 61
Disperse phase, 61
Dispersed-air flotation, 347
Dispersion, 5–7
　coefficients, 7, 139
　　figure, 6
　　methods for predicting value of, 7
　longitudinal dispersion coefficient, 5
Dissociation constants, 41
Dissolved-air flotation, 347, 404
Dissolved-materials removal trains, 319
Dissolved oxygen:
　influence of BOD on, 144–145
　influence of bottom deposits on, 145
　influence of photosynthesis on, 145–146
　influence of respiration on, 146
　influence of runoff on, 145
　limiting concentration of, 142–143
　molecular diffusion coefficient as a function of temperature, 144
　reaeration coefficient as a function of temperature, 144
　saturation concentration value as a function of temperature, 144
　source and sink terms, 142–146
Distance matrix in waste collection, 270
　systems, 271
Distribution coefficients:
　for bicarbonate ion, 52
　for carbonate ion, 52
　for dissolved carbon dioxide (and carbonic acid), 52
　for orthophosphate species, 57–58
Dosimetry, 182–183
Double layer, figure, 64
Drop-down curve, 303
Drying processes, 424–425
　air drying beds, 424
　heat dryers, 425
Dynamic behavior of models, 100

Ecological systems (*see* Ecosystems)
Ecosystems, 99–134
 aquatic: figure, 116
  subsystem of, table, 118
 components of, 91
 definition of, 72n.
 nitrogen transformation in, figure, 73
 pond, figure, 92
Effective size, 375n.
Effluent standards, 176
Electrodialysis, 393–394
Electromagnetic spectrum, figure, 91
Electroneutrality expression, 52, 54
Elutriation of sludges, 419
Emission inventory, 180
Emissivity, 20
Emphysema, 176
Emulsions, 62
Endergonic processes, 78
Endogenous respiration, 78
Endrin, 111
Enthalpy, 58
 change in, 14
 levels, figure, 60
Environmental impact statements, 186–194
 factors to be considered in preparation of, tables, 189, 190
 impacts to be considered in preparation of, tables, 191, 192
 matrix preparation for, 188–193
 requirement legislation, 186
 text of, 193–194
Environmental Policy Act of 1969, 186
Epilimnion, 23
Episodes caused by air pollution, 176
Equalizing storage, 291
 table, 292
Equilibrium constant, 47
 activity, 39
 concentration, 42
Equilibrium selectivity coefficient, 390
Estuaries:
 definition of, 154
 modeling of, 154–164
Eutrophication, 113–134
 definition of, 113
 effects of, 114–115
Evaporation, 17–19
 energy loss from, 22
 thermal energy flux, 18

Exergonic processes, 78
Extended-aeration process, 361

F curves, 33
Facultative autotrophs, 77
Fall turnover, 23
Fecal coliform organisms, 173
Ferric hydroxide equilibrium, 45
Ferric iron, 44
 removal of, 386–387
Ferric phosphate equilibrium, 45
Ferrous iron, 44
 removal of, 386–387
Ferrous phosphate equilibrium, 44
Ferrous sulfide equilibrium, 44
Fick's laws of diffusion, 2–3
Film model, 8
 figure, 9
 film penetration model, 9–10
 penetration model, 9–10
Filtration:
 centrifugal, 420
 pressure, 420
 vacuum, 420–424
Final-demand projections, tables, 207, 208
Fine screens, 336
Finite-volume transport model, 138–142, 256
 use in modeling estuaries, 154–164
  one-dimensional application of, 158
  two-dimensional application of, 163–164
 use in modeling nontidal streams, 164
Flash mix, 373
Flocculation:
 basin, 371
 process, 374
Flotation processes, 346–351
 basins used in, 336
 chemical agents used in, 348
 definition, 346
 design of, 349–351
 mechanism of, 347–348
 types, 347
Flow balance, 101
Flow models:
 completely mixed, 31
 plug, 31
Food chains, 91–93

Food web, 93
Form drag, 27
Fourier analysis, 217–221
Fourier coefficients, 218, 219
Fourier cosine transform of autocovariance, 230
Frequency component of time series, 217–218
Fresh-water withdrawal coefficients, table, 207
Frothers in flotation processes, 348

Garbage grinders, 267
Gas constant, 47
Gas solubilities, 11
  Dalton's law of partial pressure, 11
  Henry's law, 11
  Henry's law constant, table, 11
Gas transfer, 7–12
Grit chambers, 343–344
Gross national product (GNP), 201
Gross particulates removal trains, 319
Gross regional product (GRP), 201
Growth of microorganisms:
  patterns of, 82
    figure, 83
  phases of, 83
Growth kinetics, 80–84
Growth yield, 84

Half-lives of radioisotopes, 48
Hardness in water, 383
  causes of, 383
  removal by: ion-exchange processes, 389–393
    lime and soda ash, 384–386
Hardy Cross method, 283–289
Hazen-Williams equation, 281–282
  coefficient values for, table, 282
Heat:
  capacity, 13, 61
  of combustion, 59, 73
  of formation, 59
  latent, 13
  of reaction, 59
  sensible, 13
  specific, 13, 61
  vaporization, 13
    of water, 14
Heat-transfer coefficient, 17

Henry's law, 11
  constant, 53
    table, 11
Heptaclor, 111
Herbivores in ecosystems, 91
Heterogeneous thermal regimes, 21
Heterotrophic organisms, 77
High-rate activated sludge processes, 361–362
Homogeneous thermal regimes, 21
Household treatment systems, 331–333
Hydraulic grade line, 281
Hydraulic jump, figure, 301
Hydraulic radius, 281
Hydrocarbons, 178
Hydrogen sulfide, 43
  in anaerobic digesters, 413
  equilibrium, 43
  species distribution, figure, 43
  in waste stabilization lagoons, 330
Hypolimnion, 23
  reactions in, 44

Incineration processes, 424–425
  use in solid waste disposal, 266–267
Industrial sector classification, table, 202
Infectious hepatitis, 174
Inorganic carbon:
  dissolved species, 51
  total, 121
Input-output analysis, 197
Interindustry accounting system, table, 199
Interindustry transaction table, tables, 203–204
Ion balance, 52
Ion-exchange processes, 389–393
  exchange capacity, table, 392
  regeneration requirements, table, 392
Ion product of water, 41
Ionic strength, 40
  effect in fresh water, 42
Ionization constants, 41
Iron:
  ferric-ferrous equilibrium, 44
  ferric hydroxide equilibrium, 45
  ferric phosphate equilibrium, 45
  ferrous phosphate equilibrium, 44
  ferrous sulfide equilibrium, 44
  removal processes, 386–387

Isoconcentration lines, 340
Isoelectric point, 63

Lagoon systems, 326–331
   effluents, 328
     figure, 329
   (*See also* Waste stabilization lagoons)
Lake Mendota, estimated nutrient sources, table, 115
Lapse rates, air-temperature, figure, 179
Latent heat, 13–14
Law of mass action, 39
Lead:
   in drinking water standards, 172
   in pesticides, 111
Light attenuation, 122
Limiting nutrient concentration, 80
Linear programming, 256, 261, 263
   figure, 264
Logarithmic standard derivation, 248
   mean, 248
Logistics curve, 196
Loucks' solution procedure, 152
   figure, 153

Management systems, 254–278
   solid waste, 264–271
   waste-water reuse, 272–278
   water quality, 254–264
Manganese removal, 386–387
Manning equation, 302, 307
Mass-transfer coefficient:
   gas phase, 18
   overall, liquid phase basis, 8, 10
Mean travel time of tracer cloud, 7
Mechanical-energy equation, 280
   for open-channel flow, 296
   for pipe flow, 280
Mercury in pesticides, 111
Metabolism, 77
Meteorology, 178
Methane bacteria, 87–88
   in anaerobic digestion, 412–413
Micronutrients, 77
Microorganism, 76–84
   elemental composition of, 76
   energy mechanism of, 78
   metabolism of, 77–78
   nutritional requirements, 76

Minimum stream flow, prediction of, 247–249
Mixed liquor in activated sludge processes, 356
Mixed-media filter, 377
Mixing coefficient, 140
   relation to coefficient of dispersion, 157
Models:
   completely mixed, 107
   plug flow, 106
   of stream flow, 237–243
   of subsystems in aquatic ecosystems, table, 118
     carbon dioxide, 121, 122
       table, 118
     consumers, 120
       table, 118
     detritus, 120
       table, 118
     glossary of terms in, table, 119
     light attenuation, 122
     nitrogen, 121
       table, 118
     organic carbon, 121
       table, 118
     phosphorus, 121
       table, 118
     phytoplankton, 117–120
       figure, 118
     values of variables and parameters used in, 128
       table, 127
   of transport in the air environment, 165–167
   types of, 99
     population balance, 100
     transport-phenomena, 99–100
Monod function, 81
Multiple-tray aerators, 380–381

National Board of Fire Underwriters, 290n.
National Research Council, 371
Net energy flux, 21
Newton's law, 27, 28
Nitrification, 86
Nitrogen:
   cycle in nature, figure, 72
   models of subsystem, 120, 178
     table, 118
   oxides of, 178

Nitrogen:
  removal by ion exchange, 393
  transformations in aquatic ecosystems, figure, 73
Nonlinear differential equations, solution of, 124
Nonuniform flow, 296
Normalized autocorrelation function, 229
Nuclear power plants, 184
Nutrient utilization, 84

Objective function in linear programming, 257, 261
Objective spectrum for standards, figure, 173
Occurrence matrix, 130
  figure, 131
Ohm's law, 16
Oligotrophic, 113
Open-channel flow, 296–306
  critical velocity in, 297
Open dumps, 266, 267
Organic carbon, model of subsystem, 121
  table, 118
Organic materials, 70–76
  composition of, 70
  decomposition products of, 74
  decomposition stoichiometry, 75
  empirical mole of, 72
  energy content, 73
  oxidation stoichiometry, 74–75
Orthophosphates:
  distribution as a function of pH, table, 58
  distribution coefficients, 57–58
  ferric and ferrous species, 44–45
  total dissolved concentration of, 57
Oxidation ponds, 326–327
Oxygen sag curve, figure, 149
Oxygen transfer in stream aeration, 10

Particle size spectrum, 319
Particulate matter, 177
Pesticide concentration, 111–113
  equation for, 113
pH, definition, 50
Phenolic compounds in water, 174
Phosphorus:
  model of subsystem, 121
    table, 118
  removal of, 387–388

Photosynthesis, 89–91, 145–146, 327
  as a function of sunlight intensity, 145–146
Phytoplankton, 91, 116
  system, 117–120
    figure, 118
Pipe:
  fluid flow through, 280–282
  Hazen-Williams coefficient for, table, 282
  head loss through, 280–282
  minimum velocities of flow in, 308
  roughness coefficient of, table, 302
  size selection of, 282, 307, 308
Pipe network analysis, 280–289
Plug-flow model, 106
Pneumonia, 176
Polyelectrolytes, 373
Pond ecosystem, figure, 92
Population-balance models, 100
Population densities, 308
Population-growth curves, 195
Population-growth models, 194
Population predictions:
  long-term estimates, 197–208
  short-term estimates, 194–197
Power spectrum, 229
Pressure flotation, 349
Primary charges on colloids, 62
Primary clarifiers, 336
Primary productivity, 113–115
Primary respiration, 78
Primary sedimentation, 336
  basins, 344–346
Primary treatment units, 335
Process loading intensity, 325
Producers in ecosystems, 91
Promoters in flotation processes, 348
Pump-characteristic curves, 293–294
Pumping-capacity requirements, 293

Radiation, 19–21, 182
  Beer's law, 19
  exposure, table, 184
  long-wave, 19, 22
  short-wave, 19, 22
  solar, 19
  sources of, 183
  Stefan-Boltzman constant, 20
Radioactive wastes, 185
Radiological health, 181–186
  definition of, 181

Radiological health:
  dosimetry, 182–183
  measurement of, 182–183
  nuclear power plants, 184–185
  radiation exposure, table, 184
  radioactive wastes, 185
  shielding, 186
  sources of radiation, 183–185
  standards for, 185–186
  types, 182
    relative biologic effectiveness, table, 183
Radioisotope decay, 48
Rainfall intensity-duration curves, figure, 312
Random component of time series, 221
Random variate, 221
Rapid-sand-filter process, 371–377
  coagulation in, 372–374
  flocculation in, 374
  mixed-media filters in, 377
  sand filters in, 375–377
  sedimentation in, 375
Rate equations, table, 46
Reaction constant, influence of temperature on, 47
Reaction kinetics, 45–49
  table, 46
Reaction orders, 45–49
Reaeration, 143–144
  coefficient, 10
    as a function of temperature, 144
    as a function of turbulence, 143
  in streams and estuaries, 143–144
Recycle in water reuse, 272, 322–323
Recycle removal increment, 322
Redox potential, 75, 80
  table, 79
Regional-growth model, 197
Relation matrix, figure, 132
Relative biologic effectiveness, table, 183
rem, 183
Residence time, 35
Residence-time distribution curves:
  normalized, figure, 34
  time domain, figure, 32
Respiration, 77, 146
  endogenous, 78
  primary, 78
Retention time, 102
  biologic solids, 111, 363–365
  hydraulic, 363–364
Reverse osmosis, 394–396

Reynolds number, 26–29
Roentgen, 182
Roentgen equivalent man, 183
Roughness coefficient n for different types of conduits, table, 302
Runoff, effect on BOD, 145

Sanitary landfills, 266–267
Saprophytic organisms, 92
Saturation constant, 80
Screening processes, 335–336
Secondary clarifier, 368
Secondary salt effect, 42
Sedimentation, 25–31, 375
  constant, 30
  definition, 25
  in streams, 29–31
Sedimentation basins:
  figure, 344
  grit chambers, 343, 344
  multiple, inclined surface, 345, 346
  primary, 344–346
  in rapid sand filter process, 371, 375
  secondary clarifier, 368
  thickening tank, 355
Sedimentation processes, 336–342
Sedimentation regimes, 336–342
  class-1, 336–339
  class-2, 336, 338–342
  compression, 337
  figure, 337
  zone settling, 337
Sensible heat, 13
Septic tanks, 331
  system, 333
  figure, 332
Sequent peak method, 246
  table, 245
Sequential reuse, 272, 322, 323
Settleable solids, 325
Settling-column analysis, 339
Settling velocity, terminal, 25
Settling velocity curves, 409
Sewer systems, 307
Shielding, 186
Silanol groups, 63
Simplex method, 264
Skin friction, 27
Sludge cakes, 421
  coefficient of compressibility, 424
  specific resistance of, 421
  weight of solids in, 423

Sludge deposits, 30
Sludge stabilization, 355
Sludge treatment trains, 319
Sludge-volume index, 358
Soil percolation test, 333
Solar energy, table, 30
Solid flux curves, 409
Solid-waste generation:
   correlated with future economic factors, 197–208
   correlated with future population estimates, 196
Solid-waste management, 256–271
Solid wastes:
   collection systems, 268–271
   disposal systems, 266–268
Sols, 62
Solubility product, 41
Solution equilibriums, 39–45
Solvation, 65
Specific decay rate, 80
Specific energy, 297
   figures, 298–300
Specific growth rate, 80
   as function of limiting nutrient concentration, figure, 81
   maximum, 80
Specific heat, 13, 61
Spray aerators, 380–381
Spring turnover, 24
Stabilization stage in activated sludge process, 356
Standard error, 238
Standard states, 60
Standards:
   air-quality, table, 181
   Atomic Energy Commission, 185, 186
   objective spectrum for water standards, figure, 173
   stream, 173
      tables, 174, 175
   U.S. Public Health Service, 171, 186
      tables, 172
   for waste water effluents, 176
   water-quality, 170–176
Standing crop, 113, 114
Steady flow, 296
Steady-state transfer function, 142, 256
   matrix, 258
Stefan-Boltzman constant, 20
Step-aeration processes, 359–360
Sterilization, 396

Stoke's law, 27
Storage-yield relationships, 243–247
Stream aeration, oxygen transfer in, 10
Stream-flow sequences, 236–243
   daily, 237–238
   seasonal, 239–242
Stream standards, 173, 254
   tables, 174, 175
Streamlines, 27
Streams:
   critical flow period of, 148
   definition of, 146
   determination of BOD rate constants in, 150–151
   flow distribution patterns of, 239
   modeling of, 146–154
      by segmentation, 150
Storage for water distribution systems, 291–293
Storm drainage systems, 307
Storm-water collection systems, 311–314
   design of, 314
   rational method for design of, 311–314
   runoff coefficient for, table, 313
Sulfur oxides, 177
Supercritical flow, 297
Surface overflow rate, 338
Surface renewal rate, 9
Surge wave, 298
Suspended particulates, 319
   removal trains, 319
Suspended solids, 325
Swine feeding operation, 267
Synthesis, 77
Synthetic stream-flow sequences, 236–243
System-head curves, 293–294
Systems:
   definition of, 99
   ecological, 99–134
      aquatic, figure, 116
   information flow diagram of: block diagram of, figures, 106, 107
      figure, 100
   microbiological, continuous flow, 106–111

Table of direct and indirect requirements, 200
   table, 206
Table of technical coefficients, 200
   table, 205

INDEX **447**

Temperature:
  characteristics, 48
  coefficient, 48
  wet bulb, 18
Temperature inversion, 178–179
Thermal conductivity, 14, 16
Thermal diffusion, 14
Thermal diffusivity, 22
Thermal energy balance, 14, 21
Thermal energy budget, 21
Thermal energy flux, 17–22
Thermal phenomena, 13–25
Thermal regimes, 21–24
Thermal stratification, 23, 44–45
Thermochemistry, 58–61
Thermocline, 23
Thickening processes, 404–412
  capacity, 407
Tidal prism, 154
Time-capacity expansion of system, 208
Time constant, 102
Time derivatives, 1–2
Time-domain simulation, 104–106
  block elements used in, table, 105
  by CSMP, 124
    figures, 126, 129
Time growth of demand and installed capacity, figure, 209
Time-of-passage curve, 30
  figure, 31
Time series:
  analysis of, 222–236
  categories of, 222
  components of, 214, 216
    variance of, 223
  definition of, 214
  population parameters of, 239
  reduction of, figure, 223
Total dynamic head, 293
Toxaphene, 111
Transfer mechanisms, 9
Transport, 1–7
  in the environment, 4–5
Transport-phenomena models, 99–100
Transport systems:
  in the air environment, 165–167
  natural, 138–169
Transportation problem, 270, 272
Transportation-time matrix, 270
Treatment process:
  efficiency of, 324
  removal parameters, 325

Treatment trains, 318–325
  definition of, 318–319
  efficiency of, 324
  notation for, 320
  for removal of dissolved material, table, 323
  for removal of gross particulate, table, 321
  for removal of suspended particulate, table, 322
  removal parameters, 325
  for sludge, table, 324
Trend component of time series, 216–217
Trickling-filter process, 366–371
  retardant effect in, 369–370
Trophic level, 92, 112–113

Ubiquity principle, 78
Uniform flow, 296
  formulas, 301
Uniform-treatment method, 255
Uniformity coefficient, $375n$.
Unit cost:
  matrix, 271
  table, 277
Unit heat of combustion, 73
Unit loading matrix, 142, 163, 258
U.S. Public Health Drinking Water Standards, 171
  relative to radiation, 186
  table, 172
Units, systems of, $3n$.
Unsteady flow, 296

Vacuum filtration, 420
  rotary-drum filter, figure, 420
Vacuum flotation, 349
Van der Waals forces, 65
Van't Hoff-Arrhenius equation, 47
Variance, 221
  total, 223
Variance spectrum, 229
  figures, 231, 232, 235, 237
Velocity:
  of deposition, 30
  of scour, 30
Viscosity of water, $28n$.
  variation with temperature, 28
Vivianite, 44

Waste stabilization lagoons, 327–331
  efficiencies of, 328
  effluents from, 328
    figure, 329
  figure, 328
  odors from, 330
  primary, 330
  temperature effects in, 330
Waste-water collection systems, 307–310
  building sewers, 307
  collector sewers, 307
  combined, 307
  design of, 309–311
  interceptor sewers, 307
  minimum velocities of flow in, 308
  pipe sizes for, 307–309
  separate systems, 307
  storm drainage, 307
Waste-water flows, 307–308
  correlated with economic factors, 197–208
  correlated with future-population estimates, 196
Waste-water reuse systems, 272–278
Waste-water treatment systems, 318
Water consumption rates, 290
  table, 291
Water demands:
  correlated with economic factors, 197–208
  correlated with future-population estimates, 196

Water distribution systems, 290–296
  elevation of storage tanks for, 293–296
  flows in, 290–291
  pressure in, 291
  pumping capacity for, 293, 294
  storage for, 291–293
  system head curves for, 293–295
Water-quality criteria, 170, 254
Water-quality differentials, 275
  for BOD, table, 275
  for TDS, table, 276
Water-quality management, 254–264
Water renovation, 318
Water softening, 383–386
Water standards, 170
Water treatment systems, 318
Water-use goals, 254
Water-use or waste-generation matrix, 201
Weak acids, 40
Weak bases, 40
Weight balance, 100, 107
Wet bulb temperature, 18

Zeolites, 391
Zeta potential, 65
Zone:
  of circulation, 23
  of stagnation, 23
  of transition, 23
  zone settling characteristics, 405
    figure, 406